Springer Proceedings in Mathematics & Statistics

Volume 72

For further volumes:
http://www.springer.com/series/10533

Springer Proceedings in Mathematics & Statistics

This book series features volumes composed of select contributions from workshops and conferences in all areas of current research in mathematics and statistics, including OR and optimization. In addition to an overall evaluation of the interest, scientific quality, and timeliness of each proposal at the hands of the publisher, individual contributions are all refereed to the high quality standards of leading journals in the field. Thus, this series provides the research community with well-edited, authoritative reports on developments in the most exciting areas of mathematical and statistical research today.

Vladimir Rovenski • Paweł Walczak
Editors

Geometry and its Applications

Springer

Editors
Vladimir Rovenski
Department of Mathematics
University of Haifa
Haifa, Israel

Paweł Walczak
Faculty of Mathematics
University of Łódź
Łódź, Poland

ISSN 2194-1009 ISSN 2194-1017 (electronic)
ISBN 978-3-319-04674-7 ISBN 978-3-319-04675-4 (eBook)
DOI 10.1007/978-3-319-04675-4
Springer Cham Heidelberg New York Dordrecht London

Library of Congress Control Number: 2014935180

Mathematics Subject Classification (2010): 53C44, 53C12, 53C40, 53C30, 53C20, 37C10, 34C05, 14P05, 35K93, 35R01, 37B05, 57R30, 53A07, 53B25, 53B21, 76D17, 76E05, 76E07, 34K02, 34C25, 92C45, 37E99, 26D10, 26C05

Printed on acid-free paper

Springer is part of Springer Science+Business Media (www.springer.com)

Preface

This volume contains a collection of articles written by the participants, and their colleagues and collaborators, of the second International workshop *Geometry and Symbolic Computation* held at the University of Haifa (Israel) between May 15 and 18, 2013. The workshop was preceded by a day of excursions: the participants could choose between sightseeing at Jerusalem, Galilea Sp., The Dead See Sp., etc. The first International Workshop in this series, named "Reconstruction of Geometrical Objects Using Symbolic Computations", was on September 2008, at the University of Haifa.

Both workshops were sponsored by the Caesarea Edmond Benjamin de Rothschild Foundation Institute for Interdisciplinary Applications of Computer Science (CRI), the Center for Computational Mathematics and Scientific Computation (CCMSC), the Faculty of Natural Sciences and the Department of Mathematics at the University of Haifa.

Materials related to these workshops can be found on the homepage of V. Rovenski http://math.haifa.ac.il/ROVENSKI/rovenski.html and on the official cite of CRI http://www.cri.haifa.ac.il/index.php/crievents/.

The participants numbered approximately 20 and came from France, Greece, Kazakhstan, Poland, Russia, Ukraine and, of course, Israel. The scientific committee comprised of the editors of this volume and V. Golubyatnikov (Novosibirsk). The list of local organizers includes one of the editors (V. Rovenski); Workshop Secretary Dr. Irina Albinsky; Workshop Coordinator Ms. Danielle Friedlander; and Technical Consultant Mr. Hananel Hazan.

The papers contained in this volume are closely related to the lectures delivered at the conference, which was designed to cover different aspects of geometry together with some applications.

Three of the articles collected in the first part (Geometry) of the volume are related to geometric flows for submanifolds and foliated Riemannian manifolds analogous, to some extent, to the classical mean curvature and Ricci flows. The study of geometric flows for foliations was introduced by the editors in *Topics in Extrinsic Geometry of Codimension-One Foliations*, Springer Briefs in Mathematics, Springer-Verlag, 2011. We are happy to see some progress in this field. Another article related to geometric flows is devoted to the study of the classical Ricci flow on some particular homogeneous spaces. The other articles in this part reflect the current interest of the authors and are devoted to laminations, integral formulae, geometry of vector fields on Lie groups, and a general notion of osculation. Among them, one can find new results concerning generic properties of minimal foliations and laminations and a survey of integral formulae showing some relations between such formulae and geometric flows.

The articles collected in the second part (Applications) concern some particular problems of the theory of dynamical systems: mathematical models of liquid flows, study of cycles for nonlinear dynamical systems and relation with entropy of some quantities which appeared in a very special inequality (called *Remez inequality*) for C^k-functions.

We express our gratitude to all the participants, the contributors to the volume, the sponsors, and everyone who helped us while we were organizing the conference and preparing the volume for publication. In particular, we would like to mention Dr. Irina Albinsky who organized all the excursions, the registration of participants, and the opening procedure.

Haifa, Israel — Vladimir Rovenski
Łódź, Poland — Paweł Walczak

Contents

Preface v

Part I Geometry

The Ricci Flow on Some Generalized Wallach Spaces 3
N.A. Abiev, A. Arvanitoyeorgos, Yu. G. Nikonorov, and P. Siasos

Gaussian Mean Curvature Flow for Submanifolds in Space Forms 39
Aleksander Borisenko and Vladimir Rovenski

Cantor Laminations and Exceptional Minimal Sets in Codimension One Foliations 51
Gilbert Hector

Integral Formulas in Foliation Theory 73
Krzysztof Andrzejewski, Vladimir Rovenski, and Paweł Walczak

Prescribing the Mixed Scalar Curvature of a Foliation 83
Vladimir Rovenski and Leonid Zelenko

The Partial Ricci Flow for Foliations 125
Vladimir Rovenski

Osculation in General: An Approach 157
Paweł G. Walczak

Stability of Left-Invariant Totally Geodesic Unit Vector Fields on Three-Dimensional Lie Groups 167
Alexander Yampolsky

Part II Applications

Rotational Liquid Film Interacted with Ambient Gaseous Media 199
Gaissinski I., Levy Y., Rovenski V., and Sherbaum V.

On Cycles and Other Geometric Phenomena in Phase Portraits of Some Nonlinear Dynamical Systems 225
Gaidov Yu. A. and Golubyatnikov V.P.

Remez-Type Inequality for Smooth Functions 235
Yosef Yomdin

Part I
Geometry

The Ricci Flow on Some Generalized Wallach Spaces

N.A. Abiev, A. Arvanitoyeorgos, Yu. G. Nikonorov, and P. Siasos

Abstract We study the asymptotic behavior of the normalized Ricci flow on generalized Wallach spaces that could be considered as a special planar dynamical system. All nonsymmetric generalized Wallach spaces can be naturally parametrized by three positive numbers a_1, a_2, a_3. Our interest is to determine the type of singularity of all singular points of the normalized Ricci flow on all such spaces. Our main result gives a qualitative answer for almost all points (a_1, a_2, a_3) in the cube $(0, 1/2] \times (0, 1/2] \times (0, 1/2]$. We also consider in detail some important partial cases.

Keywords Riemannian metric • Einstein metric • Generalized Wallach space • Ricci flow • Ricci curvature • Planar dynamical system • Real algebraic surface

Mathematics Subject Classification (2010): 201053C30, 53C44, 37C10, 34C05, 14P05

N.A. Abiev
Taraz State University named after M.Kh. Dulaty, Tole bi str., 60, Taraz 080000, Kazakhstan
e-mail: abievn@mail.ru

A. Arvanitoyeorgos • P. Siasos
Department of Mathematics, University of Patras, GR-26500 Rion, Greece
e-mail: arvanito@math.upatras.gr; siasos@master.math.upatras.gr

Yu. G. Nikonorov (✉)
South Mathematical Institute of Vladikavkaz Scientific Centre of the Russian Academy of Sciences, Markus st. 22, Vladikavkaz 362027, Russia
e-mail: nikonorov2006@mail.ru

V. Rovenski and P. Walczak (eds.), *Geometry and its Applications*, Springer Proceedings in Mathematics & Statistics 72, DOI 10.1007/978-3-319-04675-4_1,

Introduction

In [1], we started the investigation of the normalized Ricci flow on generalized Wallach spaces. In this chapter we recall our previous results and develop more detailed study of some most interesting partial cases.

The study of the normalized Ricci flow equation

$$\frac{\partial}{\partial t}\mathbf{g}(t) = -2\,\mathrm{Ric}_{\mathbf{g}} + 2\mathbf{g}(t)\frac{S_{\mathbf{g}}}{n} \tag{1}$$

for a 1-parameter family of Riemannian metrics $\mathbf{g}(t)$ in a Riemannian manifold M^n was originally used by Hamilton in [14] and since then it has attracted the interest of many mathematicians (cf. [8,26]). Recently, there has been an increasing interest towards the study of the Ricci flow (normalized or not) on homogeneous spaces and under various perspectives ([2,6,7,13,15,18,23] and references therein).

The aim of the present work is to study the normalized Ricci flow for invariant Riemannian metrics on generalized Wallach spaces. These are compact homogeneous spaces G/H whose isotropy representation decomposes into a direct sum $\mathfrak{p} = \mathfrak{p}_1 \oplus \mathfrak{p}_2 \oplus \mathfrak{p}_3$ of three $\mathrm{Ad}(H)$-invariant irreducible modules satisfying $[\mathfrak{p}_i, \mathfrak{p}_i] \subset \mathfrak{h}$ $(i \in \{1,2,3\})$ [20,22]. For a fixed bi-invariant inner product $\langle\cdot,\cdot\rangle$ on the Lie algebra $\mathfrak{g}$ of the Lie group G, any G-invariant Riemannian metric $\mathbf{g}$ on G/H is determined by an $\mathrm{Ad}(H)$-invariant inner product

$$(\cdot,\cdot) = x_1\langle\cdot,\cdot\rangle|_{\mathfrak{p}_1} + x_2\langle\cdot,\cdot\rangle|_{\mathfrak{p}_2} + x_3\langle\cdot,\cdot\rangle|_{\mathfrak{p}_3}, \tag{2}$$

where x_1, x_2, x_3 are positive real numbers. An explicit expression for the Ricci curvature of invariant metrics (2) is obtained in [22, Lemma 2]. By using expressions for the Ricci tensor and the scalar curvature in [22] the normalized Ricci flow equation (1) reduces to a system of ODEs of the form

$$\frac{dx_1}{dt} = f(x_1,x_2,x_3), \quad \frac{dx_2}{dt} = g(x_1,x_2,x_3), \quad \frac{dx_3}{dt} = h(x_1,x_2,x_3), \tag{3}$$

where $x_i = x_i(t) > 0$ $(i = 1,2,3)$ are parameters of the invariant metric (2) and

$$\begin{aligned}
f(x_1,x_2,x_3) &= -1 - \frac{A}{d_1}x_1\Big(\frac{x_1}{x_2x_3} - \frac{x_2}{x_1x_3} - \frac{x_3}{x_1x_2}\Big) + 2x_1\frac{S_{\mathbf{g}}}{n},\\
g(x_1,x_2,x_3) &= -1 - \frac{A}{d_2}x_2\Big(\frac{x_2}{x_1x_3} - \frac{x_3}{x_1x_2} - \frac{x_1}{x_2x_3}\Big) + 2x_2\frac{S_{\mathbf{g}}}{n},\\
h(x_1,x_2,x_3) &= -1 - \frac{A}{d_3}x_3\Big(\frac{x_3}{x_1x_2} - \frac{x_1}{x_2x_3} - \frac{x_2}{x_1x_3}\Big) + 2x_3\frac{S_{\mathbf{g}}}{n},\\
S_{\mathbf{g}} &= \frac{1}{2}\Big(\frac{d_1}{x_1} + \frac{d_2}{x_2} + \frac{d_3}{x_3} - A\Big(\frac{x_1}{x_2x_3} + \frac{x_2}{x_1x_3} + \frac{x_3}{x_1x_2}\Big)\Big).
\end{aligned}$$

Here d_i, $i = 1,2,3$, are the dimensions of the corresponding irreducible modules $\mathfrak{p}_i$, $n = d_1 + d_2 + d_3$, and A is some special nonnegative number (see Sect. 1). If $A \neq 0$, then by denoting $a_i := A/d_i > 0$, $i = 1,2,3$, the functions f, g, h can be expressed in a more convenient form (independent of A and d_i) as

$$f(x_1,x_2,x_3) = -1 - a_1x_1\Big(\frac{x_1}{x_2x_3} - \frac{x_2}{x_1x_3} - \frac{x_3}{x_1x_2}\Big) + x_1B,$$

$$g(x_1,x_2,x_3) = -1 - a_2x_2\Big(\frac{x_2}{x_1x_3} - \frac{x_3}{x_1x_2} - \frac{x_1}{x_2x_3}\Big) + x_2B,$$

$$h(x_1,x_2,x_3) = -1 - a_3x_3\Big(\frac{x_3}{x_1x_2} - \frac{x_1}{x_2x_3} - \frac{x_2}{x_1x_3}\Big) + x_3B,$$

where

$$B := \Big(\frac{1}{a_1x_1} + \frac{1}{a_2x_2} + \frac{1}{a_3x_3} - \Big(\frac{x_1}{x_2x_3} + \frac{x_2}{x_1x_3} + \frac{x_3}{x_1x_2}\Big)\Big)\Big(\frac{1}{a_1} + \frac{1}{a_2} + \frac{1}{a_3}\Big)^{-1}.$$

It is easy to check that the volume $V = x_1^{1/a_1}x_2^{1/a_2}x_3^{1/a_3}$ is a first integral of system (3). Therefore, on the surface

$$V \equiv 1, \tag{4}$$

we can reduce (3) to a system of two differential equations of the type

$$\frac{dx_1}{dt} = \tilde{f}(x_1,x_2), \qquad \frac{dx_2}{dt} = \tilde{g}(x_1,x_2), \tag{5}$$

where

$$\tilde{f}(x_1,x_2) \equiv f(x_1,x_2,\varphi(x_1,x_2)), \qquad \tilde{g}(x_1,x_2) \equiv g(x_1,x_2,\varphi(x_1,x_2)),$$

$$\varphi(x_1,x_2) = x_1^{-\frac{a_3}{a_1}}x_2^{-\frac{a_3}{a_2}}.$$

Remark 1. From (4) it is clear that $(x_1^0, x_2^0) = (\gamma_1 q, \gamma_2 q)$ is a singular point of (5) if and only if $(x_1^0, x_2^0, x_3^0) = (\gamma_1 q, \gamma_2 q, \gamma_3 q)$, where $\gamma_i > 0$ for $i = 1,2,3$ is a singular point of (3) corresponding to unique $q := \gamma_1^{-d/a_1}\gamma_2^{-d/a_2}\gamma_3^{-d/a_3} > 0$, where $d := \big(1/a_1 + 1/a_2 + 1/a_3\big)^{-1}$.

It is known that every generalized Wallach space admits at least one invariant Einstein metric [22]. Later in [19, 20], a detailed study of invariant Einstein metrics was developed for all generalized Wallach spaces. In particular, it was shown that there are at most four invariant Einstein metrics (up to homothety) for every such space. Recall that invariant Einstein metrics with $V = 1$ correspond to singular points of (5); therefore, (x_1^0, x_2^0, x_3^0) is a singular point of system (3), (4) if and only if (x_1^0, x_2^0) is a singular point of (5). It is our interest to determine the type of singularity of such points, and our investigation concerns this problem for some special values of the parameters a_1, a_2, and a_3. The main result in this direction is Theorem 7, which gives a qualitative answer for almost all points (in measure theoretic sense) $(a_1, a_2, a_3) \in (0, 1/2] \times (0, 1/2] \times (0, 1/2]$. Note that the latter inclusion is fulfilled for any triple (a_1, a_2, a_3) corresponding to some generalized Wallach spaces (see the next section). However, we are interested in the behavior of the dynamical system (5) for all values $a_i \in (0, 1/2]$ despite the fact that some triples may not correspond to "real" generalized Wallach spaces.

We expect to give a more detailed study of system (5) for various values of the parameters a_1, a_2, a_3, which could help towards a deeper understanding of the behavior of the Ricci flow on more general homogeneous spaces. Also, it is quite possible that system (5) is interesting not only for the parameters $a_i \in (0, 1/2]$ but also as a more general dynamical system other than the Ricci flow. *It is clear that system (3) is naturally defined for all values of* a_1, a_2, a_3 *with* $a_1a_2 + a_1a_3 + a_2a_3 \neq 0$, *but for system (5) we should assume* $a_1a_2a_3 \neq 0$ *additionally*.

1 Generalized Wallach Spaces

We recall the definition and important properties of generalized Wallach spaces (cf. [21, pp. 6346–6347] and [22]).

Consider a homogeneous almost effective compact space G/H with a (compact) semisimple connected Lie group G and its closed subgroup H. Denote by $\mathfrak{g}$ and $\mathfrak{h}$ the Lie algebras of G and H, respectively. In what follows, $[\cdot, \cdot]$ stands for the Lie bracket of $\mathfrak{g}$ and $B(\cdot, \cdot)$ stands for the Killing form of $\mathfrak{g}$. Note that $\langle \cdot, \cdot \rangle = -B(\cdot, \cdot)$ is a bi-invariant inner product on $\mathfrak{g}$.

Consider the orthogonal complement $\mathfrak{p}$ of $\mathfrak{h}$ in $\mathfrak{g}$ with respect to $\langle \cdot, \cdot \rangle$. Every G-invariant Riemannian metric on G/H generates an $\mathrm{Ad}(H)$-invariant inner product on $\mathfrak{p}$ and vice versa [5]. Therefore, it is possible to identify invariant Riemannian metrics on G/H with $\mathrm{Ad}(H)$-invariant inner products on $\mathfrak{p}$ (if H is connected then the property to be $\mathrm{Ad}(H)$-invariant is equivalent to the property to be $\mathrm{ad}(\mathfrak{h})$-invariant). Note that the Riemannian metric generated by the inner product $\langle \cdot, \cdot \rangle\big|_{\mathfrak{p}}$ is called *standard* or *Killing*.

Let G/H be a homogeneous space such that its isotropy representation $\mathfrak{p}$ is decomposed as a direct sum of three $\mathrm{Ad}(H)$-invariant irreducible modules pairwise orthogonal with respect to $\langle \cdot, \cdot \rangle$, i.e.,

$$\mathfrak{p} = \mathfrak{p}_1 \oplus \mathfrak{p}_2 \oplus \mathfrak{p}_3,$$
$$\text{with } [\mathfrak{p}_i, \mathfrak{p}_i] \subset \mathfrak{h} \text{ for } i \in \{1, 2, 3\}.$$

Since this condition on each module resembles the condition of local symmetry for homogeneous spaces (a locally symmetric homogeneous space G/H is characterized by the relation $[\mathfrak{p},\mathfrak{p}] \subset \mathfrak{h}$, where $\mathfrak{g} = \mathfrak{h} \oplus \mathfrak{p}$ and $\mathfrak{p}$ is $\mathrm{Ad}(H)$-invariant [5]), then spaces with this property were called *three-locally-symmetric* in [19, 20]. But in this chapter we prefer the term *generalized Wallach spaces*, as in [21].

There are many examples of these spaces, e.g., the flag manifolds

$$SU(3)/T_{\max}, \quad Sp(3)/Sp(1)\times Sp(1)\times Sp(1), \quad F_4/\mathrm{Spin}(8).$$

These spaces (known as *Wallach spaces*) are interesting because they admit invariant Riemannian metrics of positive sectional curvature (see [27]). The invariant Einstein metrics on $SU(3)/T_{\max}$ were classified in [9] and, on the remaining two spaces, in [24]. In each of these cases, there exist exactly four invariant Einstein metrics (up to proportionality). Other classes of generalized Wallach spaces are the various Kähler C-spaces such as

$$SU(n_1+n_2+n_3)/S\big(U(n_1)\times U(n_2)\times U(n_3)\big),$$
$$SO(2n)/U(1)\times U(n-1), \quad E_6/U(1)\times U(1)\times \mathrm{Spin}(8).$$

The invariant Einstein metrics in the above spaces were classified in [17]. Each of these spaces admits four invariant Einstein metrics (up to scalar), one of which is Kähler for an appropriate complex structure on G/H. Another approach to $SU(n_1+n_2+n_3)/S\big(U(n_1)\times U(n_2)\times U(n_3)\big)$ was used in [3]. The Lie group $SU(2)$ $\big(H = \{e\}\big)$ is another example of a generalized Wallach space. Being three dimensional, this group admits only one left-invariant Einstein metric which is a metric of constant curvature [5].

In [22], it was shown that every generalized Wallach space admits at least one invariant Einstein metric. This result could not be improved in general (since, e.g., $SU(2)$ admits exactly one invariant Einstein metric). Later in [19,20], a detailed study of invariant Einstein metrics was developed for all generalized Wallach spaces. In particular, it is proved that there are at most four Einstein metrics (up to homothety) for every such space.

Denote by d_i the dimension of $\mathfrak{p}_i$. Let $\{e_i^j\}$ be an orthonormal basis in $\mathfrak{p}_i$ with respect to $\langle\cdot,\cdot\rangle$, where $i \in \{1,2,3\}$, $1 \le j \le d_i = \dim(\mathfrak{p}_i)$. Consider the expression $[ijk]$ defined by the equality

$$[ijk] = \sum\nolimits_{\alpha,\beta,\gamma} \langle [e_i^\alpha, e_j^\beta], e_k^\gamma\rangle^2,$$

where α, β, and γ range from 1 to d_i, d_j, and d_k, respectively. The symbols $[ijk]$ are symmetric in all three indices by bi-invariance of the metric $\langle\cdot,\cdot\rangle$. Moreover, for the spaces under consideration, we have $[ijk] = 0$ if two indices coincide. Therefore, the quantity $A := [123]$ plays an important role.

By [22, Lemma 1], we get $d_i \ge 2A$ for every $i = 1,2,3$ with $d_i = 2A$ if and only if $[\mathfrak{h},\mathfrak{p}_i] = 0$.

Note that $A = 0$ if and only if the space G/H is locally a direct product of three compact irreducible symmetric spaces (see [20, Theorem 2]).

Suppose $A \neq 0$ and let

$$a_i = A/d_i, \quad i \in \{1,2,3\}. \tag{6}$$

It is clear that $a_i \in (0, 1/2]$.

Consider the value of a_i's for some special examples of generalized Wallach spaces (see [20]).

The spaces $SU(l+m+n)/S\big(U(l)\times U(m)\times U(n)\big)$ have the property $a_1+a_2+a_3 = 1/2$. In this case, it is known that

$$a_1 = \frac{n}{2(l+m+n)}, \quad a_2 = \frac{m}{2(l+m+n)}, \quad a_3 = \frac{l}{2(l+m+n)}.$$

There are two other families of generalized Wallach spaces: $SO(l+m+n)/\big(SO(l)\times SO(m)\times SO(n)\big)$ and $Sp(l+m+n)/\big(Sp(l)\times Sp(m)\times Sp(n)\big)$. It should be noted that

$$a_1 = \frac{n}{2(l+m+n-2)}, \quad a_2 = \frac{m}{2(l+m+n-2)}, \quad a_3 = \frac{l}{2(l+m+n-2)}$$

for the orthogonal case and

$$a_1 = \frac{n}{2(l+m+n+1)}, \quad a_2 = \frac{m}{2(l+m+n+1)}, \quad a_3 = \frac{l}{2(l+m+n+1)}$$

for the symplectic case [20].

There are interesting examples with $a_1 = a_2 = a_3 =: a$ among the above examples:

$$SU(3m)/S\big(U(m)\times U(m)\times U(m)\big), \quad SO(3m)/\big(SO(m)\big)^3, \quad Sp(3m)/\big(Sp(m)\big)^3.$$

Obviously, $a = 1/6$, $a = m/(6m-4)$, and $a = m/(6m+2)$, respectively, for these spaces. Note that the space $SO(6)/\big(SO(2)\big)^3$ satisfies the equality $a = 1/4$.

Note that not every triple $(a_1, a_2, a_3) \in (0, 1/2]\times(0, 1/2]\times(0, 1/2]$ corresponds to some generalized Wallach space. For example, if $a_i = 1/2$ for some i, then there is no generalized Wallach space with $a_j \neq a_k$, where $i \neq j \neq k \neq i$ (see [22, Lemma 4]). Moreover, every a_i should be a rational number for a generalized Wallach space with simple group G (see (6), [20, Lemma 1] and [10, Table 1]).

2 Description of the Singular Points of System (3)

We will first give a description of the singular points of system (3), as Einstein metrics on generalized Wallach spaces. An easy calculation shows that for $a_1a_2 + a_1a_3 + a_2a_3 \neq 0$, the singular points (x_1, x_2, x_3) of system (3) can be found from the equations

$$\begin{aligned}(a_2+a_3)(a_1x_2^2+a_1x_3^2-x_2x_3)+(a_2x_2+a_3x_3)x_1-(a_1a_2+a_1a_3+2a_2a_3)x_1^2=0,\\ (a_1+a_3)(a_2x_1^2+a_2x_3^2-x_1x_3)+(a_1x_1+a_3x_3)x_2-(a_1a_2+2a_1a_3+a_2a_3)x_2^2=0.\end{aligned} \tag{7}$$

If $a_1a_2a_3 \neq 0$ and $x_3 = \varphi(x_1, x_2)$, then we also get singular points of system (5). Recall that we are interested only in singular points with $x_i > 0$, $i = 1, 2, 3$.

Note that system (7) is homogeneous (of degree 2) with respect to x_1, x_2, x_3. It is easy to see that $x_i = x_j = 0$ implies that either $x_k = 0$ or $a_i(a_j + a_k) = a_j(a_i + a_k) = 0$, $i \neq j \neq k \neq i$. If we have a solution with $x_i = 1$ and $x_j = 0$, then we should have $(4a_j^2 - 1)(a_i + a_k)(a_1a_2 + a_1a_3 + a_2a_3) = 0$. Therefore, if $a_i \in (0, 1/2)$ for $i = 1, 2, 3$, then system (7) has no solution with zero component. If $a_i \in (0, 1/2]$, $i = 1, 2, 3$, then it is proved in [20] that this system has (up to multiplication by a constant, for example, if we put $x_3 = 1$) at least one and at most four solutions with positive components. A detailed information on these solutions can be found in [20]. We briefly review these results below.

The case where at least two of a_i's are equal. Without loss of generality we may assume that $a_1 = a_2 = b$ and $a_3 = c$. Then system (7) is equivalent to the following one:

$$\begin{aligned}(x_2 - x_1)\big(x_3 - 2b(x_1 + x_2)\big) = 0,\\ x_2(x_3 - x_1) + (b + c)(x_1^2 - x_3^2) + (c - b)x_2^2 = 0.\end{aligned} \tag{8}$$

If $x_2 = x_1$ then the second equation of (8) becomes

$$(1 - 2c)x_1^2 - x_1x_3 + (b + c)x_3^2 = 0. \tag{9}$$

Thus, we have the following singular points

$$(x_1, x_2, x_3) = \big(2(b + c)q, 2(b + c)q, \mu q\big), \tag{10}$$

where $\mu = 1 \pm \sqrt{1-4(1-2c)(b+c)}$, $q \in \mathbb{R}$, $q > 0$. We observe that for $c = 1/2$, we have only one family of singular points $(x_1, x_2, x_3) = \big((b + c)q, (b + c)q, q\big)$. Otherwise, $1 - 2c > 0$, and all depend on the sign of the discriminant $D_1 = 1 - 4(1 - 2c)(b + c)$. Indeed, there exist one family of singular points for $D_1 = 0$, two families for $D_1 > 0$, and none for $D_1 < 0$.

If $x_2 \neq x_1$ then $x_3 = 2b(x_1 + x_2)$; so, the second equation of (8) reduces to

$$(b+c)(1-4b^2)x_1^2-\big(1-2b+8b^2(b+c)\big)x_1x_2+(b+c)\big(1-4b^2\big)x_2^2 = 0. \tag{11}$$

If $b = 1/2$ then (11) has no solution. Otherwise, $1 - 4b^2 > 0$; hence, all real roots of equation (11) are positive. The discriminant D_2 of (11) has the same sign as $T := 1 - 4b - 2c + 16b^2(b + c)$. Thus, there exist one family of singular points for $T = 0$, two families for $T > 0$, and none for $T < 0$.

In particular, if $a_1 = a_2 = a_3 = a$, $a \in (0, 1/2)$, then for $a \neq 1/4$, we get exactly four singular points (x_1, x_2, x_3) up to a positive multiple, namely $(1, 1, 1)$, $(1 - 2a, 2a, 2a)$, $(2a, 1 - 2a, 2a)$, and $(2a, 2a, 1 - 2a)$. For $a = 1/4$, we get only singular points proportional to $(1, 1, 1)$.

The case of pairwise distinct a_i's. We consider two subcases here.

The case $a_1 + a_2 + a_3 = 1/2$. Then all singular points (x_1, x_2, x_3) have the form

$$\begin{aligned}
&\big((1 - 2a_1)q,\ (1 - 2a_2)q,\ 2(a_1 + a_2)q\big),\\
&\big((1 - 2a_1)q,\ (1 - 2a_2)q,\ 2(1 - a_1 - a_2)q\big),\\
&\big((1 - 2a_1)q,\ (1 + 2a_2)q,\ 2(a_1 + a_2)q\big),\\
&\big((1 + 2a_1)q,\ (1 - 2a_2)q,\ 2(a_1 + a_2)q\big),
\end{aligned} \qquad \text{where } q \in \mathbb{R},\ q > 0\,. \tag{12}$$

The case $a_1 + a_2 + a_3 \neq 1/2$. We look for singular points of the form $(x_1, x_2, x_3) = (1, t, s)$. Then system (7) can be reduced to an equation of degree 4 either in s or in t. By eliminating the summand containing t^2, we obtain the following system equivalent to (7):

$$\begin{aligned}
&\big((a_2 + a_3)s - (a_1 + a_2)\big)t = 2a_1(a_2 + a_3)s^2 + (a_3 - a_1)s - 2a_3(a_1 + a_2),\\
&(a_2 + a_3)t^2 - (a_2 + a_3)s^2 + s - t + a_3 - a_2 = 0.
\end{aligned} \tag{13}$$

It is easy to see that $(a_2+a_3)s-(a_1+a_2) \neq 0$ (see details in [20]). Expressing t from the first equation of (13) and inserting it into the second, we obtain the following equation of degree 4:

$$\begin{aligned}
&(a_2 + a_3)^2(2a_1 - 1)(2a_1 + 1)s^4 + (a_2 + a_3)(2a_2 + 4a_1a_3 + 1 - 4a_1^2)s^3\\
&\quad +\big(2a_1^2+2a_3^2-8a_1a_2^2a_3-2a_2^2-8a_1^2a_2a_3-2a_2-8a_1a_2a_3^2-2a_1a_3-a_1-a_3-8a_1^2a_3^2\big)s^2\\
&\quad +(a_1 + a_2)(4a_1a_3 + 2a_2 + 1 - 4a_3^2)s + (2a_3 - 1)(2a_3 + 1)(a_1 + a_2)^2 = 0.
\end{aligned} \tag{14}$$

Denote by D_3 the discriminant of the polynomial on the left-hand side of (14). It can be shown ([20]) that all real solutions of (14) are positive. For $D_3 \neq 0$, equation (14) has either two or four distinct real solutions. Therefore, we get two or four families of singular points determined by (14).

Notice that the condition $D_3 = 0$ holds, for example, for the homogeneous space $SO(20)/\big(SO(5) \times SO(6) \times SO(9)\big)$ ($a_1 = 5/36$, $a_2 = 1/6$, $a_3 = 1/4$). In this special case, equation (14) has one root of multiplicity 2 and the space under consideration admits exactly three pairwise nonhomothetic singular points (i.e., invariant Einstein metrics).

3 The Study of Singular Points

In this section, we recall some facts about the type of singular points of system (5). Our basic references are [12, 16]. The functions $\tilde{f}(x_1, x_2)$ and $\tilde{g}(x_1, x_2)$ [see (5)] are analytic in a neighborhood of an arbitrary point (x_1^0, x_2^0) (where $x_1^0 > 0$ and $x_2^0 > 0$), and the following representations are valid:

$$\tilde{f}(x_1, x_2) \equiv J_{11}(x_1 - x_1^0) + J_{12}(x_2 - x_2^0) + F(x_1, x_2),$$
$$\tilde{g}(x_1, x_2) \equiv J_{21}(x_1 - x_1^0) + J_{22}(x_2 - x_2^0) + G(x_1, x_2),$$

where J_{11}, J_{12}, J_{21}, and J_{22} are the elements of the Jacobian matrix

$$J := J(x_1^0, x_2^0) = \begin{pmatrix} \dfrac{\partial \tilde{f}(x_1^0, x_2^0)}{\partial x_1} & \dfrac{\partial \tilde{f}(x_1^0, x_2^0)}{\partial x_2} \\ \dfrac{\partial \tilde{g}(x_1^0, x_2^0)}{\partial x_1} & \dfrac{\partial \tilde{g}(x_1^0, x_2^0)}{\partial x_2} \end{pmatrix}. \tag{15}$$

The functions F and G are also analytic in a neighborhood of the point (x_1^0, x_2^0) and

$$F(x_1^0, x_2^0) = G(x_1^0, x_2^0) = \frac{\partial F(x_1^0, x_2^0)}{\partial x_1} = \frac{\partial F(x_1^0, x_2^0)}{\partial x_2} = \frac{\partial G(x_1^0, x_2^0)}{\partial x_1}$$
$$= \frac{\partial G(x_1^0, x_2^0)}{\partial x_2} = 0.$$

The eigenvalues of $J(x_1^0, x_2^0)$ can be found from the formula

$$\lambda_{1,2} = \frac{\rho \pm \sqrt{\sigma}}{2}, \quad (|\lambda_1| \le |\lambda_2|),$$

where

$$\sigma := \rho^2 - 4\delta, \quad \rho := \operatorname{trace}\left(J(x_1^0, x_2^0)\right), \quad \delta := \det\left(J(x_1^0, x_2^0)\right). \tag{16}$$

We will use these notations in the sequel.

In the nondegenerate case ($\delta = \lambda_1 \lambda_2 \neq 0$), we will use the following theorem.

Theorem 1 (Theorem 2.15 in [12]). *Let* $(0, 0)$ *be an isolated singular point of the system*

$$\frac{dx}{dt} = ax + by + A(x, y), \quad \frac{dy}{dt} = cx + dy + B(x, y),$$

where A and B are analytic in a neighborhood of the origin with

$$A(0,0) = B(0,0) = \frac{\partial A(0,0)}{\partial x} = \frac{\partial A(0,0)}{\partial y} = \frac{\partial B(0,0)}{\partial x} = \frac{\partial B(0,0)}{\partial y} = 0.$$

Let λ_1 and λ_2 be the eigenvalues of the matrix $\begin{pmatrix} a & b \\ c & d \end{pmatrix}$ which represents the linear part of the system at the origin. Then the following statements hold:

(i) *If λ_1 and λ_2 are real and $\lambda_1\lambda_2 < 0$, then $(0,0)$ is a saddle.*
(ii) *If λ_1 and λ_2 are real with $|\lambda_1| \leq |\lambda_2|$ and $\lambda_1\lambda_2 > 0$, then $(0,0)$ is a node. If $\lambda_1 > 0$ (respectively < 0) then it is unstable (respectively stable).*
(iii) *If $\lambda_1 = \alpha + i\beta$ and $\lambda_2 = \alpha - i\beta$ with $\alpha, \beta \neq 0$, then $(0,0)$ is a strong focus.*
(iv) *If $\lambda_1 = i\beta$ and $\lambda_2 = -i\beta$ with $\beta \neq 0$, then $(0,0)$ is a weak focus or a center.*

Cases (i)–(iii) are known as *hyperbolic singular points*.

In the semi-hyperbolic case, we will use another approach for investigation of (5) based on the following theorem.

Theorem 2 (Theorem 2.19 in [12]). *Let $(0,0)$ be an isolated singular point of the system*

$$\frac{dx}{dt} = A(x,y), \quad \frac{dy}{dt} = \lambda y + B(x,y), \quad \lambda > 0,$$

where A and B are analytic in a neighborhood of the origin with

$$A(0,0) = B(0,0) = \frac{\partial A(0,0)}{\partial x} = \frac{\partial A(0,0)}{\partial y} = \frac{\partial B(0,0)}{\partial x} = \frac{\partial B(0,0)}{\partial y} = 0.$$

Let $y = \phi(x)$ be the solution of the equation $\lambda y + B(x,y) = 0$ in a neighborhood of $(0,0)$, and suppose that the function $\psi(x) := A(x,\phi(x))$ has the expression $\psi(x) = e_m x^m + o(x^m)$, where $m \geq 2$ and $e_m \neq 0$. Then

(i) *if m is odd and $e_m < 0$, then $(0,0)$ is a topological saddle;*
(ii) *if m is odd and $e_m > 0$, then $(0,0)$ is an unstable node;*
(iii) *if m is even, then $(0,0)$ is a saddle-node.*

Note that if A and B satisfy the conditions of Theorem 2, then in a neighborhood of the origin the following formulas are valid:

$$\begin{aligned} A(x,y) &= \textstyle\sum_{n=2}^{\infty}\sum_{i=0}^{n} p_{n-i,i}x^{n-i}y^i, \quad p_{n-i,i} = \frac{1}{i!(n-i)!}\frac{\partial^n A(0,0)}{\partial x^{n-i}\partial y^i}, \\ B(x,y) &= \textstyle\sum_{n=2}^{\infty}\sum_{i=0}^{n} q_{n-i,i}x^{n-i}y^i, \quad q_{n-i,i} = \frac{1}{i!(n-i)!}\frac{\partial^n B(0,0)}{\partial x^{n-i}\partial y^i}. \end{aligned} \tag{17}$$

A direct calculation of δ and ρ is often very complicated; so, we will obtain more convenient formulas for ρ and δ in the case of *singular points* (x_1^0, x_2^0) *of system* (5). Let $\tilde{J}$ be the Jacobian matrix of the map

$$(x_1, x_2, x_3) \mapsto \big(f(x_1, x_2, x_3), g(x_1, x_2, x_3), h(x_1, x_2, x_3)\big)$$

and let $p(t) = t^3 - \tilde{\rho}t^2 + \tilde{\delta}t + \tilde{\kappa}$ be the characteristic polynomial of $\tilde{J}$.

By direct computations using the chain rule, one can easily get the following two lemmas (see details in [1]).

Lemma 1. *If* (x_1^0, x_2^0) *is a singular point of system* (5)*, then* $\rho = \tilde{\rho}$ *and* $\delta = \tilde{\delta}$*, where* $p(t) = t^3 - \tilde{\rho}t^2 + \tilde{\delta}t + \tilde{\kappa}$ *is calculated at the point* $(x_1^0, x_2^0, x_3^0 = \varphi(x_1^0, x_2^0))$.

Lemma 2. *Let* $p(t) = t^3 - \tilde{\rho}t^2 + \tilde{\delta}t + \tilde{\kappa}$ *be the characteristic polynomial of the Jacobian matrix of system* (3) *at any point* (x_1, x_2, x_3) *with* $x_1x_2x_3 \neq 0$*. Then* $\tilde{\kappa} = 0$, $\tilde{\rho} = \frac{2F_1}{\mathcal{A}x_1x_2x_3}$*, and* $\tilde{\delta} = \frac{F_2}{\mathcal{A}^2x_1^2x_2^2x_3^2}$*, where*

$$\begin{aligned} F_1 &= a_1a_2x_1x_2 + a_1a_3x_1x_3 + a_2a_3x_2x_3 \\ &\quad -(\mathcal{A} + a_2a_3)a_1x_1^2 - (\mathcal{A} + a_1a_3)a_2x_2^2 - (\mathcal{A} + a_1a_2)a_3x_3^2, \end{aligned} \tag{18}$$

$$\begin{aligned} F_2 = {} & \big(a_1^2a_2a_3(2\mathcal{A} + a_2a_3) - \mathcal{A}^3\big)x_1^4 + \big(a_1a_2^2a_3(2\mathcal{A} + a_1a_3) - \mathcal{A}^3\big)x_2^4 \\ & + \big(a_1a_2a_3^2(2\mathcal{A} + a_1a_2) - \mathcal{A}^3\big)x_3^4 - 2a_1^2(\mathcal{A} + a_2a_3)(a_2x_2 + a_3x_3)x_1^3 \\ & - 2a_2^2(\mathcal{A} + a_1a_3)(a_1x_1 + a_3x_3)x_2^3 - 2a_3^2(\mathcal{A} + a_1a_2)(a_2x_2 + a_1x_1)x_3^3 \\ & + \big(a_1a_2(2(3a_1a_2 + a_2^2 + a_1^2)a_3^2 + 2(a_1 + a_2)a_1a_2a_3 + a_1a_2) + 2\mathcal{A}^3\big)x_1^2x_2^2 \\ & + \big(a_1a_3(2(3a_1a_3 + a_3^2 + a_1^2)a_2^2 + 2(a_1 + a_3)a_1a_2a_3 + a_1a_3) + 2\mathcal{A}^3\big)x_1^2x_3^2 \\ & + \big(a_2a_3(2(3a_2a_3 + a_3^2 + a_2^2)a_1^2 + 2(a_2 + a_3)a_1a_2a_3 + a_2a_3) + 2\mathcal{A}^3\big)x_2^2x_3^2 \\ & - 2a_1a_2a_3\big((\mathcal{A}+a_2a_3-a_1)x_1+(\mathcal{A}+a_1a_3-a_2)x_2+(\mathcal{A}+a_1a_2-a_3)x_3\big)x_1x_2x_3, \end{aligned} \tag{19}$$

and $\mathcal{A} = a_1a_2 + a_1a_3 + a_2a_3$.

We will show a convenient way to deal with degenerate singular points of system (5) in Lemma 3. By Lemma 1, a singular point (x_1, x_2) of system (5) is degenerate if and only if $\delta = \tilde{\delta} = 0$. Note that the right-hand side of (19) is homogeneous in the variables x_1, x_2, x_3. Therefore, without loss of generality we may consider these variables up to a positive multiple. Obviously from Lemma 2, it follows that $\tilde{\delta} = 0$ is equivalent to $F_2 = 0$. Hence, we have

Lemma 3 ([1]). *A singular point* (x_1, x_2) *of system* (5) *is degenerate if and only if the point* $(x_1, x_2, x_3 = \varphi(x_1, x_2))$ *satisfies the equation* $F_2 = 0$*, where* F_2 *is given by* (19).

4 A Special Case

Here, we consider the case $a_1 = a_2 = a_3 = 1/4$ (corresponding to the space $SO(6)/\big(SO(2)\big)^3$) which is of special interest.

Theorem 3. *For* $a_1 = a_2 = a_3 = 1/4$*, system of ODEs* (5) *has a unique (degenerate) singular point* $(x_1^0, x_2^0) = (1,1)$ *which is a saddle with six hyperbolic sectors.*

Proof. As the calculations show, the unique singular point of (5) is indeed $(x_1^0, x_2^0) = (1,1)$ (see Fig. 1 for a phase portrait in a neighborhood of this point). Note that in this case, $J(x_1^0, x_2^0) = \begin{pmatrix} 0 & 0 \\ 0 & 0 \end{pmatrix}$. By moving $(1,1)$ to the origin and by the analyticity of $\tilde{f}$ and $\tilde{g}$ at $(1,1)$, system (5) can be reduced to the equivalent system

$$\frac{dx}{dt} = P_2(x,y) + P_3(x,y) + P_4(x,y) + \cdots,$$
$$\frac{dy}{dt} = Q_2(x,y) + Q_3(x,y) + Q_4(x,y) + \cdots,$$

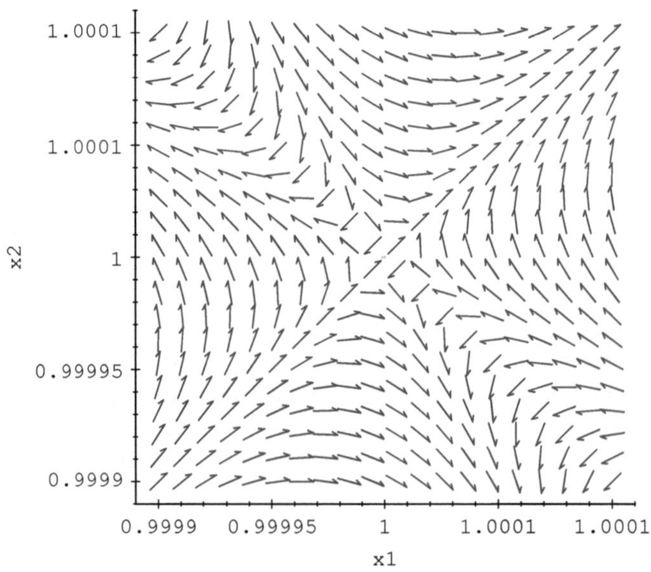

Fig. 1 The unique singular point at $a_1 = a_2 = a_3 = 1/4$

where

$$P_2(x,y) = -x^2/2 + xy + y^2, \quad Q_2(x,y) = x^2 + xy - y^2/2,$$

and $P_i(x,y), Q_i(x,y)$ $(i \geq 3)$ are some homogeneous polynomials of degree i with respect to x and y $(x = x_1 - 1,\ y = x_2 - 1)$.

Next, we use results from [16]. Using the blowing up $y = ux$, $d\tau = xdt$, we obtain the system

$$\frac{dx}{d\tau} = xP_2(1,u) + x^2P_3(1,u) + x^3P_4(1,u) + \cdots,$$
$$\frac{du}{d\tau} = \Delta(u) + x\big(Q_3(1,u) - uP_3(1,u)\big) + x^2\big(Q_4(1,u) - uP_4(1,u)\big) + \cdots,$$

where

$$P_2(1,u) = u^2 + u - 1/2, \quad \Delta(u) = -(u-u_1)(u-u_2)(u-u_3),$$
$$u_1 = -2, \quad u_2 = -1/2, \quad u_3 = 1.$$

We are in the case of [16, Sect. 6.2], where the equation $\Delta(u) = 0$ has three different real roots. So, it is obvious that the blowing-up system has the singular points $(0,u_1)$, $(0,u_2)$, and $(0,u_3)$. We show that all of these singular points are saddles.

Let $\beta_i := P_2(1,u_i)$ and let J_i be the matrix of the linear part of the blowing-up system at the point $(0,u_i)$, $i = 1,2,3$. Then

$$J_i = \begin{pmatrix} \beta_i & 0 \\ \frac{1}{2}(u_i - 1)(u_i + 1)(2u_i^2 - u_i + 2) & -3\beta_i \end{pmatrix}$$

with eigenvalues equal to β_i and $-3\beta_i$. It is clear that $\beta_i \neq 0$ for all $i = 1,2,3$. Therefore, the eigenvalues of J_i have different signs, and by Theorem 1, all the singular points $(0,u_i)$, $i = 1,2,3$, are saddles. The phase portrait of the blowing-up system is identical to the one shown in [16, Fig. 7b].

The saddles $(0,u_i)$, $i = 1,2,3$, correspond to the unique singular point $(0,0)$ of the initial system. According to the qualitative classification of singular points of degree 2 given by [16], the point $(0,0)$ is also a saddle with six hyperbolic sectors near it (see [16, Fig. 3(12)]). □

5 The "Degeneration" Set Ω

Recall that systems (3) and (5) are well defined for all $(a_1,a_2,a_3) \in \mathbb{R}^3$ with $a_1a_2 + a_1a_3 + a_2a_3 \neq 0$ [with the additional restriction $a_1a_2a_3 \neq 0$ for system (5)].

The special case considered in Sect. 4 leads us to considering the set

$$\Omega' = \{(a_1, a_2, a_3) \in \mathbb{R}^3 \mid \text{system (5) has at least one degenerate singular point}\}.$$

Now, we represent the set Ω' as a part of an algebraic surface Ω in $\mathbb{R}^3$. Under the assumptions $a_1a_2 + a_1a_3 + a_2a_3 \neq 0$, $a_1a_2a_3 \neq 0$, and $x_3 = \varphi(x_1, x_2)$, the singular points (x_1, x_2) of system (5) can be found from equation (7), and according to Lemma 3, they are degenerate if and only if $F_2 = 0$. Note that (7) and $F_2 = 0$ are homogeneous with respect to x_1, x_2, x_3. Setting $x_3 = 1$ and eliminating x_1 and x_2 from these three equations (e.g., using *Maple* or *Mathematica*), we obtain the equation

$$(4a_1^2-1)(4a_2^2-1)(a_1+a_3)(a_2+a_3)(a_1a_2+a_1a_3+a_2a_3)^2 \cdot Q(a_1, a_2, a_3) = 0, \quad (20)$$

where

$$\begin{aligned} Q(a_1, a_2, a_3) = {} & (2s_1 + 4s_3 - 1)(64s_1^5 - 64s_1^4 + 8s_1^3 + 12s_1^2 - 6s_1 + 1 \\ & +240s_3s_1^2 - 240s_3s_1 - 1536s_3^2s_1 - 4096s_3^3 + 60s_3 + 768s_3^2) \\ & -8s_1(2s_1 + 4s_3 - 1)(2s_1 - 32s_3 - 1)(10s_1 + 32s_3 - 5)s_2 \qquad (21) \\ & -16s_1^2(13 - 52s_1 + 640s_3s_1 + 1024s_3^2 - 320s_3 + 52s_1^2)s_2^2 \\ & +64(2s_1 - 1)(2s_1 - 32s_3 - 1)s_2^3 + 2048s_1(2s_1 - 1)s_2^4 \end{aligned}$$

and

$$s_1 = a_1 + a_2 + a_3, \quad s_2 = a_1a_2 + a_1a_3 + a_2a_3, \quad s_3 = a_1a_2a_3.$$

Note that for $a_1 = \pm 1/2$, $a_2 = \pm 1/2$, $a_3 = -a_1$, and $a_3 = -a_2$, we have no additional triples (a_1, a_2, a_3) of "degenerate" parameters. Therefore, the first four factors in (20) can be ignored. Another reason to ignore them is the symmetry of the problem under the permutation $a_1 \to a_2 \to a_3 \to a_1$. All these arguments imply

Lemma 4. *If a point (a_1, a_2, a_3) with $a_1a_2 + a_1a_3 + a_2a_3 \neq 0$ and $a_1a_2a_3 \neq 0$ lies in the set Ω', then $Q(a_1, a_2, a_3) = 0$, where Q is defined by* (21).

It is easy to see that $Q(a_1, a_2, a_3)$ is a symmetric polynomial in a_1, a_2, a_3 of degree 12. Therefore, **the equation** $Q(a_1, a_2, a_3) = 0$ (without the restrictions $a_1a_2 + a_1a_3 + a_2a_3 \neq 0$ and $a_1a_2a_3 \neq 0$) **defines an algebraic surface in** $\mathbb{R}^3$ **that we will denote by** Ω. From Lemma 4 we see that $\Omega' \subset \Omega$.

In the rest of this section, we consider only points $(a_1, a_2, a_3) \in (0, 1/2] \times (0, 1/2] \times (0, 1/2]$. It is very important to describe in detail the set

$$\Omega \cap (0, 1/2]^3 = \{(a_1, a_2, a_3) \in (0, 1/2] \times (0, 1/2] \times (0, 1/2] \,:\, Q(a_1, a_2, a_3) = 0\}.$$

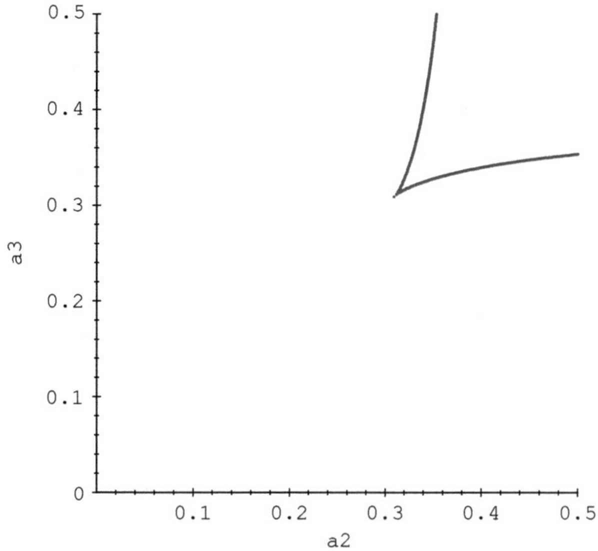

Fig. 2 The set (curve) $\Omega \cap (0, 1/2]^3$ for $a_1 = 1/2$ containing a cusp at $a_2 = a_3$

As usual, the most complicated and interesting problem is the study of this surface in neighborhoods of singular points of Ω determined by $\nabla Q(a_1, a_2, a_3) = 0$.

For $a_1 = 1/2$, the equation $Q = 0$ is equivalent to

$$4\tilde{s}_2(4\tilde{s}_2+1)^2 - 4(4\tilde{s}_2-1)(4\tilde{s}_2+1)^2\tilde{s}_1 - 13(4\tilde{s}_2+1)^2\tilde{s}_1^2 + 4(4\tilde{s}_2-1)\tilde{s}_1^3 + 44\tilde{s}_1^4 = 0,$$

where $\tilde{s}_1 = a_2 + a_3$ and $\tilde{s}_2 = a_2 a_3$. If $a_2, a_3 \in (0, 1/2]$ then this set is a curve homeomorphic to the interval $[0, 1]$ with endpoints $(1/2, 1/2, \sqrt{2}/2)$ and $(1/2, \sqrt{2}/4, 1/2)$ and with the singular point (a cusp) at the point $a_3 = a_2 = (\sqrt{5} - 1)/4 \approx 0.3090169942$ (see Fig. 2). The same is also valid under the permutation $a_1 \to a_2 \to a_3 \to a_1$.

Note that for $s_1 = a_1 + a_2 + a_3 = 1/2$, the equation $Q = 0$ is equivalent to $s_3^2(s_2 - 2s_3)^2 = \frac{1}{2}a_1a_2a_3(-5a_1a_2 + a_1 - 2a_1^2 + a_2 - 2a_2^2 + 6a_1a_2(a_1 + a_2)) = 0$. It is easy to check that for $a_1, a_2 \in [0, 1/2]$, the equality $-5a_1a_2 + a_1 - 2a_1^2 + a_2 - 2a_2^2 + 6a_1a_2(a_1 + a_2) = 0$ holds only when (a_1, a_2) is one of the points $(0, 0)$, $(0, 1/2)$, and $(1/2, 0)$.

Therefore, $s_1 = a_1 + a_2 + a_3 = 1/2$ in the set $\Omega \cap [0, 1/2]^3$ only for points in the boundary of the triangle with vertices $(0, 0, 1/2)$, $(0, 1/2, 0)$, and $(1/2, 0, 0)$. For all other points in $\Omega \cap (0, 1/2]^3$, we have the inequality $s_1 = a_1 + a_2 + a_3 > 1/2$.

It is clear that $(1/4, 1/4, 1/4) \in \Omega$. Note that for $s_1 = a_1 + a_2 + a_3 = 3/4$, the equation $Q = 0$ is equivalent to

$$(24s_2^2 + 8s_2 + 64s_2s_3 - 8s_3 - 128s_3^2 + 1)(32s_2 - 64s_3 - 5)^2 = 0.$$

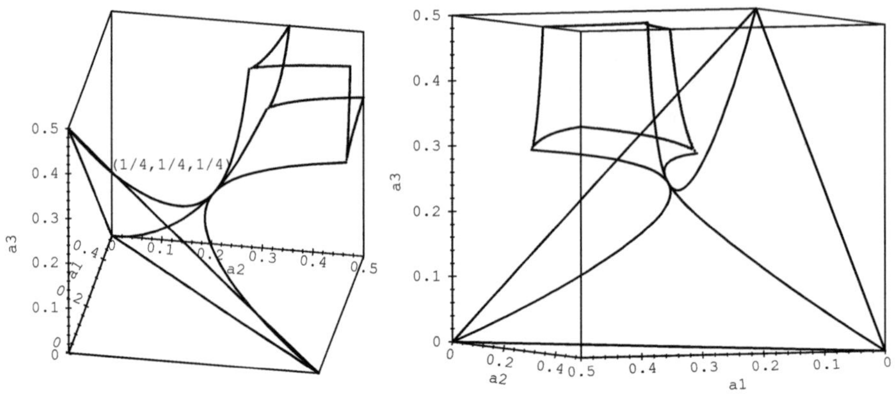

Fig. 3 Three curves of singular points on Ω. The common point $(1/4, 1/4, 1/4)$ is an elliptic umbilic

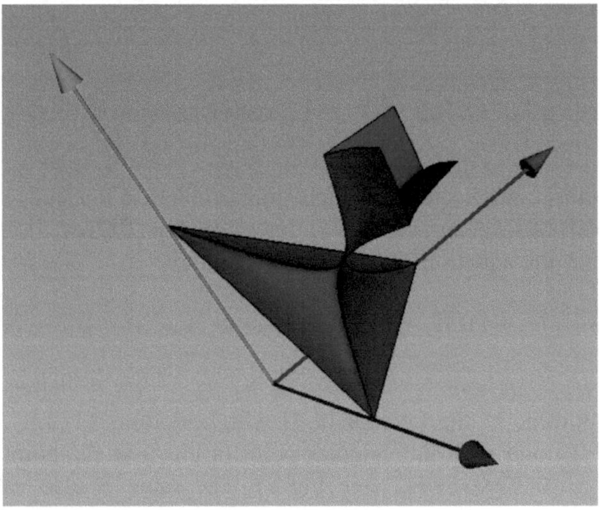

Fig. 4 The set (surface) $\Omega \cap (0, 1/2]^3$

It is not difficult to show that $(1/4, 1/4, 1/4)$ is the only point in $\Omega \cap [0, 1/2]^3$ satisfying the additional condition $s_1 = a_1 + a_2 + a_3 = 3/4$.

It turns out that the point $(1/4, 1/4, 1/4)$ is a singular point of degree 3 of the algebraic surface Ω (see Figs. 3 and 4). The type of this point is **elliptic umbilic** in the sense of Darboux (see [11, pp. 448–464] and [25, p. 320]) or of type D_4^- in other terminology (see, e.g., [4, Chap. III, Sects. 21.3, 22.3]).

6 On the Signs of σ and ρ for Singular Points of System (5)

In this section, we study the singular points of system (5) according to the signs of σ and ρ [see (16)]. Results depend on conditions on the parameters a_1, a_2, a_3.

Lemma 5. *The quadratic form*

$$G(x,y,z) := -2(a_1a_3+a_2a_3)xy - 2(a_1a_2+a_1a_3)yz - 2(a_2a_3+a_1a_2)xz \\ +(\mathcal{A}+a_1^2)x^2 + (\mathcal{A}+a_2^2)y^2 + (\mathcal{A}+a_3^2)z^2,$$

where $\mathcal{A} = a_1a_2+a_1a_3+a_2a_3$, is nonnegative if $\mathcal{A} > 0$ (in particular, if $a_i > 0$ for $i = 1,2,3$) and achieves its absolute minimum (equal to zero) exactly at the points

$$(x,y,z) = \big((a_2+a_3)t, (a_1+a_3)t, (a_1+a_2)t\big), \quad t \in \mathbb{R}.$$

Proof. It is easy to show that the matrix of the form G has nonnegative eigenvalues 0, $2\mathcal{A}$, and $a_1^2+a_2^2+a_3^2+\mathcal{A}$. Obviously, the last two numbers are positive for $\mathcal{A} > 0$. Therefore, G is a nonnegative form. Note that the equation $G = 0$ has the solutions given in the statement of the lemma. □

Theorem 4. *For $a_1a_2+a_1a_3+a_2a_3 > 0$, all singular points of system (5) are such that $\sigma \equiv \rho^2 - 4\delta \geq 0$. In particular, a nondegenerate singular point (i.e. $\delta \neq 0$) of (5) is either a node (if $\delta > 0$) or a saddle (if $\delta < 0$).*

Proof. Let $\mathcal{A} = a_1a_2+a_1a_3+a_2a_3 > 0$. By Lemmas 1 and 2, we get

$$\begin{aligned}\sigma &= \rho^2 - 4\delta = \tilde{\rho}^2 - 4\tilde{\delta} \\ &= \big(-2(a_1a_3+a_2a_3)x_1^2x_2^2 - 2(a_1a_2+a_1a_3)x_2^2x_3^2 - 2(a_2a_3+a_1a_2)x_1^2x_3^2 \\ &\quad +(\mathcal{A}+a_1^2)x_1^4 + (\mathcal{A}+a_2^2)x_2^4 + (\mathcal{A}+a_3^2)x_3^4\big)x_1^{-2}x_2^{-2}x_3^{-2}.\end{aligned}$$

Using Lemma 5 for $x = x_1^2$, $y = x_2^2$, and $z = x_3^2$, we get that $\sigma \geq 0$ for all $x_1, x_2, x_3 > 0$, $x_3 = \varphi(x_1,x_2)$. So, in particular, $\sigma \geq 0$ holds for all singular points of system (5), and by Theorem 1, its nondegenerate singular points can be only either a node or a saddle. □

Remark 2. From Theorem 4 and Lemma 5, we get the following. If (x_1,x_2) is a singular point of system (5) with $\sigma = 0$ then

$$(x_1,x_2,x_3) = \big(q\sqrt{a_2+a_3}, q\sqrt{a_1+a_3}, q\sqrt{a_1+a_2}\big) \tag{22}$$

for a unique $q \in \mathbb{R}$, $q > 0$, determined by the equality $x_3 = \varphi(x_1,x_2)$.

Next, we are interested in the values $(a_1,a_2,a_3) \in (0,1/2]\times(0,1/2]\times(0,1/2]$ for which system (5) has at least one singular point with $\sigma = 0$. By direct calculation, we have got

Theorem 5 ([1]). *The only two families of the parameters a_i, $i = 1,2,3$, satisfying the conditions $a_i \in (0, 1/2]$ and which can give singular points of system (5) with the property $\sigma = 0$, are the following:*

$$a_1 = a_2 = a_3 = s, \quad s \in (0, 1/2], \tag{23}$$

$$a_i = a_j = \frac{(2s^2-1)^2}{8s^2}, \quad a_k = \frac{4s^4+4s^2-1}{8s^2}, \quad s \in (s_1, s_2), \tag{24}$$

where $s_1 := \sqrt{2\sqrt{2}-2}/2$, $s_2 := \sqrt{2}/2$ $(i, j, k \in \{1,2,3\}, i \neq j \neq k \neq i)$.

Remark 3. Now we can find all singular points of system (5) corresponding to the families (23), (24) and having $\sigma = 0$.

According to Remark 2, family (23) provides a unique singular point $(x_1, x_2) = (1, 1)$ of (5) satisfying $\sigma = 0$ for all $s \in (0, 1/2]$.

Analogously, by Remark 2, it follows that family (24) provides only the following singular points of (5) satisfying $\sigma = 0$ for all $s \in (s_1, s_2)$:

$$(2s^2q, 2s^2q), \quad \left(2s^2q, (1-2s^2)q\right), \quad \left((1-2s^2)q, 2s^2q\right),$$

where $q = (2s^2)^{\frac{-2(4s^4+4s^2-1)}{(6s^2-1)(2s^2+1)}}(1-2s^2)^{\frac{-(2s^2-1)^2}{(6s^2-1)(2s^2+1)}} > 0$ is determined by the condition $V \equiv 1$ [see (4)].

Remark 4. Using Theorem 5 and Lemma 4, we can detect all values of $a_i \in (0, 1/2]$, $i = 1,2,3$, such that system (5) has at least one degenerate singular point with $\sigma = 0$. According to Remark 3 for all $s \in (0, 1/2]$, family (23) provides a unique singular point $(x_1, x_2) = (1, 1)$ of (5) with $\sigma = 0$. In this case, Q [see (21)] takes the form

$$Q = -(2s+1)^4(4s-1)^8,$$

and the equation $Q = 0$ implies that $s = 1/4$. Then according to Lemma 4, the point $(1, 1)$ is a degenerate singular point ($\delta = 0$) only for $s = 1/4$ (and its type has been determined in Theorem 3). For $s \in (0, 1/4) \cup (1/4, 1/2]$ the point $(1, 1)$ is a node.

Analogously, for family (24), we have that

$$Q = s^8(1-8s^2-4s^4)(1-2s^2)^3(3-2s^2)^3,$$

and the equation $Q = 0$ has only three positive roots $\sqrt{2\sqrt{5}-4}/2$, $\sqrt{2}/2$, and $\sqrt{6}/2$, but none of these values belong to the interval (s_1, s_2). Therefore, (24) cannot give singular points of (5) with $\sigma = \delta = 0$.

Next, we denote by S the set of points (a_1, a_2, a_3) such that there is a singular point (x_1^0, x_2^0) of system (5) with $\rho = 0$. Recall that for points with $a_1a_2 + a_1a_3 + a_2a_3 = 0$, system (5) is undefined.

Theorem 6 ([1]). *A point (a_1, a_2, a_3) with $a_1a_2+a_1a_3+a_2a_3 \neq 0$ and $a_1a_2a_3 \neq 0$ lies on the surface S if and only if*

$$Q_1(a_1, a_2, a_3) := 4(a_1+a_2)(a_1+a_3)(a_2+a_3)-2a_1-2a_2-2a_3+1 = 0. \qquad (25)$$

The proof of this theorem is straightforward. Indeed, equations (7) and (18) are homogeneous with respect to x_1, x_2, x_3. Now setting $x_3 = 1$ and eliminating x_1 and x_2 from the above three equations (using, e.g., *Maple* or *Mathematica*), we get the equation

$$(a_1+a_3)(a_2+a_3)(a_1a_2+a_1a_3+a_2a_3) \cdot \big(4(a_1+a_2)(a_1+a_3)(a_2+a_3) \\ -2(a_1+a_2+a_3)+1\big)=0.$$

The proof finishes by noting that for $a_3 = -a_1$ and $a_3 = -a_2$, we have no additional sets of the parameters (a_1, a_2, a_3).

Remark 5. It should also be noted that the point $(a_1, a_2, a_3) = (1/4, 1/4, 1/4)$ is the unique singular point of the surface S ($\nabla Q_1(1/4, 1/4, 1/4) = 0$). This is clear by reducing (25) to the simpler equation $4z_1z_2z_3 - z_1 - z_2 - z_3 + 1 = 0$ using the substitutions $z_1 = a_1 + a_2, z_2 = a_1 + a_3, z_3 = a_2 + a_3$. It is easy to see that S divides the cube $[0, 1/2]^3$ into three domains $\tilde{O}_1$, $\tilde{O}_2$, and $\tilde{O}_3$ containing the points $(0, 0, 0)$, $(1/2, 1/2, 1/2)$, and $(1/8, 1/4, 3/8)$, respectively.

Remark 6. It is easy to show that system (5) has a singular point with $\rho = \delta = 0$ if and only if $(a_1, a_2, a_3) = (1/4, 1/4, 1/4)$ (in this case system (5) has exactly one degenerate singular point $(x_1, x_2) = (1, 1)$, see Sect. 4). Indeed, from the equations $\rho = 0$ and $\delta = 0$, we have $\sigma = \rho^2 - 4\delta = 0$. According to Remark 4, the system of equations $\delta = 0$, $\sigma = 0$ has a unique solution $(a_1, a_2, a_3) = (1/4, 1/4, 1/4)$. It is easy to check that this solution satisfies (25) as well. Therefore, Theorem 6 implies that $\rho = 0$.

Remark 7. There are no values of the parameters $a_i \in (0, 1/2]$, $i = 1, 2, 3$, for which the singular points of (5) are nilpotent ($\lambda_1 = \lambda_2 = 0$, $J \neq 0$). In fact, assume that $\lambda_1 = \lambda_2 = 0$ and $J \neq 0$. Then $\delta = \lambda_1\lambda_2 = 0$ and $\rho = \lambda_1 + \lambda_2 = 0$. By Remark 6, the equalities $\rho = \delta = 0$ are possible only at $a_1 = a_2 = a_3 = 1/4$, hence we have the case of Theorem 3 again (i.e., the singular point is $(1, 1)$ with $J = 0$).

7 Singular Points for Parameters in the Set $(0, 1/2)^3 \setminus \Omega$

We first discuss a part of the surface Ω (see Sect. 5) in the cube $(0, 1/2)^3$. Recall that Ω is invariant under the permutation $a_1 \to a_2 \to a_3 \to a_1$. It should be noted that the set $(0, 1/2)^3 \cap \Omega$ is connected (it can be shown by lengthy computations using suitable geometric tools). There are three curves ("edges") of *singular points* on Ω (i.e., points where $\nabla Q = 0$): one of them has parametric representation $a_1 = -\frac{1}{2}\frac{16t^3-4t+1}{8t^2-1}, a_2 = a_3 = t$, and the others are defined by permutations of a_i. These curves have a common point $(1/4, 1/4, 1/4)$ which is an elliptic umbilic on the surface Ω (see Figs. 3 and 4). The part of Ω in $(0, 1/2)^3$ consists of three (pairwise isometric) "bubbles" spanned on every pair of "edges" (cf. pictures of elliptic umbilics at [28, pp. 64–91]). The Gaussian curvature at every nonsingular point of the surface $\Omega \cap (0, 1/2)^3$ is negative, as it could be checked by direct calculations.

From the above discussion and some geometric considerations (that could be rigorous but very lengthy), we see that the set $(0, 1/2)^3 \setminus \Omega$ has exactly three connected components. Denote by O_1, O_2, and O_3 the components containing the points $(1/6, 1/6, 1/6)$, $(7/15, 7/15, 7/15)$, and $(1/6, 1/4, 1/3)$, respectively.

Let us fix $j \in \{1, 2, 3\}$. By the definition of Ω, for all points $(a_1, a_2, a_3) \in O_j$, system (5) has only nondegenerate singular points. The number of these points and their corresponding types are the same on each component O_j (under some suitable identification for various values of the parameters a_1, a_2, a_3). Therefore, it suffices to check *only one point in the set O_j*.

One of the main results of this chapter is the following theorem which clarifies the above observation and provides a general result about the type of the nondegenerate singular points of system (5).

Theorem 7. *For $(a_1, a_2, a_3) \in O_j$, the following possibilities for singular points of system (5) can occur:*

(i) *If $j = 1$, then there is one singular point with $\delta > 0$ (an unstable node) and three singular points with $\delta < 0$ (saddles)*
(ii) *If $j = 2$, then there is one singular point with $\delta > 0$ (a stable node) and three singular points with $\delta < 0$ (saddles)*
(iii) *If $j = 3$, then there are two singular points with $\delta < 0$ (saddles).*

Proof. By Theorem 4, a nondegenerate singular point is either a node (if $\delta > 0$) or a saddle (if $\delta < 0$).

Recall that for $(a_1, a_2, a_3) \in O_j$, all singular points of system (5) are not degenerate. Moreover, there are no singular points (x_1, x_2, x_3) with some zero component. Therefore, **the number of singular points** and **the set of signs** of $\delta = \tilde{\delta}$ for these points are constant on each component O_j. This is easy to be checked in a small neighborhood of any point $(a_1, a_2, a_3) \in O_j$ (it follows from *the stability of nondegenerate singular points*), and then the proof can be spread to all points of the connected set O_j via continuous paths (as in standard analytical monodromy theorems).

Consider the component O_1 containing the representative point $(1/6, 1/6, 1/6)$. Then as the calculations show, under $x_3 = 1$, system (7) in the variables (x_1, x_2) has four solutions, given by $(1, 1)$, $(2, 1)$, $(1/2, 1/2)$, and $(1, 2)$. By Lemma 2, it follows that the point $(x_1, x_2) = (1, 1)$ corresponds to the value $\delta = 1/9$ (an unstable node with $\rho = 2/3$ and $\sigma = 0$). If (x_1, x_2) is one of the solutions $(2, 1)$, $(1/2, 1/2)$, and $(1, 2)$, then δ equals to $-2/9$, $-8/9$, and $-2/9$, respectively (so these points are saddles).

Consider now the component O_2 containing the representative point $(7/15, 7/15, 7/15)$. By the same manner using Lemmas 1 and 2, we get the following four solutions (x_1, x_2) of system (7): $(1, 1)$ with $\delta = 169/25$ (a stable node with $\rho = -26/25$ and $\sigma = 0$) and $(1/14, 1)$, $(1, 1/14)$, $(14, 14)$ with δ equal to $-4901/225$, $-4901/225$, and $-4901/44100$, respectively (three saddles).

Finally, consider the component O_3 containing the point $(1/6, 1/4, 1/3)$. In this case, we get the following two solutions of system (7): $(x_1, x_2) = (4/5, 3/5)$ with $\delta = -35/72$ (a saddle) and $(x_1, x_2) \approx (2.284185494, 2.372799295)$ with $\delta = -0.0982$ (a saddle). □

Remark 8. The following is a natural (and practical) question: Let (a_1^0, a_2^0, a_3^0) be any triple in $(0, 1/2)^3 \setminus \Omega$. *Is there a way to decide on which connected component O_1, O_2, or O_3 does this triple belong to?* The answer is affirmative.

Indeed, consider first the simplest case where $a_1^0 = a_2^0 = a_3^0 =: a^0$. Then obviously $(a_1^0, a_2^0, a_3^0) \in O_1$ for $a^0 < 1/4$ and $(a_1^0, a_2^0, a_3^0) \in O_2$ for $a^0 > 1/4$ (recall that $(1/4, /1/4, 1/4)$ is a very special point of Ω).

Assume now that $a_1^0 : a_2^0 : a_3^0 \neq 1 : 1 : 1$. Then we find (solving approximately an algebraic equation of degree at most 12 with respect to t) the intersection of Ω with the interval I containing points of the form $(a_1, a_2, a_3) = (a_1^0 t, a_2^0 t, a_3^0 t)$, where $0 < t < 1$. This means that we give numerical values to (a_1^0, a_2^0, a_3^0) and then solve the corresponding equation with respect to t (it could be done by Maple® or by Mathematica®).

From simple geometric arguments we have the following: If the number of intersection points is 0, 1, and 2, then (a_1^0, a_2^0, a_3^0) belongs to O_1, O_3, and O_2, respectively. For instance, if all solutions of the corresponding equation are complex, then the number of intersection points is 0 and $(a_1^0, a_2^0, a_3^0) \in O_1$.

Now, we consider an important partial case: $a_1 + a_2 + a_3 = 1/2$, $a_i \in (0, 1/2)$. It is easy to see that all these points are in the component O_1 from Theorem 7 (see the discussion in Sect. 5). On the other hand, we know the explicit form of all four singular points of system (5) for this case, see (12). Hence, we can get a refinement of Theorem 7 for this special case.

It is clear that for fixed $a_i \in (0, 1/2)$, $i = 1, 2, 3$, every family in (12) gives a singular point (x_1^0, x_2^0) of system (5), with q uniquely determined by Remark 1.

Theorem 8. *For $a_1 + a_2 + a_3 = 1/2$, $a_i \in (0, 1/2)$, all singular points of system (5) are nondegenerate. Moreover, the points from the first family in (12) are unstable nodes, and the points from the other families in (12) are saddles.*

Proof. At first, we prove that (12) cannot provide degenerate singular points of (5). In fact, at $s_1 = 1/2$ the function $Q(a_1, a_2, a_3)$ takes the form (see Lemma 4) $Q = -4096 s_3^2(2s_3 - s_2)$. It is easy to check that the equalities $s_1 = 1/2$ and $Q = 0$ are fulfilled if and only if

$$a_i = t, \quad a_j = -a_i, \quad a_k = 1/2, \quad t \in \mathbb{R},$$

where $i, j, k \in \{1, 2, 3\}$, $i \neq j \neq k \neq i$. Therefore, according to Lemma 4, every family in (12) can provide only nondegenerate ($\delta \neq 0$) singular points of (5). Moreover, by Theorem 4 such nondegenerate points should be either nodes or saddles. Using Lemmas 1 and 2, we get $\delta > 0$ exactly for the first family. Therefore, according to Theorem 7 only the first family in (12) can provide nodes of (5), and the other families give saddles. Since $\rho = 1/q > 0$ then all nodes are unstable. □

Remark 9. At $a_1 = a_2 = a_3 = 1/6$ and $q = 3/2$ from the first family in (12), we get a singular point $(x_1^0, x_2^0) = (1, 1)$ with the property $\sigma = 0$. Note that the general result about singular points with $\sigma = 0$, obtained in Remark 3, implies that (5) has a unique singular point $(1, 1)$ with $\sigma = 0$ for all $a_1 = a_2 = a_3 = s$, $s \in (0, 1/2]$.

8 Tools for Semi-Hyperbolic Cases

In Sect. 9, we will study semi-hyperbolic singular points (x_1^0, x_2^0) of system (5) where the matrix $J(x_1^0, x_2^0)$ has one of the following special forms

$$k \begin{pmatrix} 1 & 1 \\ 1 & 1 \end{pmatrix}, \quad k \neq 0, \tag{26}$$

or

$$k \begin{pmatrix} 1 & -1 \\ -1 & 1 \end{pmatrix}, \quad k \neq 0. \tag{27}$$

Lemma 6. *Let $J(x_1^0, x_2^0)$ has form* (26) *with $k > 0$. Then the following statements hold.*

(i) *There is a nondegenerate linear transformation of variables $(x_1, x_2) \mapsto (x, y)$ reducing (5) to the canonical form*

$$\frac{dx}{dt} = X(x, y), \quad \frac{dy}{dt} = 2ky + Y(x, y), \quad k > 0, \tag{28}$$

with

$$X(x, y) \equiv F\left(x_1(x, y), x_2(x, y)\right) - G\left(x_1(x, y), x_2(x, y)\right),$$

$$Y(x, y) \equiv F\left(x_1(x, y), x_2(x, y)\right) + G\left(x_1(x, y), x_2(x, y)\right).$$

(ii) *The equation* $2ky + Y(x,y) = 0$ *has a unique solution* $y = \phi(x)$*, and there is* $e_m \neq 0$ *with* $m \geq 2$ *in the power series* $X(x,\phi(x)) = \sum_{n=2}^{\infty} e_n x^n$*.*
(iii) *All possible types of singularities of the point* (x_1^0, x_2^0) *are exactly the types of singularities given in items (i)–(iii) of Theorem 2.*

Proof. (i) It is easy to check that the linear transformation

$$x_1 = x_1(x,y) := (x+y)/2 + x_1^0, \quad x_2 = x_2(x,y) := (y-x)/2 + x_2^0$$

moves (x_1^0, x_2^0) to the origin $(0,0)$ and reduces (5) to the required canonical form. Since F and G are analytic in a neighborhood of (x_1^0, x_2^0), then the functions X and Y satisfy the conditions of Theorem 2.

(ii) Let $\bar{Y}(x,y) := 2ky + Y(x,y)$. Since $\bar{Y}_y(0,0) = 2k \neq 0$, then by the implicit function theorem in a sufficiently small neighborhood of $(0,0)$, the equation $2ky + Y = 0$ has a unique analytic solution $y = \phi(x)$, $\phi(0) = 0$, $\phi'(0) = 0$. Since Y is represented by the Taylor series analogous to (17) [put $B := Y$ in (17)], then $y = \phi(x)$ is represented by the power series $y = \phi(x) = \sum_{n=2}^{\infty} \nu_n x^n$, where

$$\nu_2 = -q_{2,0}/(2k),$$
$$\nu_3 = -(q_{1,1}\nu_2 + q_{3,0})/(2k),$$
$$\nu_4 = -(q_{1,1}\nu_3 + q_{0,2}\nu_2^2 + q_{2,1}\nu_2 + q_{4,0})/(2k),$$
$$\nu_5 = -(q_{1,1}\nu_4 + 2q_{0,2}\nu_2\nu_3 + q_{2,1}\nu_3 + q_{1,2}\nu_2^2 + q_{3,1}\nu_2 + q_{5,0})/(2k), \ \ldots$$

Since X can be represented by the Taylor series analogous to (17) (put $A := X$ in (17)), then the function $\psi(x) := X(x,\phi(x))$ is also analytic in a neighborhood of 0 and can be represented by the power series $\psi(x) = \sum_{n=2}^{\infty} e_n x^n$, $\psi(0) = 0$, $\psi'(0) = 0$, where

$$e_2 = p_{2,0},$$
$$e_3 = p_{1,1}\nu_2 + p_{3,0},$$
$$e_4 = p_{1,1}\nu_3 + p_{0,2}\nu_2^2 + p_{2,1}\nu_2 + p_{4,0},$$
$$e_5 = p_{1,1}\nu_4 + 2p_{0,2}\nu_2\nu_3 + p_{2,1}\nu_3 + p_{1,2}\nu_2^2 + p_{3,1}\nu_2 + p_{5,0}, \ \ldots$$

There exists a first nonzero term $e_m \neq 0$ in the sequence $\{e_n\}, n \geq 2$. Otherwise (i.e., $\psi(x) \equiv 0$), we have the family of non-isolated singular points of (28) along the line $y = \phi(x)$ in spite of our conditions.

(iii) Since $k > 0$, then we can apply Theorem 2 to system (28). Depending on the values of m and $e_m \neq 0$, the point $(0,0)$ takes one of the types of singularities given in Theorem 2. Now, we return to system (5) in variables x_1, x_2. Then $(0,0)$ corresponds to the singular point (x_1^0, x_2^0) of (5). □

Remark 10. If $J(x_1^0, x_2^0)$ has form (27) with $k > 0$, then system (5) can be reduced to the canonical form (28) with

$$X(x,y) \equiv F\left(x_1(x,y), x_2(x,y)\right) + G\left(x_1(x,y), x_2(x,y)\right),$$

$$Y(x,y) \equiv F\left(x_1(x,y), x_2(x,y)\right) - G\left(x_1(x,y), x_2(x,y)\right).$$

The corresponding transformation is

$$x_1 = x_1(x,y) := (x+y)/2 + x_1^0, \quad x_2 = x_2(x,y) := (x-y)/2 + x_2^0.$$

Remark 11. For $k < 0$, we will assume (without loss of generality) that

$$e_2 = -p_{2,0}, \quad e_3 = -p_{1,1}\nu_2 - p_{3,0}, \quad \dots.$$

for both cases (26) and (27). Indeed, we may use the transformation $t \mapsto -t$ reducing systems of form (28) with $k < 0$ to the form $x'(t) = -X(x,y)$, $y'(t) = -2ky - Y(x,y)$, with $-k > 0$.

Remark 12. To determine the first nonzero coefficient e_m, $m \geq 2$, it is sufficient to know only $p_{i,j}$ and $q_{i,j}$, where $2 \leq i + j \leq m$. Therefore, we do not need the terms of order $i + j > m$ in calculations using the power series of X and Y. Note also that we do not need the value of k if $p_{2,0} \neq 0$.

We will use all the above results in the next section in order to study singular points of (3) for an interesting special case.

9 Singular Points in the Case $a_1 = a_2$

In this section, we are interested in the case of coincident a_i and a_j with different indexes. By permuting the indexes we may assume without loss of generality that $i = 1$ and $j = 2$. Recall that the singular points of (3) corresponding to the case $a_1 = a_2 = b$, $a_3 = c$ were described in Sect. 2. There are two types of such singular points (x_1^0, x_2^0, x_3^0): those with the property $x_1^0 = x_2^0$ and those with the property $x_3^0 = 2b(x_1^0 + x_2^0)$.

The explicit form of *singular points of the type* $x_1^0 = x_2^0$ is given by (10).

Now, consider more closely *the singular points of the type* $x_3^0 = 2b(x_1^0 + x_2^0)$. For $b \neq 1/2$, we can easily find the following singular points of (3) as solutions of (11):

$$(x_1^0, x_2^0, x_3^0) := (\gamma_1 q, \gamma_2 q, \gamma_3 q), \quad q \in \mathbb{R}, q > 0, \tag{29}$$

Table 1 Special values for b and c

Values of b	Values of c_1	Values of c_2	Values of c_3	Values of c_4	Relations
$\in (0, b_1)$	–	–	$\in (0, 1/2)$	$> 1/2$	$c_3 < c_4$
b_1	–	–	$\in (0, 1/2)$	$1/2$	$c_3 < c_4$
$\in (b_1, b_2)$	–	–	$\in (0, 1/2)$	$\in (0, 1/2)$	$c_3 < c_4$
b_2	$\in (0, 1/2)$	$\in (0, 1/2)$	$\in (0, 1/2)$	$\in (0, 1/2)$	$c_1 = c_2 < c_3 < c_4$
$\in (b_2, 1/4)$	$\in (0, 1/2)$	$\in (0, 1/2)$	$\in (0, 1/2)$	$\in (0, 1/2)$	$c_1 < c_2 < c_3 < c_4$
$1/4$	0	$1/4$	$1/4$	$1/4$	$c_1 < c_2 = c_3 = c_4$
$\in (1/4, b_3)$	< 0	$\in (0, 1/2)$	$\in (0, 1/2)$	$\in (0, 1/2)$	$c_4 < c_2 < c_3$
b_3	< 0	$\in (0, 1/2)$	$1/2$	$\in (0, 1/2)$	$c_4 < c_2 < c_3$
$\in (b_3, b_4)$	< 0	$\in (0, 1/2)$	$> 1/2$	$\in (0, 1/2)$	$c_4 < c_2 < c_3$
b_4	< 0	$\in (0, 1/2)$	–	0	$c_4 < c_2$
$\in (b_4, 1/2]$	< 0	$\in (0, 1/2)$	< 0	< 0	

where

$$\gamma_1 := 2(b+c)(1-4b^2),\ \gamma_2 := -2b + 8b^3 + 8cb^2 + 1 \pm \sqrt{\Delta},$$
$$\gamma_3 := 2b\big(2c + 1 \pm \sqrt{\Delta}\big),$$
$$\Delta := (2c+1)\big(1 - 4b - 2c + 16b^2(b+c)\big).$$

It is easy to check that $\gamma_1, \gamma_2, \gamma_3 > 0$ for $\Delta \geq 0$. We will need the following special values for b and c:

$$b_1 = (\sqrt{3}-1)/4, \quad b_2 = (\sqrt{2}-1)/2,\ b_3 = (\sqrt{5}-1)/4, \quad b_4 = \sqrt{2}/4,$$
$$c_1 := \big(1 - 2b - \sqrt{4b^2 + 4b - 1}\big)/4,\ c_2 := \big(1 - 2b + \sqrt{4b^2 + 4b - 1}\big)/4,$$
$$c_3 := \big(16b^3 - 4b + 1\big)/\big(2 - 16b^2\big),\ c_4 := \big(1 - 8b^2\big)/(8b).$$

Remark 13. In Table 1, we write down all values of b such that every c_i, $i = 1, \dots, 4$, is well defined and satisfies the condition $c_i \in (0, 1/2]$. Moreover, we show some relations among c_i, $i = 1, \dots, 4$. It should also be noted that $\lim\limits_{b \to b_4 - 0} c_3 = +\infty$ and $\lim\limits_{b \to b_4 + 0} c_3 = -\infty$.

Further, we determine values of $a_1 = a_2 = b$ and $a_3 = c$ with $(a_1, a_2, a_3) \in \Omega$.

Lemma 7. *Given $b, c > 0$, there exists a degenerate point of system* (5) *if and only if one of the following three conditions hold:*

(i) $c = c_1,\ \ b \in [b_2, 1/4)$;
(ii) $c = c_2,\ \ b \in [b_2, 1/2]$;
(iii) $c = c_3,\ \ b \in (0, b_3]$.

Proof. This follows from Lemma 4. In the case $a_1 = a_2 := b$, $a_3 := c$, the function $Q(a_1, a_2, a_3)$ takes the form

$$Q = (c + 1/2)(c - c_1)(c - c_2)(c - c_3)^3.$$

Using Remark 13, it is easy to determine all values of b with $c_i \in (0, 1/2]$, $i = 1, 2, 3$. □

Remark 14. It is easy to see that all points of curves (i) and (ii) in Lemma 7 (except the point $(1/4, 1/4, 1/4)$) are regular points of the surface Ω. On the other hand, curve (iii) is a part of an "edge" of Ω in $(0, 1/2]^3$ (see also Figs. 3 and 4).

9.1 Classification of Singular Points of the Type $x_3^0 = 2b(x_1^0 + x_2^0)$

At first, we consider

Lemma 8. *Let b, c be such that $\Delta \geq 0$. Then, depending on the choice of $\pm\sqrt{\Delta}$, formula* (29) *gives singular points of* (5) *of the form*

$$(x_1^0, x_2^0) = (\gamma_1 q, \gamma_2 q), \tag{30}$$

where q is a unique positive real number satisfying (4). Moreover, the following formula is valid for $J(x_1^0, x_2^0)$:

$$\delta = \frac{-\Delta}{2b(1 - 4b^2)(b + c)\left(2c + 1 \pm \sqrt{\Delta}\right)^2 q^2}.$$

Proof. Use Lemmas 1 and 2 where in (19) we substitute $a_1 = a_2 = b$, $a_3 = c$ and (x_1^0, x_2^0, x_3^0) as in (29). The existence and the uniqueness of a suitable q follows from Remark 1. □

Using Lemmas 7 and 8, we can find all possible *degenerate* singular points (30) of system (5).

Lemma 9. *For singular points* (30) *of system* (5)*, the following assertions hold:*

(i) *There are no degenerate singular points for $b \in [b_2, 1/4)$, $c = c_1$ and for $b \in [b_2, 1/2] \setminus \{1/4\}$, $c = c_2$*
(ii) *All singular points are degenerate (semi-hyperbolic) saddles for $b \in (0, b_3] \setminus \{1/4\}$, $c = c_3$.*

Proof. For singular points (30), we get $\Delta = 2(2c + 1)(8b^2 - 1)(c - c_3)$.

(i) Recall that $c_2 = c_3$ only for $b = 1/4$ by Remark 13; hence, $\Delta \neq 0$ for $c = c_2$, $b \in [b_2, 1/2] \setminus \{1/4\}$. By Remark 13, there are no values of b such that $c_1 = c_3$. Then, by the same reason as above, $\Delta \neq 0$ for $c = c_1$, $b \in [b_2, 1/4)$. Therefore, $\delta \neq 0$ by Lemma 8.

Table 2 Types of singularities in the case $\Delta \geq 0$

Values of b	Values of c with hyperbolic saddles $\Delta > 0$	Values of c with semi-hyperbolic saddles $\Delta = 0$	Values of c giving linearly zero saddles $\Delta = 0$
$b \in (0, 1/4)$	$c \in (0, c_3)$	$c = c_3$	–
$b = 1/4$	$c \in (0, 1/4)$	–	$c = c_3 = 1/4$
$b \in (1/4, b_3)$	$c \in (0, c_3)$	$c = c_3$	–
$b = b_3$	$c \in (0, c_3)$	$c = c_3 = 1/2$	–
$b \in (b_3, 1/2)$	$c \in (0, 1/2]$	–	–

(ii) In this case, $\Delta = 0$ for singular points of (5) given by (30). Then, $\delta = 0$ by Lemma 8. As the calculations show the matrix of linear parts $J = J(x_1^0, x_2^0)$ has the form $J = k_0\begin{pmatrix} 1 & 1 \\ 1 & 1 \end{pmatrix}$, where

$$k_0 := -\frac{1}{2}\frac{4b-1}{(2b+1)(2b-1)^2 q}.$$

By Lemma 6, we can reduce (5) to the canonical form (28). Note that the expressions for corresponding X and Y in this case are too big to be included here. Taking into account Remark 12, we have

$$p_{2,0} = 0, \quad p_{1,1}\nu_2 + p_{3,0} = -\frac{1}{8}\frac{(8b^2-1)^3}{b(4b-1)(2b-1)^4(2b+1)^2 q^3}.$$

The case $b \in (0, 1/4)$. Note that $p_{1,1}\nu_2 + p_{3,0} < 0$ in this case. Since $k_0 > 0$, then $e_3 = p_{1,1}\nu_2 + p_{3,0} < 0$, and by Lemma 6 we have saddles.

The case $b \in (1/4, b_3]$. In this case, $p_{1,1}\nu_2 + p_{3,0} > 0$. Since $k_0 < 0$, then system (5) satisfies the conditions of Remark 11; so, $e_3 = -p_{1,1}\nu_2 - p_{3,0} < 0$, and we have saddles in this case too.

Note that in the above formulas, we assumed that the corresponding exact values of q are the same as in Remark 1. □

Theorem 9. *Let* $\Delta \geq 0$. *Then singular point* (30) *of system* (5) *has one of the types of singularities depicted in Table 2.*

Proof. Note that $\Delta > 0$ for $b \in (0, b_3]$, $c \in (0, c_3)$ and for $b \in (b_3, 1/2)$, $c \in (0, 1/2]$. Lemma 8 implies $\delta < 0$ for $\Delta > 0$. Now, it suffices to take into account Lemma 9 and Theorem 3. □

Table 3 The sign of the value D

Values of b	Values of c for $D < 0$	Values of c for $D = 0$	Values of c for $D > 0$
$b \in (0, b_2)$	–	–	$c \in (0, 1/2]$
$b = b_2$	–	$c = c_1 = c_2$	$c \neq c_1 = c_2$
$b \in (b_2, 1/4)$	$c \in (c_1, c_2)$,	$c = c_1, c = c_2$	$c \in (0, c_1) \cup (c_2, 1/2]$
$b \in [1/4, 1/2]$	$c \in (0, c_2)$	$c = c_2$	$c \in (c_2, 1/2]$

Table 4 The sign of the value $8b(b+c) - 1$

Values of b	Values of c for $8b(b+c) < 1$	Values of c for $8b(b+c) = 1$	Values of c for $8b(b+c) > 1$
$b \in (0, b_1)$	$c \in (0, 1/2]$	–	–
$b = b_1$	$c \in (0, 1/2)$	$c = c_4 = 1/2$	–
$b \in (b_1, b_4)$	$c \in (0, c_4)$	$c = c_4$	$c \in (c_4, 1/2]$
$b \in [b_4, 1/2]$	–	–	$c \in (0, 1/2]$

9.2 Classification of Singular Points of the Type $x_1^0 = x_2^0$

We need the following

Lemma 10. *Let b, c be such that either $D := 1 - 4(1-2c)(b+c) \geq 0$ for $\mu = 1 + \sqrt{D}$, or $0 \leq D < 1$ for $\mu = 1 - \sqrt{D}$. Then, depending on the choice of $\mu = 1 \pm \sqrt{D}$, formula* (10) *gives singular points of* (5) *of the form*

$$(x_1^0, x_2^0) = (2(b+c)q, 2(b+c)q), \tag{31}$$

where q is a unique positive real number satisfying (4). Moreover, the following formulas are valid for $J(x_1^0, x_2^0)$:

$$\rho = \frac{2(1-2(b+c))}{\mu q}, \quad \delta = \frac{D \pm \sqrt{D}}{4(b+c)^2\mu^2 q^2}\big(8b(b+c) - \mu\big). \tag{32}$$

Proof. This follows from Lemmas 1 and 2 where in (19) we substitute $a_1 = a_2 = b$, $a_3 = c$, and (x_1^0, x_2^0, x_3^0) as in (10). Remark 1 guarantees the existence and uniqueness of a suitable q. □

Formulas (32) show that we need to analyze the signs of D, $8b(b+c) - 1$, and $D_1 := D - \big(8b(b+c) - 1\big)^2$.

Lemma 11. *The signs of the values D, $8b(b+c) - 1$, and D_1 are analyzed in Tables 3, 4, and 5 respectively, where the sign "–" means that such a corresponding combination does not occur.*

Proof. The sign of D. D can be represented as a quadratic polynomial $D = 8c^2 + 4(2b-1)c + 1 - 4b$ with respect to c, having discriminant $4b^2 + 4b - 1$.

Table 5 The sign of the value D_1

Values of b	Values of c for $D_1 < 0$	Values of c for $D_1 = 0$	Values of c for $D_1 > 0$
$b \in (0, b_3)$	$c \in (0, c_3)$	$c = c_3$	$c \in (c_3, 1/2]$
$b = b_3$	$c \in (0, 1/2)$	$c = c_3 = 1/2$	–
$b \in (b_3, 1/2]$	$c \in (0, 1/2]$	–	–

For $b \in (0, b_2)$, this discriminant is negative, and we have $D > 0$ for all $c \in (0, 1/2]$. For $b = b_2$ we have $D \geq 0$, where $D = 0$ only for $c = (2 - \sqrt{2})/4$. It is clear that for all $b \in (b_2, 1/2]$, the equation $D = 0$ has real roots c_1, c_2. Recall that the signs of c_1 and c_2 were studied in Remark 13; hence, Table 3 is confirmed.

The signs of $8b(b + c) - 1$ *and* D_1. It is clear that equations $8b(b + c) - 1 = 0$ and $D_1 = 0$ are equivalent to $c = c_4$ and $c = c_3$, respectively. Using Remark 13, we complete the proof of the lemma. □

Using Lemmas 7 and 10, we will find in Lemma 12 and Theorem 10 all possible degenerate singular points of system (5) of type (31).

Lemma 12. *All singular points* (31) *of system* (5) *are degenerate (semi-hyperbolic) saddles for* $b \in (0, b_3] \setminus \{1/4\}$, $c = c_3$ *(case* (iii)*) of Lemma 7).*

Proof. Recall that in this case singular points (31) of (5) correspond to family (10) for $\mu = 1 \pm \sqrt{D}$. Since $D = \frac{(4b-1)^2}{(8b^2-1)^2} > 0$ for $b \neq 1/4$ and $c = c_3$, then according to Lemma 10 the equality $\delta = 0$ implies $8b(b + c_3) - \mu = 0$, i.e., $\mu = \dfrac{4b(2b-1)}{8b^2-1} > 0$. For this μ, the matrix J has the form $J = k_3 \begin{pmatrix} 1 & 1 \\ 1 & 1 \end{pmatrix}$, where $k_3 := \frac{4b-1}{(2b-1)q}$.

Using Remark 12, we have

$$p_{2,0} = 0, \quad p_{1,1}\nu_2 + p_{3,0} = \frac{1}{8} \frac{(8b^2-1)^3(2b+1)}{b(4b-1)(2b-1)q^3}.$$

By the same manner as in Lemma 9, it is easy to prove that all singular points are saddles for $b \in (0, b_3] \setminus \{1/4\}$, $c = c_3$.

Note that in the above formulas we assumed that the corresponding exact values of q are determined as in Remark 1. □

Remark 15. A more detailed analysis of Lemma 12 gives in addition the following: For $\mu = 1 - \sqrt{D}$ (resp. $\mu = 1 + \sqrt{D}$), the semi-hyperbolic saddle case occurs for $b \in (0, 1/4)$ (resp. $b \in (1/4, b_3]$), $c = c_3$.

Theorem 10. *Let* $D = 0$. *Then singular points* (31) *of system* (5) *are of types depicted in Table 6.*

Table 6 Types of singularities in the case $D = 0$

Values of b	Values of c with semi-hyperbolic saddle-nodes, $D = 0$	Values of c with linearly zero saddles, $D = 0$
$b \in [b_2, 1/4)$	$c = c_1, c = c_2$	–
$b = 1/4$	–	$c = c_2 = 1/4$
$b \in (1/4, 1/2]$	$c = c_2$	–

Proof. **The case** $b \neq 1/4$. By Lemmas 7 and 9, we conclude that the equality $\delta = 0$ holds only for singular points of type (31). In fact, as the calculations show $D = 0$ at $c = c_i$, $i = 1, 2$, hence $\delta = 0$ by Lemma 10.

Now we will determine the type of such degenerate singular points of (5). For $c = c_i$, $i = 1, 2$, we easily get that $J = k_i \begin{pmatrix} 1 & -1 \\ -1 & 1 \end{pmatrix}$, where

$$k_1 := \frac{1 - 2b + \sqrt{4b^2 + 4b - 1}}{2q}, \quad k_2 := \frac{1 - 2b - \sqrt{4b^2 + 4b - 1}}{2q}.$$

Using Remark 10, we can reduce system (5) to the canonical form (28), and taking into account Remark 12, we can determine the first coefficient $p_{2,0}$ in the Taylor series (17) of the function X corresponding to this canonical form.

The case $b \in [b_2, 1/4)$, $c = c_1$ (case (i) of Lemma 7). By a series of calculations, we find

$$p_{2,0} = \frac{1 + 4c_1 - 8c_1^2}{16b(b + c_1)q^2}.$$

It is clear that $p_{2,0} > 0$ because of $0 < c_1 < 1/2$ by Remark 13 in this case. Since $k_1 > 0$, then $e_2 = p_{2,0} > 0$, by Lemma 6 we have saddle-nodes.

The case $b \in [b_2, 1/2] \setminus \{1/4\}$, $c = c_2$ (case (ii) of Lemma 7). The first coefficient $p_{2,0}$ is

$$p_{2,0} = \frac{1 + 4c_2 - 8c_2^2}{16b(b + c_2)q^2}.$$

Since $0 < c_2 < 1/2$ by Remark 13, then $p_{2,0} > 0$. If $b \in [b_2, 1/4)$, then $k_2 > 0$ and $e_2 = p_{2,0} > 0$. Hence by Lemma 6, we have saddle-nodes. If $b \in (1/4, 1/2]$, then $k_2 < 0$ and $e_2 = -p_{2,0} < 0$ by Remark 11. Therefore, such b also gives saddle-nodes.

The case $b = 1/4$ is covered by Theorem 3. □

Theorem 11. *Let* $D > 0$, $\mu = 1 + \sqrt{D}$. *Then singular points* (31) *of system* (5) *are of types depicted in Table 7.*

Table 7 Types of singularities in the case $D > 0$, $\mu = 1 + \sqrt{D}$

Values of b	Values of c with hyperbolic stable nodes, $\delta > 0, \rho < 0$	Values of c with hyperbolic saddles, $\delta < 0$	Values of c with semi-hyperbolic saddles, $\delta = 0$
$b \in (0, b_2)$	–	$c \in (0, 1/2]$	–
$b \in [b_2, 1/4)$	–	$c \in (0, c_1) \cup (c_2, 1/2]$	–
$b = 1/4$	–	$c \in (1/4, 1/2]$	–
$b \in (1/4, b_3)$	$c \in (c_2, c_3)$	$c \in (c_3, 1/2]$	$c = c_3$
$b = b_3$	$c \in (c_2, c_3)$	–	$c = c_3 = 1/2$
$b \in (b_3, 1/2]$	$c \in (c_2, 1/2]$	–	–

Table 8 Types of singularities in the case $D > 0$, $\mu = 1 - \sqrt{D}$

Values of b	Values of c with hyperbolic unstable nodes, $\delta > 0, \rho > 0$	Values of c with hyperbolic saddles, $\delta < 0$	Values of c with semi-hyperbolic saddles, $\delta = 0$
$b \in (0, b_2)$	$c \in (0, c_3)$	$c \in (c_3, 1/2)$	$c = c_3$
$b \in [b_2, 1/4)$	$c \in (0, c_1) \cup (c_2, c_3)$	$c \in (c_3, 1/2)$	$c = c_3$
$b = 1/4$	–	$c \in (1/4, 1/2)$	–
$b \in (1/4, 1/2]$	–	$c \in (c_2, 1/2)$	–

Proof. According to Lemma 10 for $\mu = 1 + \sqrt{D}$ the singular points (x_1^0, x_2^0) of system (5) have form (31) with $q > 0$ determined by Remark 1.

By Theorem 1, the nondegenerate node case occurs only when $\delta > 0$ in (32). Note that the inequality $D + \sqrt{D} < 0$ has no solutions. Hence, $\delta > 0$ implies

$$D > 0, \quad D < (8b(b + c) - 1)^2, \quad 8b(b + c) - 1 > 0. \tag{33}$$

The solutions of these inequalities were obtained in Lemma 11; so, we easily find the solutions of system (33), depicted in the first and the second columns of Table 7.

Now, we analyze the sign of ρ in (32). Since $2(b+c_2)-1 > 0$ at $b \in (1/4, 1/2]$, then for b, c ensuring nodes we obtain $2(b+c)-1 > 2(b+c_2)-1 > 0$. Therefore, $\rho < 0$; hence, all nodes are stable by Theorem 1.

The contents of the fourth column is known from Lemma 12 and Remark 15. By Theorem 1, the nondegenerate saddle case occurs only if $\delta < 0$. By Lemma 11 (excepting the values of b and c with $D \leq 0$), we easily get the third column of Table 7. □

Theorem 12. *Let $D > 0$, $\mu = 1 - \sqrt{D}$. Then singular points* (31) *of system* (5) *are of types depicted in Table 8.*

Proof. According to Lemma 10 for $\mu = 1 - \sqrt{D}$, the singular points (x_1^0, x_2^0) of system (5) have form (31) with $q > 0$ determined by Remark 1. It is easy to show that $\mu = 1 - \sqrt{D} > 0$ only for $D > 0$ and $c < 1/2$. Therefore, we may suppose that $0 < c < 1/2$.

Since the inequality $D - \sqrt{D} > 0$ has no solutions, then $\delta > 0$ is possible in (32) only if $D - \sqrt{D} < 0$ and $8b(b + c) - 1 + \sqrt{D} < 0$. Note that the inequality $D - \sqrt{D} < 0$ is equivalent to $D > 0$ at $0 < c < 1/2$. Therefore the inequality $\delta > 0$ is equivalent to system of the first and the second inequalities of (33) and $8b(b+c)-1 < 0$. Using Lemma 11, we can find that $\delta > 0$ has the solutions given in the first and the second columns of Table 8.

Now, we analyze the sign of ρ in (32). Since $2(b + c_3) - 1 < 0$ at $b \in (0, 1/4)$, then for b, c ensuring nodes we have $2(b+c)-1 < 2(b+c_3)-1 < 0$. Therefore, $\rho > 0$ and all nodes are unstable according to Theorem 1.

The contents of the fourth column are known from Lemma 12 and Remark 15. By Theorem 1, the nondegenerate saddle case occurs if $\delta < 0$. By Lemma 11 (excepting the values of b and c with $D \leq 0$), we easily get the third column of Table 8. □

9.3 *Singular Points for* $(a_1, a_2, a_3) \in \Omega \cap (0, 1/2]^3$ *with* $a_1 = a_2$

It is easy to see that in the nondegenerate cases ($(a_1, a_2, a_3) \notin \Omega$), Theorems 9, 11, and 12 are consistent with Theorem 7 stating that for every $(a_1, a_2, a_3) \in (0, 1/2)^3 \setminus \Omega$, system (5) has either one node and three saddles or two saddles.

Now, we consider singular points for $(a_1, a_2, a_3) \in \Omega \cap (0, 1/2]^3$ with $a_1 = a_2$. From Lemma 7 and Remark 14, we see that ($a_1 = a_2 = b$ and $a_3 = c$) all points of the curves $c = c_1 := \left(1 - 2b - \sqrt{4b^2 + 4b - 1}\right)/4$, $b \in [b_2, 1/4)$ and $c = c_2 := \left(1 - 2b + \sqrt{4b^2 + 4b - 1}\right)/4$, $b \in [b_2, 1/2]$ (cases (i) and (ii) of Lemma 7) are regular points of the surface Ω (except the point $(1/4, 1/4, 1/4)$), but the curve $c = c_3 := \left(16b^3 - 4b + 1\right)/\left(2 - 16b^2\right)$, $b \in (0, b_3]$ (case (iii) of Lemma 7) is one of the three "edges" of Ω (see Figs. 3 and 4).

Theorem 13. *If* $(a_1, a_2, a_3) = (b, b, c) \in \Omega \cap (0, 1/2]^3$, *then the following assertions about the singular points of system* (5) *hold:*

- (i) *There exists a nondegenerate singular point of system (5) if and only if* $b \in [b_2, 1/4)$, $c = c_1$ *or* $b \in [b_2, 1/4) \cup (1/4, 1/2]$, $c = c_2$. *Moreover, in both these cases* (5) *has exactly two nondegenerate saddles of form* (30).
- (ii) *If* $b \in [b_2, 1/4)$, $c = c_1$ *or* $b \in [b_2, 1/4) \cup (1/4, 1/2]$, $c = c_2$, *then system* (5) *has exactly one degenerate singular point that is a saddle-node of form* (31).
- (iii) *For every fixed* $b \in (0, b_3] \setminus \{1/4\}$ *and* $c = c_3$, *system* (5) *has exactly two (degenerate) singular points that are semi-hyperbolic saddles of forms* (30) *and* (31).
- (iv) *There is exactly one (degenerate) singular point* $(1, 1)$ *for* $(a_1, a_2, a_3) = (1/4, 1/4, 1/4)$, *that is a linearly zero saddle.*

Proof. Recall that by Lemma 7, the equalities $c = c_i$, $i = 1, 2, 3$, are necessary for $(b, b, c) \in \Omega \cap (0, 1/2]^3$. Recall that $c_1 \neq c_3$ and $c_2 \neq c_3$ for all $b \neq 1/4$.

(i) By Theorem 9, system (5) has two nondegenerate saddles of form (30) in both cases $b \in [b_2, 1/4)$, $c = c_1$ or $b \in [b_2, 1/4) \cup (1/4, 1/2]$, $c = c_2$. Now, we prove that $\delta = 0$ for all other admissible cases of the parameters b, c. Consider the case $c = c_3$. Then, we have $\Delta = 0$ (see Table 2) and $8b(b + c) - \mu = 0$ (see the proof of Lemma 12). Therefore, by Lemmas 8 and 10, we have $\delta = 0$ for both cases (30) and (31) of singular points. Next, consider the cases $c = c_1$ and $c = c_2$. Since $D = 0$, then by Lemma 10 we have $\delta = 0$ for singular points of form (31). This proves (i).

Let us consider case (ii). By Theorem 10, there exists exactly one saddle-node for $b \in [b_2, 1/4)$, $c = c_1$, and also for $b \in [b_2, 1/4) \cup (1/4, 1/2]$, $c = c_2$.

Cases (iii) and (iv) are covered by Lemmas 9, 12 and Theorem 3. □

Conclusion

Theorem 7 gives a general picture for types of singular points of system (5) with $(a_1, a_2, a_3) \in (0, 1/2) \times (0, 1/2) \times (0, 1/2)$. Nevertheless, it would be interesting to study "degenerate" sets of parameters (a_1, a_2, a_3) from the set $\Omega \cap (0, 1/2]^3$. For the point $(a_1, a_2, a_3) = (1/4, 1/4, 1/4)$, we obtained suitable results in Sect. 4. It should be noted that the point $(a_1, a_2, a_3) = (1/4, 1/4, 1/4)$ is a very special one on the algebraic surface

$$\Omega = \{(a_1, a_2, a_3) \in \mathbb{R}^3 \mid Q(a_1, a_2, a_3) = 0\}$$

(see Sects. 5 and 7). The following questions are worth for further investigation.

Question 1. *Find a tool to study points (a_1, a_2, a_3) of Ω for determining the type of singular points (x_1, x_2) of system (5).*

Note that this question is completely solved (see Theorem 13) for the case $a_i = a_j, i \neq j$, for $(a_1, a_2, a_3) \in \Omega \cap (0, 1/2]^3$. We specify the above question for some other partial cases.

Question 2. *Study in detail the case $a_k = 1/2$, $a_i, a_j \in (0, 1/2]$, $i \neq j \neq k \neq i$.*

Question 3. *What is the number and corresponding types of singular points of system (5) for regular points (i.e., points (a_1, a_2, a_3) with $\nabla Q(a_1, a_2, a_3) \neq 0$) on the surface $\Omega \cap (0, 1/2)^3$?*

Question 4. *Are there regular points (a_1, a_2, a_3) of the surface $\Omega \cap (0, 1/2)^3$ with at least two degenerate singular points of (5)?*

Acknowledgements The authors are indebted to Prof. Yusuke Sakane for useful discussions concerning computational aspects of this project and to Tanya Nikonorova for her help with the graphics. The project was supported in part by the State Maintenance Program for the Leading Scientific Schools of the Russian Federation (grant NSh-921.2012.1) and by Federal Target Grant "Scientific and educational personnel of innovative Russia" for 2009–2013 (agreement no. 8206, application no. 2012-1.1-12-000-1003-014).

References

1. *Abiev N.A., Arvanitoyeorgos A., Nikonorov Yu.G., Siasos P.* The dynamics of the Ricci flow on generalized Wallach spaces // Differential Geom. Appl. (2014), http://dx.doi.org/10.1016/j.difgeo.2014.02.002
2. *Anastassiou S., Chrysikos I.* The Ricci flow approach to homogeneous Einstein metrics on flag manifolds // J. Geom. Phys. (2011), V. 61, No. 8, P. 1587–1600.
3. *Arvanitoyeorgos A.* New invariant Einstein metrics on generalized flag manifolds // Trans. Amer. Math. Soc. (1993), V. 337, No. 2, P. 981–995.
4. *Arnold V. I., Gusein-Zade S. M. and Varchenko A. N.* Singularities of differentiable maps. Vol. I, Monogr. Math. 82, Birkhäuser Boston, Inc., Boston, MA, 1986.
5. *Besse A. L.* Einstein Manifolds. Springer-Verlag. Berlin, etc., 1987.
6. *Böhm C., Wilking B.* Nonnegatively curved manifolds with finite fundamental groups admit metrics with positive Ricci curvature // GAFA Geom. Func. Anal. (2007), v. 17, P. 665–681.
7. *Buzano M.* Ricci flow on homogeneous spaces with two isotropy summands // Ann. Glob. Anal. Geom. (2014), V. 45, No. 1, P. 25–45.
8. *Chow B., Knopf D.* The Ricci Flow: an Introduction. Mathematical Surveys and Monographs, AMS, Providence, RI, 2004.
9. *D'Atri J. E., Nickerson N.* Geodesic symmetries in space with special curvature tensors // J. Different. Geom. (1974), V. 9, P. 251–262.
10. *D'Atri J. E., Ziller W.* Naturally reductive metrics and Einstein metrics on compact Lie groups // Mem. Amer. Math. Soc. (1979), V. 215, P. 1–72.
11. *Darboux G.* Leçons sur la Théorie générale des Surfaces, IV, Gauthier-Villars, Paris, 1896.
12. *Dumortier F., Llibre J., Artes J.* Qualitative theory of planar differential systems. Universitext. Springer-Verlag, Berlin, 2006. xvi+298 pp.
13. *Glickenstein D. Payne T. L.* Ricci flow on three-dimensional, unimodular metric Lie algebras // Commun. Anal. Geom. (2010), V. 18, No. 5, P. 927–961.
14. *Hamilton R. S.* Three-manifolds with positive Ricci curvature // J. Differential Geom. (1982), V. 17, P. 255–306.
15. *Isenberg J., Jackson M., Lu P.* Ricci flow on locally homogeneous closed 4-manifolds // Comm. Anal. Geom. (2006), V. 14, No. 2, P. 345–386.
16. *Jiang Q., Llibre J.* Qualitative classification of singular points // Qual. Theory Dyn. Syst. (2005), V. 6, No. 1, P. 87–167.
17. *Kimura M.* Homogeneous Einstein metrics on certain Kähler C-spaces // Adv. Stud. Pure Math. (1990), V. 18, No. 1, P. 303–320.
18. *Lauret J.* Ricci flow of homogeneous manifolds // Math. Z. (2013), V. 274, No. 1–2, P. 373–403.
19. *Lomshakov A. M., Nikonorov Yu. G., Firsov E. V.* On invariant Einstein metrics on three-locally-symmetric spaces // Doklady Mathematics (2002), V. 66, No. 2, P. 224–227.
20. *Lomshakov A. M., Nikonorov Yu. G., Firsov E. V.* Invariant Einstein Metrics on Three-Locally-Symmetric Spaces // Matem. tr. (2003), V. 6, No. 2. P. 80–101 (Russian); English translation in: Siberian Adv. Math. (2004), V. 14, No. 3, P. 43–62.
21. *Nikonorov Yu. G., Rodionov E. D., Slavskii V. V.* Geometry of homogeneous Riemannian manifolds // Journal of Mathematical Sciences (New York) (2007), V. 146, No. 7, P. 6313–6390.
22. *Nikonorov Yu. G.* On a class of homogeneous compact Einstein manifolds // Sibirsk. Mat. Zh. (2000), V. 41, No. 1, P. 200–205 (Russian); English translation in: Siberian Math. J. (2000), V. 41, No. 1, P. 168–172.
23. *Payne T. L.* The Ricci flow for nilmanifolds // J. Mod. Dyn. (2010), V. 4, No. 1, P. 65–90.
24. *Rodionov E. D.* Einstein metrics on even-dimensional homogeneous spaces admitting a homogeneous Riemannian metric of positive sectional curvature // Sibirsk. Mat. Zh. (1991), V. 32, No. 3, P. 126–131 (Russian); English translation in: Siberian Math. J. (1991), V. 32, No. 3, P. 455–459.
25. *Thom R.* Topological models in biology // Topology (1969), V. 8, P. 313–335.

26. *Topping P.* Lectures on the Ricci flow, London Mathematical Society Lecture Note Series, vol. 325, Cambridge University Press, Cambridge, 2006.
27. *Wallach N.* Compact homogeneous Riemannian manifolds with strictly positive curvature // Annals of Mathematics, Second Series. (1972), V. 96, P. 277–295.
28. *Woodcock A. E. R., Poston T.* A geometrical study of the elementary catastrophes, Lecture Notes in Mathematics, vol. 373, Springer-Verlag, Berlin-New York, 1974.

Gaussian Mean Curvature Flow for Submanifolds in Space Forms

Aleksander Borisenko and Vladimir Rovenski

Abstract In this chapter we investigate the convergence of the mean curvature flow of submanifolds in Euclidean and hyperbolic spaces with Gaussian density. For Euclidean case, we prove that the flow deforms a closed submanifold with pinching condition to a "round point" in finite time.

Keywords Riemannian metric • Mean curvature flow • Density • Conformal transformation

Mathematics Subject Classifications (2010): Primary 53C44, Secondary 35K93, 35R01

1 Introduction

The *mean curvature flow* (MCF) was proposed by W. Mullins (1956) to describe the formation of grain boundaries in annealing metals. Brakke [5] introduced the motion of a submanifold by its MCF in arbitrary codimension and constructed a generalized varifold solution for all time. There are many works for the classical solution of MCF on hypersurfaces. Huisken [7] showed that if the initial hypersurface in the Euclidean space is compact and uniformly convex, then MCF converges to a "round point" in a finite time. He also studied MCF of hypersurfaces in a Riemannian manifold satisfying a pinching condition in a sphere, see [1].

A. Borisenko (✉)
Department MA and MO, Sumy State University, Sumy, Ukraine
e-mail: aborisenk@gmail.com

V. Rovenski
Mathematical Department, University of Haifa, Haifa, Israel
e-mail: rovenski@math.haifa.ac.il.

V. Rovenski and P. Walczak (eds.), *Geometry and its Applications*, Springer Proceedings in Mathematics & Statistics 72, DOI 10.1007/978-3-319-04675-4_2,

For MCF of submanifolds with higher codimension, fruitful results were obtained for submanifolds with low dimension or admitting some special structures, see survey [11, 12]. Andrews and Baker [2] proved a convergence theorem for MCF of closed submanifolds satisfying a suitable pinching condition in the Euclidean space. Baker [3] and Liu–Xu–Ye–Zhao [8, 9] generalized Andrews-Baker's convergence theorem [2] for MCF of submanifolds in the Euclidean space to the case of MCF of arbitrary codimension in spherical and hyperbolic space forms and Riemannian manifolds.

Morgan [10] introduced manifolds with density, which provides a new concept of curvature. A. Borisenko and V. Miquel considered MCF with density for hypersurfaces in Euclidean space.

In this chapter we study the convergence of the MCF of submanifolds in Euclidean and hyperbolic spaces with Gaussian density. For Euclidean case, we prove that the flow deforms a closed submanifold satisfying pinching condition to a "round point" in finite time. For hyperbolic case, we find maximal radius (or minimal normal curvature) of central hypersphere in a hyperbolic space that shrinks to the origin under the MCF with Gaussian density; moreover, for central spheres of smaller radius we estimate the collapsing time.

2 The MCF in Riemannian Manifolds and Space Forms

Consider immersions of a closed manifold M^n into a space form:

$$F_t : M^n \to \bar{M}^{n+p}(c), \quad F_t(q) = F(q,t), \quad q \in M^n,\ t \in [0,T).$$

Denote by h_t the second fundamental tensor, and by $H_t = \operatorname{Tr}_g h_t$ the mean curvature vector field of the immersions (g is the induced metric on M). The *MCF* is the evolution equation (see [2, 11])

$$\partial_t F = H, \tag{1}$$

where $F_0 : M^n \to \bar{M}^{n+p}(c)$ provides initial data.

Remark 1. The *general form of the MCF* is

$$(\partial_t F)^{\perp} = H, \tag{2}$$

where $^{\perp}$ denotes the projection onto the normal space of $F_t(M)$. This equation is equivalent to (1) up to diffeomorphisms of M (see [12]; the proof is the same as for $p = 1$ in [6]).

Let $\bar{M}^{n+p}(c)$ be endowed with a continuous *density function* $f = e^{\psi}$, where $\psi \in C^2(\bar{M}^{n+p}(c))$. The generalization of the mean curvature of submanifolds in such spaces, obtained by the first variation of the volume, is given in [10] as

$$H_{\psi} = H - (\nabla\psi)^{\perp}.$$

It is natural to study flows governed by H_ψ instead of H:

$$\partial_t F = H - (\nabla \psi)^\perp. \tag{3}$$

Any n-dimensional submanifold satisfies $|h|^2 \geq \frac{1}{n}|H|^2$ (where $|H|$ and $|h|$ are norms), and totally umbilical submanifolds give the equality.

Lemma 1 ([13]). *Let M^n be an n-dimensional submanifold in an $(n+p)$-dimensional Riemannian manifold $\bar{M}^{n+p}$ and π a tangent two-plane on $T_q(M)$ at a point $q \in M$. Choose an orthonormal two-frame $\{e_1, e_2\}$ at q such that $\pi = \operatorname{span}\{e_1, e_2\}$. Then*

$$K(\pi) \geq \frac{1}{2}\left(2\bar{K}_{\min} + \frac{H^2}{n-1} - |h|^2\right) + \sum\nolimits_{a=n+1,\, j>i}^{n+p} \sum\nolimits_{(i,j)\neq(1,2)} (h_{ij}^a)^2 .$$

Recently, Andrews–Baker [2] proved convergence theorem for the MCF of closed submanifolds satisfying a pinching condition in the Euclidean space.

Theorem A ([2]). *Let $n \geq 2$, and suppose that $F_0(M^n)$ is a closed submanifold smoothly immersed in $\mathbb{R}^{n+p}$. If $F_0(M^n)$ has $H \neq 0$ everywhere and satisfies*

$$|h|^2 \leq \begin{cases} \frac{4}{3n}|H|^2, & \text{if } n = 2, 3, \\ \frac{1}{n-1}|H|^2, & \text{if } n \geq 4, \end{cases} \tag{4}$$

then MCF (1) *has a unique smooth solution $F_t : M^n \times [0, T) \to \mathbb{R}^{n+p}$ on a finite maximal time interval, and F_t converges uniformly to a point $q \in \mathbb{R}^{n+p}$ as $t \to T$. The rescaled maps $\tilde{F}_t = \frac{F_t - q}{\sqrt{2n(T-t)}}$ converge in C^∞ as $t \to T$ to an embedding $\tilde{F}_T$ with image equal to a regular unit n-sphere in some $(n+1)$-dimensional subspace of $\mathbb{R}^{n+p}$. If $n \geq 4$, pinching ratio* (4) *is optimal.*

Liu–Wei–Zghao [8] extended Theorem A to submanifolds in hyperbolic spaces.

Theorem A′ ([8]). *Let $F_0(M^n)$ $(n \geq 2)$ be a closed submanifold smoothly immersed in hyperbolic space $\mathbb{H}^{n+p}(c)$ of constant curvature $c < 0$. If $F_0(M^n)$ satisfies*

$$|h|^2 \leq \begin{cases} \frac{4}{3n}|H|^2 + \frac{n}{2}c, & \text{if } n = 2, 3, \\ \frac{1}{n-1}|H|^2 + 2c, & \text{if } n \geq 4, \end{cases} \tag{5}$$

then MCF (1) *with F_0 as initial value has a unique smooth solution $F_t : M^n \times [0, T) \to \mathbb{H}^{n+p}(c)$ on a finite maximal time interval, and $F_t(M^n)$ converges uniformly to a "round point" as $t \to T$.*

3 Gaussian MCF in Euclidean Space

The *Gaussian density* $e^{-\frac{n}{2}\mu^2|x|^2}$ (for some $\mu > 0$) in $\mathbb{R}^{n+p}$ is rotational invariant and corresponds to the radial function

$$\psi(x) = -\frac{n}{2}\,\mu^2|x|^2. \tag{6}$$

In this case, $\nabla\psi(x) = -n\mu^2 x$ for all $x \in \mathbb{R}^{n+p}$. Along the submanifold $F(M)$ we have $(\nabla\psi)^\perp = -n\mu^2 F^\perp$. Since $H_\psi = H - (\nabla\psi)^\perp$, see [10], the *MCF in* $\mathbb{R}^{n+p}$ *with Gaussian density* is defined by

$$\partial_t F = H + n\mu^2 F^\perp. \tag{7}$$

Lemma 2 (see [4]). *Let ψ be a radial function on $\mathbb{R}^{n+p}$. The vector field $\nabla\psi$ is conformal if and only if*

$$\psi(x) = \pm\frac{n}{2}\,\mu^2|x|^2 \qquad \text{(hence, } \nabla\psi = \pm n\mu^2 x) \qquad \text{for some } \mu > 0.$$

Borisenko–Miquel [4] proved convergence theorem for the MCF with Gaussian density on a hypersurface in $\mathbb{R}^{n+1}$.

Theorem B ([4]). *Let $F_0 : M \to \mathbb{R}^{n+1}$ be a convex hypersurface with a chosen unit normal vector N, which evolves under MCF with Gaussian density (see (7) with $p = 1$)*

$$\partial_t F = (H + n\mu^2\langle F, N\rangle)N. \tag{8}$$

Then its evolution F_t remains convex for all time $t \in [0, T)$ where it is defined.

If $h \ge \mu g$ and $h(v, v) > \mu g(v, v)$ in some vector at some point v, then there is a point q_0 inside the convex domain $F_0(M)$ such that $F_0(M)$ lies in the ball B with center q_0 of radius $1/\mu$. Moreover,

1. *$T < \infty$ and $h > \mu g$ for $t \in (0, T)$,*
2. *$F_t(M)$ belongs to a ball of radius $1/\mu$ all time and shrinks to a "round point" when $t \to T$.*

Lemma 3 ([2]). *If a solution $F_t : M^n \to \mathbb{R}^{n+p}$ $(0 \le t < T)$ of MCF (1) satisfies $|h|^2 + a < C|H|^2$ for some constants $C \le \frac{1}{n} + \frac{1}{3n}$ and $a > 0$ at $t = 0$, then this remains true for all $0 \le t < T$.*

Using Theorem A, we extend Theorem B for submanifolds in Euclidean space.

Theorem 1. *Let $F_0 : M^n \to \mathbb{R}^{n+p}$ be a complete smoothly immersed submanifold with the condition*

$$|h|^2 + \beta^2 \le C\,|H|^2 := \begin{cases} \frac{4}{3n}\,|H|^2, & \text{if } n = 2, 3, \\ \frac{1}{n-1}\,|H|^2, & \text{if } n \ge 4, \end{cases} \tag{9}$$

where

$$\beta^2 \geq (\pi\mu)^2 \frac{n+p}{n+p+1} - \Big(\frac{1}{n-1} - C\Big)|H|^2. \tag{10}$$

Then the MCF with the Gaussian density in $\mathbb{R}^{n+p}$, (7), *has a unique smooth solution* $F_t : M^n \times [0, T) \to \mathbb{R}^{n+p}$ *on a finite maximal time interval, and* F_t *converges uniformly to a "round point" when* $t \to T$.

Proof. Its main steps coincide with ones in the proof of Theorem B.

By Lemma 1, at each point $q \in M^n$ the smallest sectional curvature $K_{\min}$ satisfies

$$K_{\min}(q) \geq \frac{1}{2}\Big(\frac{1}{n-1}|H(q)|^2 - |h(q)|^2\Big). \tag{11}$$

Substituting $|h|^2$ from our assumption (9) into inequality (11), for $q \in M$ we obtain

$$K_{\min}(q) \geq \frac{1}{2}\left(\left(\frac{1}{n-1} - C\right)|H|^2 + \beta^2\right).$$

Note that $\frac{1}{n-1} - C \geq 0$. By Theorem of Bonnet, Hopf-Rinow and Myers for $t = 0$, we have

$$\operatorname{diam} M \leq \pi\sqrt{2}\,\tilde{d}, \qquad \tilde{d} = \left[\left(\frac{1}{n-1} - C\right)|H|^2 + \beta^2\right]^{-1/2}.$$

Note that the inner diameter of M is greater than or equal to diameter d of $F_0(M)$. The Yung's Theorem (1901) tells us that every set $K \subset \mathbb{R}^{n+p}$ of diameter d is contained in a ball in $\mathbb{R}^{n+p}$ of radius $r_0(K) = \sqrt{\frac{n+p}{2(n+p+1)}}\,d$. Thus, $F_0(M)$ is contained in a ball in $\mathbb{R}^{n+p}$ of radius

$$r_0 \leq \pi\sqrt{\frac{n+p}{n+p+1}}\,\tilde{d}. \tag{12}$$

Recall that if $F_0(M)$ is contained in a ball $B(r_0)$ of radius $r_0 > 0$, then flow (1) must develop singularity (collapsing to a point) before the time $T = r_0^2/(2n)$, see [2].

Condition (10) for β yields the inequality $r_0^2/(2n) < 1/(2n\mu^2)$.

By Proposition 1, the MCF $\hat{F}_{\hat{t}}$ of (14) is equivalent to the flow F_t of (7) for all $\hat{t} \in [0, \hat{T}]$.

The submanifold $\hat{F}_0(M) = F_0(M)$ satisfies the conditions of Theorem A. Then (1) has a unique smooth solution $\hat{F}_{\hat{t}} : M^n \times [0, \hat{T}) \to \mathbb{R}^{n+p}$ on a finite maximal time interval, and it converges uniformly to a point $\hat{q} \in \mathbb{R}^{n+p}$ as $\hat{t} \to \hat{T}$. The rescaled maps converge in C^∞ as $\hat{t} \to \hat{T}$ to an embedding with image equal to a regular n-sphere in some $(n+1)$-dimensional subspace of $\mathbb{R}^{n+p}$.

From equivalence of flows (14) and (7) we conclude that F_t converges in a finite time uniformly to a point $q \in \mathbb{R}^{n+p}$. Since submanifolds $\hat{F}_{\hat{t}}(M)$ and $F_t(M)$ are homothetic, we obtain that F_t converges to a "round point" $q \in \mathbb{R}^{n+p}$. □

By the next proposition, one may transfer any result on MCF (1) to a result on flow (3) with ψ given in (6).

Proposition 1 (For $p = 1$, see [4]). *MCF (7) in $\mathbb{R}^{n+p}$ with Gaussian density is equivalent, up to tangential diffeomorphisms, with the parameter change*

$$\hat{t} = -\frac{1}{2n\mu^2}(e^{-2n\mu^2 t} - 1) \tag{13}$$

to the MCF in $\mathbb{R}^{n+p}$

$$\frac{\partial \hat{F}}{\partial \hat{t}} = \hat{H} \qquad \Big(\text{for } \hat{t} < \frac{1}{2n\mu^2}\Big). \tag{14}$$

Proof. The one-parameter family of diffeomorphisms $\phi_t(x) = e^{-n\mu^2 t}x$ is the solution of the ODE

$$\frac{d}{dt}\phi_t(x) = -n\mu^2\phi_t(x)$$

with the initial condition $\phi_0(x) = x$ and is associated with the vector field $X(x) = -n\mu^2 x$ on $\mathbb{R}^{n+p}$. If F flows by the mean curvature with density $f = e^{-\frac{1}{2}n\mu^2|x|^2}$, then the flow $\hat{F}_t = \phi_t \circ F_t$ has the form

$$\hat{F} = e^{-n\mu^2 t}F. \tag{15}$$

To check this and to find the corresponding reparametrization of time, we compute

$$\begin{aligned}\partial_t \hat{F} &= -n\mu^2 e^{-n\mu^2 t}F + e^{-n\mu^2 t}(H + n\mu^2 F^{\perp})\\ &= -n\mu^2 e^{-n\mu^2 t}F^{\top} + e^{-n\mu^2 t}H = -n\mu^2\hat{F}^{\top} + e^{-n\mu^2 t}H.\end{aligned}$$

By (15), the second fundamental tensors of $\hat{F}$ and F are related by $\hat{h} = e^{-n\mu^2 t}h$; hence, $\hat{H} = e^{n\mu^2 t}H$. Therefore, the evolution for $\hat{F}$ is

$$\partial_t\hat{F} = -n\mu^2\hat{F}^{\top} + e^{-2n\mu^2 t}\hat{H}. \tag{16}$$

If we define $\hat{t}$ by (13), we get $dt/d\hat{t} = (d\hat{t}/dt)^{-1} = e^{2n\mu^2 t}$, and

$$\frac{\partial\hat{F}}{\partial\hat{t}} = \frac{\partial\hat{F}}{\partial t}\cdot\frac{dt}{d\hat{t}} = -n\mu^2 e^{2n\mu^2 t}\hat{F}^{\top} + \hat{H} = \frac{1}{2}\Big(\hat{t} - \frac{1}{2n\mu^2}\Big)^{-1}\hat{F}^{\top} + \hat{H}. \tag{17}$$

Flow (17) is, up to a tangential diffeomorphism (see Remark 1), equivalent to the MCF equation $\partial_t \hat{F} = \hat{H}$ for $\hat{t} < \hat{T} = \frac{1}{2} n\mu^{-2}$ (because at $\hat{T}$ the tangential diffeomorphism giving the equivalence is not well defined: the time $\hat{t} = \hat{T}$ corresponds in (13) to $t = \infty$). □

Remark 2. For Euclidean case, we find $t = -\frac{1}{2n\mu^2} \log(1 - 2n\mu^2 \hat{t})$, and the converse of (15) is

$$F = e^{n\mu^2 t} \hat{F} = (1 - 2n\mu^2 t)^{-1/2} \hat{F}.$$

In [3] Baker proved a convergence result for the MCF of submanifolds in a sphere $S^{n+p}(c)$ of constant curvature $c > 0$. Using this, one may deduce the convergence theorem for the MCF for closed submanifolds satisfying a pinching condition in the sphere with Gaussian density.

4 Gaussian MCF in Hyperbolic Space

Let r be the *distance function* from a fixed point q (the origin) on $\mathbb{H}^{n+p} := \mathbb{H}^{n+p}(-1)$.

The *Gaussian density* $e^{n\mu^2(1-\cosh r)}$ (for some $\mu > 0$) in a hyperbolic space $\mathbb{H}^{n+p}$ is rotational invariant and corresponds to the radial function

$$\psi(x) = -n\mu^2(\cosh r(x) - 1). \tag{18}$$

In this case, $\nabla\psi(x) = -n\mu^2(\sinh r(x))\partial_r$ for all $x \in \mathbb{H}^{n+p}$.

The *MCF with Gaussian density* for a submanifold $F_0 : M^n \to \mathbb{H}^{n+p}$ is

$$\partial_t F = H + n\mu^2(\sinh r(F))\partial_r^{\perp}. \tag{19}$$

For a hypersurface $F_0 : M^n \to \mathbb{H}^{n+1}$ with a chosen unit normal vector N this reads

$$\partial_t F = \big(H + n\mu^2 \sinh r(F)\langle\partial_r, N\rangle\big)\, N. \tag{20}$$

Lemma 4. *(i) Let $\psi = \varphi \circ r$ be a radial function on $\mathbb{H}^{n+p}$ (for a function $\varphi : \mathbb{R}_+ \to \mathbb{R}$ of class C^1). Then the vector field $\nabla\psi$ is conformal if and only if $\varphi(r) = \pm n\mu^2(\cosh r - 1)$ for some $\mu \in \mathbb{R}_+$.*

(ii) In spherical coordinates $(r, \tilde{x})$ in $\mathbb{H}^{n+p}$ the conformal diffeomorphisms belonging to $X(x) = -n\mu^2(\sinh r(x))\partial_r$ have a form $\tilde{\phi}_t(r, \tilde{x}) = (\phi_t(r), \tilde{x})$, where

$$\phi_t(r) = 2\,\text{arctanh}\big(\tanh(r/2)\, e^{-n\mu^2 t}\big). \tag{21}$$

Proof. (i) We have $\nabla\psi = \varphi'\,\nabla r$. The condition for the vector field $\nabla\psi$ being conformal, that is, $\mathrm{Hess}_\psi = \lambda\, g$, translates into

$$\varphi''\nabla r \otimes \nabla r + \varphi'\,\mathrm{Hess}_r = \lambda\, g. \tag{22}$$

The hessian is defined as a symmetric $(0,2)$-tensor such that $\mathrm{Hess}_\psi(X,Y) = g(S(X),Y)$, where $S(X) = \nabla_X\nabla\psi$ is a self-adjoint $(1,1)$-tensor.

The normal curvature of a sphere of radius r in $\mathbb{H}^{n+p}$ is $\coth r$. Hence,

$$\mathrm{Hess}_r = (\coth r)(g - \nabla r \otimes \nabla r).$$

Collecting terms with g and $\nabla r \otimes \nabla r$ in (22), we obtain the system

$$\varphi'' = (\coth r)\,\varphi', \qquad \lambda = (\coth r)\,\varphi'.$$

The solution of the first ODE with the initial condition $\varphi(0) = 0$ has the required form. Notice that $\varphi \approx \mp\frac{1}{2}\,n\mu^2 r^2$ for $r \approx 0$, see Lemma 2.

(ii) The one-parameter family $\phi_t(r)$ of conformal radial diffeomorphisms belonging to $\tilde{X}(r) = -n\mu^2(\sinh r)\partial_r$ is the solution of the Cauchy's problem

$$\frac{d}{dt}\,\phi_t(r) = -n\mu^2 \sin h\phi_t(r), \qquad \phi_0(r) = r.$$

The unique solution has form (21). □

Remark 3. One may represent $\mathbb{H}^{n+p}$ as a unit ball $B(0,1) \subset \mathbb{R}^{n+p}$ with the metric

$$\mathrm{ds}^2 = \frac{4\,\mathrm{dx}^2}{(1-x^2)^2}, \quad \text{where} \quad x = (x_1,\dots,x_{n+p}), \quad x^2 = \sum\nolimits_i x_i^2.$$

For the hyperbolic radial distance r we have $\mathrm{dr} = \frac{2\,\mathrm{d}|x|}{1-x^2}$ and

$$r = 2\,\mathrm{arctanh}(|x|) \iff |x| = \tanh(r/2).$$

Hence, $\sinh r = \frac{2|x|}{1-x^2}$, and the unit radial vector is $\partial_r = \frac{1-x^2}{2|x|}\,F$.

If F flows by the mean curvature with density $f = e^{n\mu^2(1-\cosh r)}$, for the density we obtain

$$\nabla\psi = -n\mu^2(\sinh r(F))\,\partial_r = -n\mu^2 F\,.$$

Then the flow $\hat{F}_t = \phi_t(F_t)$, where $\phi_t(x) = e^{-n\mu^2 t}x$, has the form, see (15),

$$\hat{F} = e^{-n\mu^2 t}\,F\,. \tag{23}$$

The derivation in t yields

$$\partial_t \hat{F} = \partial_t(e^{-n\mu^2 t} F) = e^{-n\mu^2 t}\big((H+n\mu^2 F^{\perp})-n\mu^2 F\big) = e^{-n\mu^2 t} H - n\mu^2 e^{-n\mu^2 t} F^{\top} .$$

Note that $\coth r = \frac{1+x^2}{2|x|}$ and $\coth \hat{r} = \frac{1+x^2 e^{-2n\mu^2 t}}{2|x|\,e^{-n\mu^2 t}}$, where $\hat{r} = 2\,\mathrm{arctanh}(e^{-n\mu^2 t}|x|)$ due to (23). Since the mapping of $\mathbb{H}^{n+p}$ into itself given in (23) is conformal, for the mean curvature vectors H and $\hat{H}$ of submanifolds F and $\hat{F}$ we have

$$\hat{H} = \lambda H, \quad \text{where} \quad \lambda = \frac{\coth \hat{r}}{\coth r} = \frac{1+x^2 e^{-2n\mu^2 t}}{(1+x^2)\,e^{-n\mu^2 t}}.$$

Thus, $e^{-n\mu^2 t} H = \frac{(1+x^2)\,e^{-2n\mu^2 t}}{1+x^2 e^{-2n\mu^2 t}} \hat{H}$ and the PDE above reduces to

$$\partial_t \hat{F} = \frac{(1+x^2)\,e^{-2n\mu^2 t}}{1+x^2 e^{-2n\mu^2 t}} \hat{H} - n\mu^2 \hat{F}^{\top}.$$

After suitable tangential transformation of M^n, we obtain the PDE that generalizes (1):

$$\partial_t \hat{F} = \frac{1+x^2}{e^{2n\mu^2 t} + x^2}\, \hat{H}\, . \tag{24}$$

Note that (24) reduces to MCF (14) when $\mu \to 0$.

In the next proposition we find maximal radius (or minimal normal curvature) of central hypersphere in a hyperbolic space that shrinks to the origin under the MCF with Gaussian density; for central spheres of smaller radius we estimate the collapsing time.

Proposition 2. *Let either the radius r_0 of the central hypersphere $S^n(r_0) \subset \mathbb{H}^{n+1}$ or its normal curvature k satisfy the certain of inequalities*

$$\cosh r_0 < \sigma_1 := \frac{1+\sqrt{1+4\mu^4}}{2\,\mu^2}, \qquad k > \mu\sqrt{\sigma_1}. \tag{25}$$

Then $S^n(r_0)$ shrinks to the origin under MCF (20) *with Gaussian density by the time*

$$\begin{aligned} T &= \frac{1}{n\sqrt{1+4\mu^4}} \ln \frac{(1-2\mu^2+\sqrt{1+4\mu^4})(2\mu^2\cosh r_0 - 1 + \sqrt{1+4\mu^4})}{(2\mu^2-1+\sqrt{1+4\mu^4})(1-2\mu^2\cosh r_0+\sqrt{1+4\mu^4})} \\ &< \frac{\sigma_1}{n\mu^2(\sigma_1+1)} \cdot \frac{\cosh r_0 - 1}{\sigma_1 - \cosh r_0}\, . \end{aligned} \tag{26}$$

The central sphere of radius $r_1 = \mathrm{arccosh}(\sigma_1)$ is a fixed point of the flow. The central sphere of radius $r > r_1$ expands without limit.

Proof. The mean curvature of the central hypersphere $S^n(r)$ of radius r is $H = -n\coth r$; hence, $N = \partial_r$ and (20) reads as the ODE for the radius $r(t) > 0$,

$$\frac{d}{dt}r = -n\coth r + n\mu^2\sinh r, \qquad r(0) = r_0. \tag{27}$$

The sphere shrinks to a point when

$$\coth r - \mu^2\sinh r > 0 \quad \Leftrightarrow \quad \mu^2\cosh^2 r - \cosh r - \mu^2 < 0.$$

The roots of quadratic equation $\mu^2\sigma^2 - \sigma - \mu^2 = 0$ are $\sigma_{1,2} = \frac{1\pm\sqrt{1+4\mu^4}}{2\mu^2}$. The positive root $\sigma_1 \geq 1$, and the negative root $\sigma_2 \in (-1,0)$. Hence, the central sphere of radius $r_1 = \operatorname{arccosh}(\sigma_1)$ is a fixed point of the flow, the central sphere of radius $r > r_1$ expands without limit, and the central sphere of radius $r < r_1$ shrinks to the origin. The normal curvature of the r_1-sphere is $k_1 = \coth r_1 = \cosh r_1/\sqrt{\cosh^2 r_1 - 1} = \sqrt{\frac{1+\sqrt{1+4\mu^4}}{2}} > \mu$.

Assuming $\sigma(t) = \cosh r(t) > 1$ and $\sigma_o = \cosh r_0$, we reduce (27) to

$$d\sigma/dt = n(\mu^2\sigma^2 - \sigma - \mu^2) = n\mu^2(\sigma-\sigma_1)(\sigma-\sigma_2), \qquad \sigma(0) = \sigma_o.$$

We have

$$\frac{1}{(\sigma-\sigma_1)(\sigma-\sigma_2)} = -\frac{1}{\sigma_1-\sigma_2}\Big(\frac{1}{\sigma_1-y} + \frac{1}{y-\sigma_2}\Big),$$

$$\int_{\sigma_o}^{\sigma}\frac{\mathrm{d}y}{(y-\sigma_1)(y-\sigma_2)} = n\mu^2 t.$$

If the initial value satisfies $\sigma_o \in (1,\sigma_1)$, then the integral above is $\log\frac{y-\sigma_2}{\sigma_1-y}\big|_{\sigma_o}^{\sigma} = n\mu^2(\sigma_1-\sigma_2)\,t$; hence, the solution $\sigma(t)$ is a decreasing function

$$\sigma(t) = \frac{\sigma_2\,\alpha+\sigma_1}{\alpha+1}, \qquad \text{where} \quad \alpha = \frac{\sigma_1-\sigma_o}{\sigma_o-\sigma_2}\,e^{\,n\mu^2(\sigma_1-\sigma_2)\,t}.$$

Note that $\lim_{t\to\infty}\sigma(t) = \sigma_2 < 0 < 1 < \sigma_1 = \lim_{t\to-\infty}\sigma(t)$. The collapse $r(T) = 0$ at $t = T$ (i.e., $\sigma(T) = 1$) appears at

$$T = \frac{1}{n\mu^2(\sigma_1-\sigma_2)}\log\frac{(\sigma_o-\sigma_2)(\sigma_1-1)}{(\sigma_1-\sigma_o)(1-\sigma_2)} > 0,$$

that is, (26). Using the inequality $\log(1+y) < y$ for $y > 0$ and relation $\sigma_2 = -1/\sigma_1$, we obtain

$$T < \frac{1}{n\mu^2(\sigma_1 - \sigma_2)}\Big(\frac{(\sigma_o - \sigma_2)(\sigma_1 - 1)}{(\sigma_1 - \sigma_o)(1 - \sigma_2)} - 1\Big)$$
$$= \frac{\sigma_o - 1}{n\mu^2(\sigma_1 - \sigma_o)(1 - \sigma_2)} = \frac{\sigma_1}{n\mu^2(\sigma_1 + 1)} \cdot \frac{\sigma_o - 1}{\sigma_1 - \sigma_o}.$$

Certainly, for initial value $\sigma_o > \sigma_1$, the solution $\sigma(t)$ is a monotone increasing function. □

Remark 4. For the MCF of a hypersphere in $\mathbb{H}^{n+1}$, the radius obeys the PDE $\frac{d}{dt}r = -n \coth r$; hence, $\cosh r(t) = e^{-nt}\cosh r_0$ and the existence time is $\tilde{T} = \frac{1}{n}\log(\cosh r_0)$, i.e., $r(\tilde{T}) = 0$. For $\mu = 0$, flow (20) reduces to the MCF, and in this case we have $\lim_{\mu\to 0} T = \tilde{T}$. For the MCF of a submanifold M^n in $\mathbb{H}^{n+p}$ $(n, p > 1)$, we have the course estimate $\tilde{T} < \frac{1}{n-1} r_0$, see [8]. We conjecture that Theorem A′ can be extended to the convergence theorem (like Theorem 1) for the MCF of closed submanifolds satisfying a pinching condition in the hyperbolic space with Gaussian density.

Acknowledgements Supported by the Marie-Curie actions grant EU-FP7-P-2010-RG, No. 276919.

References

1. Andrews B. *Contraction of convex hypersurfaces in Riemannian spaces*. J. Diff. Geom. v. 39 (1994) 407–431.
2. Andrews B., and Baker C. *Mean curvature flow of pinched submanifolds to spheres*. J. Diff. Geom. v. 85 (2010) 357–395.
3. Baker C. *The mean curvature flow of submanifolds of high codimension*, 113 pp. arXiv.org/math.DG/1104.4409 v1.
4. Borisenko A., and Miquel v. *Gaussian mean curvature flow*. J. of Evolution Equations, v. 10 (2010) 413–423.
5. Brakke K. *The motion of a surface by its mean curvature*. Prinston, New Jersey: Prinston University Press, 1978.
6. Ecker K. *Regularity theory for mean curvature flow*. Birkhäuser Boston, 2004.
7. Huisken G. *Flow by mean curvature of convex surfaces into spheres*, J. Differ. Geom. v. 20 (1984) 237–266.
8. Liu K., Xu H., and Zhao E. *Mean curvature flow of higher codimension in hyperbolic spaces*, 14 pp. arXiv.org/math.DG/1105.5686 v1.
9. Liu K., Xu H., and Zhao E. *Mean curvature flow of higher codimension in Riemannian manifolds*, 28 pp. arXiv.org/math.DG/1204.0107 v1.
10. Morgan F. *Manifolds with density*, Notices AMS, 52 (2005) 853–858.
11. Smoszyk K. *Mean curvature flow in higher codimension – Introduction and survey*, 37 pp. arXiv.org/math.DG/1104.3222 v2.

12. Wang M.-T. *Lectures on mean curvature flow in higher codimensions*, Handbook of geometric analysis. No. 1, 525–543, Adv. Lect. Math. (ALM), 7, Int. Press, Somerville, MA, 2008 (see also arXiv.org/math.DG/1104.3354 v1).
13. Xu H. W., and Gu J. R.: *An optimal differentiable sphere theorem for complete manifolds*, Math. Res. Lett. 17 (2010), 1111–1124.

Cantor Laminations and Exceptional Minimal Sets in Codimension One Foliations

Gilbert Hector

Abstract In this paper we deal with two types of questions concerning the structure of foliations (or laminations) on compact spaces:

1. Describe generic properties of foliations and laminations and refine the known ones,
2. Discuss the embeddability of n-dimensional minimal Cantor laminations as minimal sets in codimension one foliations on compact $(n + 1)$-manifolds or as closed sets in $\mathbb{R}^{n+1}$ (or any simply connected $(n + 1)$-manifold).

The two questions are related by the fact that exceptional minimal sets in codimension one present stronger generic constraints.

Keywords Minimal • Generic • Residual • Endset • End-rigid • Embeddability

Mathematics Subject Classification (2000): 37B05, 57R30

Introduction

A codimension m foliation is transversely modeled on $\mathbb{R}^m$; when replacing the latter by a topological space $\mathbb{K}$ one obtains a $\mathbb{K}$-*lamination*; it is a *Cantor lamination* if $\mathbb{K}$ is the standard Cantor set. Now our goal in this paper is twofold:

1. Refine the known generic properties of foliations and laminations and provide simplified proofs for them,

G. Hector (✉)
Institut C. Jordan UMR CNRS 5208, Université C. Bernard bât 101, 43 Boulevard du 11 novembre 1918, 69622 Villeurbanne Cedex, France
e-mail: hector@math.univ-lyon1.fr

V. Rovenski and P. Walczak (eds.), *Geometry and its Applications*, Springer Proceedings in Mathematics & Statistics 72, DOI 10.1007/978-3-319-04675-4_3,

2. Discuss the embeddability of n-dimensional minimal Cantor laminations as minimal sets in codimension one foliations on compact $(n+1)$-manifolds (exceptional minimal sets) or as closed sets in $\mathbb{R}^{n+1}$ (or any simply connected $(n+1)$-manifold).

The two questions are related by the fact that exceptional minimal sets in codimension one present stronger generic constraints.

(1) Concerning the first item, recall that some property is *generic* for a foliation or lamination if the union of all leaves sharing it is a residual set: the intersection of a countable sequence of open dense sets. Historically, the first generic property appeared in the literature states that for any foliation or lamination $(M, \mathcal{F})$ the union of all leaves with trivial holonomy is a residual set M_* (see [11] or [6]). Another obvious property is that if a foliation admits a dense (or recurrent) leaf, then its leaves are generically recurrent, the foliation being called *totally recurrent.* Following a pioneering work of Ghys [8], Cantwell and Conlon showed in [4] (see also [2]) that the leaves of a minimal foliation or lamination have generically the same endset; moreover, there are only three possibilities, namely: they have 1, 2, or a Cantor set of ends (situation similar to that of finitely generated infinite groups).

Our first goal will be to refine the results of [2, 4] and propose a new proof of them. Indeed, given $(M, \mathcal{F})$ we will search for a residual set $M_\bullet$ contained in M_* such that all leaves contained in $M_\bullet$ have the same endset, and in a second step we will describe the subset $M_* \backslash M_\bullet$ when it is nonempty.

More precisely, we provide a very short proof for the following:

Theorem A. *For any minimal lamination $(M, \mathcal{F})$ one of the following holds:*

(i) there exists a residual subset $M_\bullet \subset M_*$ such that all leaves of $M_\bullet$ have one end or all have two ends,
(ii) all leaves of M_* have a Cantor set of ends.

With similar arguments we also get an unexpected result of topological genericity:

Theorem B. *For any compact minimal lamination, the union of planar leaves is residual when it is nonempty.*

We will say that $(M, \mathcal{F})$ is *end-rigid* if all leaves without holonomy have the same endset or equivalently if $M_\bullet = M_*$. By Theorem A, this is the case for any foliation or lamination having generically a Cantor set of ends, and a result of Blanc in [3] asserts that it is also true for any foliation or lamination with generically two ends.

In order to describe the case of laminations with generically one end, we will use the notion of vanishing separatrix, similar to the well-known notion of vanishing cycle introduced by Novikov in [17]. Roughly speaking, a *nontrivially vanishing separatrix* in an n-dimensional lamination is a closed connected $(n-1)$-manifold Σ embedded in a leaf L of $\mathcal{F}$ such that:

(i) Σ separates L into two unbounded connected components,
(ii) Σ lifts homeomorphically to the nearby leaves, these lifts being generically null homologous in the corresponding leaf.

The precise definition will be given in Definition 2.12. Our second result concerns the structure of a particular class of laminations with generically one end: the so-called *generically tree-like* laminations (see Definition 2.9):

Theorem C. *For any compact minimal lamination $(M, \mathcal{F})$ which is generically tree-like and with generically one end, one of the following holds:*

(i) either $\mathcal{F}$ is end-rigid, i.e., $M_{\bullet} = M_{*}$,
(ii) or $\mathcal{F}$ supports a nontrivially vanishing separatrix.

There exist examples of both types: laminations by planes of Denjoy type for (i) (see 4.1) and the Ghys-Kenyon Cantor laminations for (ii): they have generically one end but are not end-rigid (see Examples 2.7).

(2) It is a natural question in foliation theory to ask whether any open manifold can be a leaf in some codimension one compact foliated manifold? The answer is no: a first example was provided by Ghys in [7]. Other authors asked the same question for isometric types of manifolds (see, for example, [20]). Here we raise the analogous question for compact Cantor laminations $(\mathcal{M}, \mathcal{L})$: can we embed them as an exceptional minimal set in a compact codimension one foliated manifold or as a compact subset in $\mathbb{R}^{n+1}$.

Vanishing separatrices will be the appropriate technical tool to be used for dealing with this question, but one should note that a vanishing separatrix in the minimal set $\mathcal{M}$ may be nonvanishing in M; therefore, we say that it is *sporadically vanishing* in M.

Our main result in this context is the following:

Theorem D. *Any sporadically vanishing separatrix in an exceptional minimal set $(\mathcal{M}, \mathcal{L})$ of a codimension one foliation $(M, \mathcal{F})$ is trivially vanishing; thus, if $\mathcal{L}$ is generically tree-like, it is also end-rigid. The same result holds for $(\mathcal{M}, \mathcal{L})$ embedded in $\mathbb{R}^{n+1}$.*

Theorem E. *No Ghys-Kenyon lamination embeds as an exceptional minimal set in a codimension one compact foliated manifold neither as a closed subset into $\mathbb{R}^{n+1}$.*

In contrast with this result, we will notice that Cantor laminations embed in codimension two.

All over the paper, we will deal with minimal foliations or laminations but it is clear that all statements and results extend to totally recurrent foliations. Also for the sake of simplicity, we will assume that all manifolds and structures considered here are orientable and transversely orientable but we will not make any differentiability assumption.

1 Preliminaries on Foliations and Laminations

In order to facilitate our descriptions below, we fix here some notations and recall some definitions, including very standard ones, for general foliations and laminations.

We consider first foliated manifolds and indicate briefly how to adapt our descriptions to laminations.

A local chart $\varphi : U \to \mathbb{R}^p$ of a p-manifold will be called a *nice open cube* if it extends as a homeomorphism $\bar{\varphi} : \overline{U} \to [0,1]^p \subset \mathbb{R}^p$.

Notations and definitions 1.1. *Holonomy pseudogroup of a foliation $\mathcal{F}$.*

1. Let $p = m + n$. A foliation $\mathcal{F}$ of dimension n and codimension m on the p-manifold M may be defined by a *nice foliated cocycle* $\mathcal{C} = (\{(U_i, f_i)\}, \{g_{ij}\})$, where

   (i) $\mathcal{U}$ is a locally finite cover of M (finite when M is compact),
   (ii) each U_i is a *a nice open p-cube* in M and if $U_i \cap U_j \neq \emptyset$, there exists a nice open cube U_{ij} (not necessarily belonging to $\mathcal{U}$) such that $\overline{U_i \cup U_j} \subset U_{ij}$,
   (iii) the *distinguished map* $f_i : U_i \to \mathbb{R}^m$ is a submersion whose fibers (the *plaques* of $\mathcal{F}$ in U_i) are nice n-cubes and the image $Q_i = f_i(U_i)$ is a nice open cube in $\mathbb{R}^m$ which represents the set of plaques of U_i,
   (iv) the local homeomorphisms $g_{ij} : \mathbb{R}^m \to \mathbb{R}^m$ verify the cocycle condition and relate the distinguished maps by

   $$f_i = g_{ij} \circ f_j.$$

   For each i, we identify Q_i with a section of f_i contained in U_i in such a way that all these subsets are pairwise disjoint. We call Q_i the *axis* of U_i and $Q = \coprod_j Q_i$ the *axis* of $\mathcal{F}$. Moreover for any plaque P in U_i, $x = P \cap Q_i$ is called the *center* of P.

2. Due to (ii) above, any plaque of U_i cuts at most one plaque of U_j; thus, the change g_{ij} of transverse coordinates determines a local homeomorphism generally defined on a proper open subset of Q_j (and denoted by the same symbol):

   $$g_{ij} : Q_j \to Q_i.$$

   The set $\Gamma = \{g_{ij}\}$ generates the *holonomy pseudogroup* of $\mathcal{F}$ (with respect to $\mathcal{C}$), a pseudogroup of local homeomorphisms of Q denoted by $(Q, \mathcal{P})$. This pseudogroup depends on $\mathcal{C}$, but for a suitable notion of isomorphism of pseudogroups its isomorphism class becomes independent of $\mathcal{C}$ and depends only on the foliation $\mathcal{F}$ (see [11]).

   It is important to note that, in general, Q is not compact even when M is compact.

3. For a $\mathbb{K}$-lamination, a nice open cube is by definition homeomorphic to $]0,1[^n \times T$ with T open in $\mathbb{K}$. In case of Cantor laminations, we require the distinguished maps $f_i : U_i \to \mathbb{K}$ to have clopen images so that the axis Q_i of each U_i is clopen (in particular compact) as well as the global axis Q of $\mathcal{F}$ when M is compact.

Notations and definitions 1.2. *Associated graphed structure and essential skeleton of* $(M, \mathcal{F})$.

1. Let $(Q, \mathcal{P})$ be the holonomy pseudogroup of $(M, \mathcal{F})$ as above and let $\Gamma = \{g_{ij}\}$ be the canonical generating set of $\mathcal{P}$. Denote by $\text{dom}(g_{ij}) \subset Q_j$ and $\text{im}(g_{ij}) \subset Q_i$ the domain and image of g_{ij}, respectively. We construct a topological space M^e as follows:

   (i) the basic piece is the axis Q of the foliated atlas $\mathcal{U}$,
   (ii) for any ij, we glue on Q the tube of edges $\mathcal{T}_{ij} = \text{dom}(g_{ij}) \times [0,1]$ by identifying $(x, 0) \in \text{dom}(g_{ij}) \times [0,1]$ with $x \in \text{dom}(g_{ij})$ and $(x, 1) \in \text{dom}(g_{ij}) \times [0,1]$ with $g_{ij}(x) \in \text{im}(g_{ij})$.

2. Observe that M^e inherits a "fine" or "foliated" topology when we endow Q and all the sets $\text{dom}(g_{ij})$ with the discrete topology. Thus, M^e becomes a foliated space with a lamination $\mathcal{F}^e$, whose leaves are locally finite graphs representing the orbits of $\mathcal{P}$ on Q. By construction the holonomy pseudogroup of $\mathcal{F}^e$ identifies with $(Q, \mathcal{P})$.

   Also embedding each tube $\mathcal{T}_{ij}$ in M, we can view M^e as immersed in M so that the leaves of $\mathcal{F}^e$ are the traces on M^e of the leaves of $\mathcal{F}$.

   We call $(M^e, \mathcal{F}^e)$ the *graphed structure* associated to $(M, \mathcal{F})$ (also called the Schreier continuum of $(M, \mathcal{F})$ in [12]). By means of the immersion $M^e \to M$ we can consider it as a kind of a 1-skeleton of $(M, \mathcal{F})$ and we call it also the *essential skeleton* of $(M, \mathcal{F})$. As we will see below, $(M^e, \mathcal{F}^e)$ inherits many dynamical properties of the foliation $\mathcal{F}$.
3. In case $(M, \mathcal{F})$ is a compact Cantor lamination, M^e will be compact and $\mathcal{F}^e$ will be a compact Cantor lamination by graphs.

Example 1.3. Let B and F be closed manifolds of dimension n and m, respectively. A group representation $h : \pi_1(B) \to \text{homeo}(F)$ defines by suspension an n-dimensional foliation $\mathcal{F}$ on a locally trivial bundle

$$F \longrightarrow M \xrightarrow{\pi} B$$

whose monodromy identifies with h. The holonomy pseudogroup of $\mathcal{F}$ is just generated by the group action $(F, \pi_1(B))$.

Let k be the cardinal of a finite set Γ of generators of $\pi_1(B)$ and let B^e be a join of k circles. Then h induces a representation $h^e : \pi_1(B^e) \to \text{homeo}(F)$ whose suspension will be the graphed structure $(M^e, \mathcal{F}^e)$ associated to $(M, \mathcal{F})$. The leaves of $\mathcal{F}$ are coverings of B and those of $\mathcal{F}^e$ are coverings of B^e.

In the general case, we obtain the following immediate but fundamental result.

Lemma 1.4. *If M is compact, any two corresponding leaves $L \in \mathcal{F}$ and $L^e = L \cap M^e$ of $\mathcal{F}^e$ have isomorphic endsets.*

Proof. This is trivial if $\mathcal{F}$ admits a leaf-wise Riemannian metric implying that L and L^e are quasi-isometric. It extends easily to the general topological case. □

2 Structure of Minimal Foliations and Laminations

In this section we seek the first goal of the paper: refine the genericity results of Cantwell and Conlon (see [2,4]) and propose a drastically simplified proof for them. After that we will be in a position to apply these results to the "embeddability" problem in Sect. 4. Here, we work in the context of laminations.

So let $(M^e, \mathcal{F}^e)$ be the graphed structure associated to the compact laminated space $(M, \mathcal{F})$ with holonomy pseudogroup $(Q, \mathcal{P})$. The leaf $L^e \in \mathcal{F}^e$ corresponding to $L \in \mathcal{F}$ is the complete connected subgraph generated by the set $L \cap Q$ of vertices. It may be equipped with its natural "graph metric" d and for $x \in Q$ and $r \in N$, we will denote by $\beta(x, r)$ the closed ball of center x and radius r in L^e.

Also for any subset $X \subset L \cap Q$, we denote by X^e the complete subgraph of L^e generated by the set of vertices X. Its Γ-*boundary* (*or boundary* for short) $\partial^\Gamma X^e$ is by definition the subset of all elements $x \in X$ for which there exists a generator $\gamma \in \Gamma$ such that $x \in \mathrm{dom}(\gamma)$ and $\gamma(x) \notin X$.

We will also use the $\mathcal{U}$-*fattening* $\hat{X}$ of X defined as the union of all $\mathcal{U}$-plaques P whose center $x = P \cap Q$ belongs to X.

We need one more notion.

Notations and definitions 2.1. 1. Let $\Sigma \subset L$ be a codimension one closed submanifold of a leaf $L \in \mathcal{F}$. We will say that Σ is *nice* (with respect to $\mathcal{C}$) if

(i) for any $\mathcal{F}$-plaque $P \subset U_i$ with center x, $P \cap \Sigma \neq \emptyset \Leftrightarrow x \in \Sigma$,
(ii) the complete subgraph $(\Sigma \cap Q)^e$ is contained in Σ.

It will be denoted by Σ^e and called the *essential skeleton* of Σ.

2. Next, a compact n-domain $D \subset L$ will be called *nice* if its boundary ∂D is nice. Its *essential skeleton* is given by $D^e = (D \cap Q)^e$. The skeletons of D and ∂D are related by $\partial D^e = (\partial D)^e$.

It is not difficult to verify that any codimension one submanifold Σ, and thus also any compact n-domain D, is isotopic to a nice one, so we will deal only with nice domains in the sequel.

2.1 General Genericity Properties

From now on, we assume that $(M, \mathcal{F})$ is minimal. For $r \in \mathbb{N}$, we denote by $W_r \subset Q$ the set of all points $x \in Q$ such that there exists a compact connected domain $D_r(x)$ in the leaf $L_x \in \mathcal{F}$ through x verifying the following conditions:

(i) $D_r(x)$ is nice and $\partial D_r(x)$ has one or two connected components,
(ii) $D_r(x) \supset \beta(x, r)$,
(iii) $D_r(x)$ is a submanifold with trivial holonomy.

For each r, $W_{r+1} \subset W_r$ and we set $W_\infty = \bigcap_r W_r$. Similar inclusions hold for the $\mathcal{U}$-fattenings: $\hat{W}_r$ and we set $\hat{W}_\infty = \bigcap_r \hat{W}_r$.

Recall that the union M_* of all leaves without holonomy of $\mathcal{F}$ is residual.

Lemma 2.2. *If not empty, $\hat{W}_\infty$ is a saturated residual subset of M_* and thus of M. Moreover either all leaves of $\hat{W}_\infty$ have one end or they all have two ends.*

Proof.

1. Indeed, suppose that W_∞ is not empty and take $x \in W_\infty$. The leaf $L_x^e \in \mathcal{F}^e$ through x verifies $\bigcup_r \beta(x, r) = L_x^e$; thus, by definition, we immediately get

$$L_x^e \subset \bigcup_r D_r(x) \quad \text{and consequently} \quad L_x = \hat{L}_x^e = \bigcup_r D_r(x).$$

   The latter relation implies immediately that $\hat{W}_\infty$ is nonempty, saturated, and dense in M; it is also contained in M_* by condition (ii) above. Moreover each $\hat{W}_r$ containing $\hat{W}_\infty$ is dense; it is open by condition (iii) and Reeb's local stability theorem Reeb's theorem (see [21]). It follows that $\hat{W}_\infty$ is residual.
2. Condition (ii) also implies that any leaf in $\hat{W}_\infty$ has at most two ends. If there exists one with only one end, we may repeat the previous argument by restricting to the subsets $W_r' \subset W_r$ defined by condition
   i') $D_r(x)$ is nice with connected boundary,
   instead of (i). We will conclude that $\hat{W}_\infty' = \bigcap_r \hat{W}_r'$ is residual and all its leaves have exactly one end. This finishes the proof. □

Recall that an end ϵ of a leaf L is defined, up to equivalence, by a decreasing sequence $\{V_q\}_{q \in \mathbb{N}}$ of unbounded domains $V_q \subset L$ with compact connected boundary ∂V_q and empty intersection. We say that this end *has trivial holonomy* if there exists q such that the domain V_q has trivial holonomy. We may assume that all V_q are nice and we get the following.

Lemma 2.3. *If there exists a leaf L_0 of $\mathcal{F}$ having an isolated end ϵ with trivial holonomy, then $\hat{W}_\infty = \bigcap_r \hat{W}_r$ is residual and $\mathcal{F}$ has generically one or two ends.*

Proof. Indeed, the fact that ϵ is isolated means that for any $q > 0$, the submanifold $V_{0q} = V_0 \backslash \mathrm{int}(V_q)$ is a nice compact connected domain whose boundary has exactly two connected components ∂V_0 and ∂V_q.

Fix $r \in N$; we may assume without loss of generality that the distance of the two sets ∂V_r^e and ∂V_0^e verifies $\mathrm{dist}(\partial V_r^e, \partial V_0^e) > r$ so that there exists $q(r)$ verifying

$$\beta(x, r) \subset V_{0q(r)}$$

for any $x \in V_r \cap Q$. By minimality of $\mathcal{F}$, V_r is dense in M and assuming that V_0 has trivial holonomy, we conclude that $(V_r \cap Q) \subset W_r$; thus, $\hat{W}_r$ is open dense and finally $\hat{W}_\infty$ is residual. □

We reach to the wanted result (Theorem A of the introduction):

Theorem 2.4. *For any minimal lamination $(M, \mathcal{F})$, one of the following holds:*

(i) there exists a residual subset $M_\bullet \subset M_*$ such that all leaves of $M_\bullet$ have one end or all have two ends,
(ii) all leaves of M_* have a Cantor set of ends.

Proof. From Lemma 2.3, we deduce the following alternative:

(a) either there exists a leaf in $\mathcal{F}$ which has an isolated end with trivial holonomy and we are in case (i) with $M_\bullet = \hat{W}_\infty$,
(b) or no leaf in M_* has an isolated end; any such leaf has a Cantor set of ends and we are in case (ii). □

The formulation of the previous theorem suggests the question whether, in case (i), the two sets $M_\bullet$ and M_* may differ or not. To state it in a precise way, we introduce the following definition:

Definition 2.5. We will say that the minimal lamination $(M, \mathcal{F})$ is *end-rigid* if all leaves without holonomy have the same endset that is $M_\bullet = M_*$.

According to Theorem 2.4, a minimal lamination whose leaves have generically a Cantor set of ends is end-rigid. On the other hand, E. Blanc proved the following in his thesis (see [3]):

Proposition 2.6. *If the leaves of a minimal lamination $(M, \mathcal{F})$ have generically two ends, then $\mathcal{F}$ is end-rigid, all leaves without holonomy have two ends, and any leaf in $M \backslash M_*$ has one end and nontrivial but finite holonomy.*

We will also propose a simplified proof of this claim in [13]. Now remains the question whether laminations with generically one end are end-rigid or not. Indeed, the following family of examples will show that they are not.

Examples 2.7. The Ghys-Kenyon laminations.

We present here a large family of compact Cantor laminations whose leaves have generically one end. All are based on a primary example of a compact Cantor lamination $(\mathcal{M}, \mathcal{L})$ foliated by trees and obtained as a subspace of the space of all subgraphs of the Cayley graph of the group $\mathbb{Z}^2$ with the Gromov-Hausdorff topology (see [9]). The construction is rather involved and we refer to [9] or [1] for a precise description. The relevant properties of $(\mathcal{M}, \mathcal{L})$ are the following:

(i) all leaves of $\mathcal{L}$ are trees and therefore have trivial holonomy,
(ii) the leaves of $\mathcal{L}$ have generically one end but some leaves have more than two ends; $\mathcal{L}$ is not end-rigid.

Next, for any n one associates with this primary example a Cantor lamination of dimension n with the same transverse structure, thus sharing the previous two properties. This is done in two steps: first thicken the graphed space $\mathcal{M}$ by replacing each vertex by the $(n+1)$-ball $\mathbb{B}^{n+1}$ and each edge $[0,1]$ by the thickened edge $[0,1] \times \mathbb{B}^n$; second, take the boundary of the space obtained this way: it will be foliated by n-manifolds.

All these examples are called *Ghys-Kenyon laminations.*

It is worth noticing that our method of proof for the main Theorem 2.4 extends to the topological setting (see also Proposition 2.11). This is Theorem B of "Introduction."

Theorem 2.8. *For any minimal lamination $(M, \mathcal{F})$ the union of all planar leaves is residual if not empty.*

Proof. Replacing condition (i) in the definition of W_r by (i″), each $D_r(x)$ is homeomorphic to the closed disk of dimension n, we define sets W''_r such that for any $x \in W''_\infty = \bigcap_r W''_r$, we get $L_x = \bigcup_r D_r(x)$ showing that L_x is a plane and the leaves of $\mathcal{F}$ are generically planes. □

2.2 End-Rigidity and Vanishing Separatrices

Our next goal is to investigate more precisely the notion of end-rigidity for minimal laminations with generically one end at least in the particular case of foliations which are "generically tree-like" (see Definition 2.9) for which we also get a result of topological genericity similar to Theorem 2.8.

For a space X we denote by $H_*^\infty(X)$ the homology group of locally finite simplicial chains with real coefficients. We need new technical tools.

Notations and definitions 2.9. *Separatrices and tree-like manifolds.*

Let Σ be a closed $(n-1)$-submanifold of an open connected n-manifold L. According to our general orientability assumption, it is two-sided, thus disconnects a fundamental family of neighborhoods.

We will say that Σ is a *separatrix* of L if it disconnects L into two connected components L^+ and L^- and it is *null-homologous* if one of these two components is compact. Note that an open manifold L has one end if and only if any separatrix in L is null homologous.

A $(n-1)$-manifold Σ is a separatrix if its homology class $[\Sigma] \in H_{n-1}^\infty(L)$ is zero; it is null homologous if $[\Sigma] \in H_{n-1}(L)$ is zero. Further, if Σ does not separate L, then $0 \neq [\Sigma] \in H_{n-1}(L)$ and Σ admits a dual loop γ which cuts Σ in exactly one point; Σ and γ cut transversely and both being oriented, their algebraic intersection $\theta \wedge \Sigma = \pm 1$.

Note also that if $H^{\infty}_{n-1}(L) = 0$, any closed connected $(n-1)$-submanifold is a separatrix whether null homologous or not. By analogy with trees, we will say that such a manifold is *tree-like*. Of course a tree is a tree-like complex, a surface will be tree-like if and only if its genus is zero, and a manifold with trivial first Betti number is tree-like.

Notations and definitions 2.10. *Transverse cylinders and towers in laminations.*

An open set or manifold embedded in a leaf of $\mathcal{F}$ will be called *horizontal*. Now it is a standard fact that any relatively compact open horizontal subset $X \subset L$ admits an open "tubular neighborhood" in M that is a disk bundle $\lambda : \Lambda \to X$ over X whose fibers are transverse to $\mathcal{F}$. For example, in the differentiable case, it can be constructed by local integration of the normal bundle.

(a) Next, suppose that X is a neighborhood in a leaf L of an embedded closed horizontal submanifold $Y \subset L$ of dimension $q < n$. The restriction of λ to $C(Y) = \lambda^{-1}(Y)$ is still a disk bundle with an induced foliation $\mathcal{C}(Y)$ of dimension q having Y as a proper compact leaf; in general, $\mathcal{C}$ has both compact and noncompact leaves. We call $C(Y)$ a *transverse cylinder over* Y.
(b) In case Y has trivial holonomy in $\mathcal{C}(Y)$, the transverse cylinder $C(Y)$ contains a transverse cylinder $T(Y)$ trivially foliated by the induced foliation $\mathcal{T}(Y)$ whose leaves are level sets Y_t homeomorphic to Y and indexed by a parameter $t \in T_y = \lambda^{-1}(y)$, the local transverse disk at a base-point $y \in Y$, Y itself being identified with Y_y. We call $T(Y)$ a *transverse tower over* Y.
(c) In case of a Cantor lamination, we can always assume that T_y identifies with a clopen subset of the transverse Cantor set and so $T(Y)$ is compact.

For example, if L has trivial holonomy in $\mathcal{F}$, we can always restrict a transverse cylinder to a transverse tower.

We leave it to the reader to transfer the previous notions and definitions to the setting of graphed structures.

As an immediate by-product we get a new genericity result:

Proposition 2.11. *For a minimal compact lamination* $(M, \mathcal{F})$ *the following are equivalent:*

(i) $\mathcal{F}$ *admits one leaf* $L_0 \subset M_*$ *which is tree-like,*
(ii) all leaves in M_* *are tree-like,*
(iii) $\mathcal{F}$ *is generically tree-like.*

Proof. Implications (ii) $\Rightarrow$ (iii) $\Rightarrow$ (i) are trivial so let us show that (i) $\Rightarrow$ (ii). We proceed by contradiction and suppose that there exists $L_1 \subset M_*$ admitting a closed $(n-1)$-submanifold Σ with nontrivial homology class $[\Sigma] \in H_{n-1}(L_1)$. Let $\gamma \subset L_1$ be a dual loop with algebraic intersection $\gamma \wedge \Sigma = +1$. As L_1 is without holonomy by assumption, we have two foliated towers $T(\Sigma)$ and $T(\gamma)$ foliated by level surfaces indexed by a common fiber T_u with $u = \gamma \cap \Sigma$. There exists an open set $T'_u \subset T_u$ such that for any $t \in T'_u$, we get

$$\gamma_t \wedge \Sigma_t = +1.$$

As $\mathcal{F}$ is minimal, $L_0 \cap T'_u \neq \emptyset$ and for any $t_0 \in L_0 \cap T'_u$ the submanifold $\Sigma_{t_0} \subset L_0$ is not a separatrix contradicting the fact that L_0 is tree-like. Our claim follows. □

We introduce now a technical tool which will prove essential for the description of laminations with generically one end. It is similar to the vanishing cycles introduced by Novikov for the study of foliations on three-manifolds (see [17]).

Definition 2.12. Consider a tower $T(\Sigma_u)$ over a separatrix Σ_u parametrized by a transverse disk T_u. We will say that Σ_u is *vanishing* if there exists an open set $T' \subset T_u$ such that $u \in \overline{T}'$ and the level surface Σ_t is null homologous for any $t \in T'$. It is *trivially vanishing* if Σ_u itself is null homologous.

We are now in a position to prove Theorem C:

Theorem 2.13. *Let $(M, \mathcal{F})$ be a minimal lamination which is generically tree-like. If $\mathcal{F}$ has generically one end, one of the following holds:*

(i) all leaves in M_* are tree-like with one end,
(ii) $\mathcal{F}$ admits a nontrivially vanishing separatrix.

Proof. Any leaf L_0 without holonomy and with more than one end supports a separatrix Σ_0 with a transverse tower $T(\Sigma_0) \cong \Sigma_0 \times T_0$. Then take $L_1 \subset M_*$ which is tree-like. The subset $T'_0 \subset T_0$ of all values t such that Σ_t is null homologous contains $T_0 \cap L_1$ and thus is dense in T_0. It is open by Reeb's stability. and different from T_0 because Σ_0 is a separatrix. We get a nontrivially vanishing separatrix Σ_u for any $u \in \overline{T}'_0 \backslash T'_0$. □

Of course, it would be much more satisfactory to have the result of Proposition 2.13 without restricting to tree-like laminations. Indeed, we conjecture that this more general statement holds (see 4.5).

2.3 Vanishing Separatrices and Transverse Invariant Measures

To finish the section, we associate to a nontrivially vanishing separatrix a transverse invariant measure for $(M, \mathcal{F})$. We refer to [14] for details on *averaging sequences* and associated *transverse invariant measures* as defined in [10, 19].

For any compact connected horizontal domain X in the graphed structure $(M^e, \mathcal{F}^e)$, we denote by $\sharp(X)$ the cardinality of $X \cap Q$, the number of vertices of X. A sequence $\{X_p\}_{p \in \mathbb{N}}$ of such domains is *a strong averaging sequence* if

$$\sharp(\partial X_p) \quad \text{is bounded while} \lim_{p \to \infty} \sharp(X_p) = \infty.$$

Then up to extracting an appropriate subsequence, we may assume that this sequence defines a measure μ on Q by setting

$$\mu(Y) = \lim_{p\to\infty} \frac{1}{\sharp(X_p)} \sharp(Y \cap X_p)$$

for any Borel set $Y \subset Q$. This measure is invariant by the holonomy pseudogroup $\mathcal{P}$ and thus extends to a transverse invariant measure for $\mathcal{F}$.

In the topological setting, a sequence $\{Y_p\}_{p\in\mathbb{N}}$ of nice compact horizontal domains in $(M,\mathcal{F})$ is *a strong averaging sequence* if

$$\sharp(\partial Y_p) \quad \text{is bounded while } \lim_{p\to\infty} \sharp(Y_p) = \infty,$$

where $\sharp(Y_p) = \sharp(Y_p^e)$ and $\partial\sharp(Y_p) = \partial\sharp(Y_p^e)$.

Now given a tower $T(\Sigma_u)$ as in Definition 2.12, choose a sequence $\{t_p\}_{p\in\mathbb{N}} \subset T'$ converging to u. For each p, Σ_{t_p} is null homologous; thus, there exists a compact domain $K_{t_p} \subset L_{t_p}$ such that $\Sigma_{t_p} = \partial K_{t_p}$. We suppose that all Σ_{t_p} and K_{t_p} are nice (in the sense of Notations and Definitions 2.1).

Proposition 2.14. *Let Σ_u be a nontrivially vanishing separatrix of $(M,\mathcal{F})$. With the previous notations, we get*

(i) $\sharp(\Sigma_{t_p}^e)$ is uniformly bounded, *(ii) $\lim_{p\to\infty} \sharp(K_{t_p}^e) = +\infty$.*

The sequence $\{K_{t_p}^e\}_{p\in\mathbb{N}}$ is a strong averaging sequence defining a transverse invariant measure μ for $(M,\mathcal{F})$.

Proof. Claim (i) is obvious and if (ii) does not hold, then, up to extracting convenient subsequences, we may assume successively that

(a) $\sharp(K_{t_p}^e)$ is constant independent of p,
(b) the finite complexes $K_{t_p}^e$ are all isomorphic,
(c) $\{K_{t_p}^e\}_{p\in\mathbb{N}}$ is a convergent sequence of compact subsets in M^e.

By continuity, the limit K_*^e of this sequence will be such that $\partial K_*^e = \Sigma_u^e$ implying that our separatrix Σ_u is trivially vanishing, a contradiction which proves the claim. □

3 Vanishing Separatrices in Codimension One Exceptional Minimal Sets

Here, we come to the second part of our study: we consider minimal Cantor laminations $(\mathcal{M},\mathcal{L})$ of dimension n embedded as exceptional minimal sets in transversely orientable codimension one compact foliated manifolds $(M,\mathcal{F})$. Our goal is to show that

(i) any vanishing separatrix of $(\mathcal{M},\mathcal{L})$ is trivially vanishing in $\mathcal{M}$,
(ii) these Cantor laminations are end-rigid in the sense of Definition 2.5 provided that they are generically tree-like.

We also show that the same result is valid for minimal codimension one foliations.

3.1 Cohomology Class Associated to a Transverse Invariant Measure

The appropriate tool to be used for the study of end-rigidity in the context of codimension one foliations will be the cohomology class naturally defined by a transverse invariant measure. This cohomology class was also considered by Levitt in [15] and is a particular case of Sullivan's "foliation cycles" (see [22]).

Construction 3.1. *The cohomology class $\chi_\mu \in H_1(M)$ for foliated manifolds.*

So let μ be a transverse invariant measure of a compact codimension one transversely orientable foliated manifold $(M, \mathcal{F})$. Suppose that $\mathcal{F}$ is defined by some nice atlas $\mathcal{U} = \{U_i, \varphi_i)\}$ with oriented axis $Q = \coprod Q_i$.

1. For any continuous path $c : [0,1] \to M$, there exists a finite sequence $0 = t_0 < t_1 < \cdots < t_s = 1$ such that each $c_j = c_{|[t_{j-1},t_j]}$, $j = 1, 2, \ldots, s$, is contained in some $U_{i_j} \in \mathcal{U}$. We denote by $\bar{c}_j$ the natural projection of c_j to the axis Q_{i_j} and define a function $\bar{\mu} : M^{[0,1]} \to \mathbb{R}$ by setting
   (i) $\bar{\mu}(c_j) = \pm\mu[\bar{c}_j(t_{j-1}), \bar{c}_j(t_j)]$ depending on the orientation of the interval $[\bar{c}_j(t_{j-1}), \bar{c}_j(t_j)]$ in Q_{i_j},
   (ii) $\bar{\mu}(c) = \sum_{j=1}^{s} \bar{\mu}(c_j)$.

   Then using the fact that any two nice coverings have a common refinement, one shows that the definition of $\bar{\mu}$ is independent of all special choices involved and does not depend on path homotopies, thus defining a *period homomorphism*

   $$\mathrm{Per}_\mu : \pi_1(M) \to \mathbb{R}.$$

   The latter factorizes through $H_1(M)$, thus defining the *associated cohomology class* $\chi_\mu \in H^1(M)$.
2. According to [14, vol. B, Chap. X, Theorem 2.3.2)], the support of any ergodic transverse invariant measure of a codimension one foliation reduces to exactly one minimal set $\mathcal{M}$. Thus, for any positively oriented closed transversal θ to $\mathcal{F}$ which cuts $\mathcal{M}$, we get

   $$\chi_\mu(\theta) > 0,$$

   showing that $H^1(M) \neq 0$ if $\mathcal{M}$ admits such a transversal.

A classical result of Sacksteder (see [19]) shows the converse and we get the following.

Proposition 3.2. *For any minimal set $\mathcal{M}$ of a codimension one transversely orientable foliation $(M, \mathcal{F})$, the two following conditions are equivalent:*

(i) $\mathcal{M}$ supports an invariant measure μ,
(ii) The intrinsic holonomy of the restriction $\mathcal{L}$ of $\mathcal{F}$ to $\mathcal{M}$ is trivial.

Here the intrinsic holonomy group of a leaf L in a minimal set $\mathcal{M}$ is the holonomy group of L with respect to the induced lamination $\mathcal{L}$: it is in general a proper subgroup of the holonomy group of L with respect to $\mathcal{F}$. Vanishing of the intrinsic holonomy of some leaf $L \in \mathcal{L}$ does not imply vanishing of the "global" holonomy group of L. We illustrate this fact with the following example:

Example 3.3. Consider the group $\mathcal{G}$ of orientation preserving homeomorphisms of $\mathbb{S}^1$ generated by two elements:

(a) a Denjoy-type homeomorphism D: it is fixed point free, preserves globally a Cantor subset $\mathbb{K} \subset \mathbb{S}^1$ and acts minimally on $\mathbb{K}$,
(b) a homeomorphism φ whose fixed points set coincides with $\mathbb{K}$.

The suspension of $\mathcal{G}$ defines a foliation $(M, \mathcal{F})$ on a three-manifold which is an $\mathbb{S}^1$-bundle over some closed surface. This foliation admits an exceptional minimal set $\mathcal{M}$ which is a $\mathbb{K}$-subbundle whose transverse structure is essentially generated by D; for any leaf $L \subset \mathcal{M}$, the intrinsic holonomy group is trivial while its "global" holonomy group is infinite cyclic generated by the germ of φ.

3.2 Minimal Foliated Compact Manifolds

Our first application concerns codimension one minimal foliations.

Proposition 3.4. *Let $(M, \mathcal{F})$ be a transversely orientable codimension one foliation on a closed manifold M. If $\mathcal{F}$ is minimal, any vanishing separatrix is trivially vanishing.*

Proof. Indeed, by Proposition 2.13, the existence of a nontrivially vanishing separatrix implies the existence of a nontrivial transverse invariant measure μ which in turn implies the triviality of holonomy (see Proposition 3.2). Now by a result of R. Sacksteder (see [19]), the foliation $\mathcal{F}$ is conjugate to a foliation defined by a closed one form. In this case all leaves are homeomorphic and they have either one or two ends (see [14]). In particular, any vanishing separatrix is trivially vanishing. □

As a consequence we get the following.

Theorem 3.5. *Any transversely orientable codimension one minimal foliation $(M, \mathcal{F})$ on a closed manifold M which is tree-like is end-rigid.*

3.3 Exceptional Minimal Sets and Sporadically Vanishing Separatrices

Next we focus on exceptional minimal sets $(\mathcal{M}, \mathcal{L})$ in codimension one compact foliated manifolds $(M, \mathcal{F})$; our goal is to establish a result similar to Theorem 3.5 for such minimal sets. A more precise description of vanishing separatrices in this context will be helpful. We adapt the general description of Notations and definitions 2.10; in particular, it will be convenient to consider transverse cylinders and towers which are compact.

Notations 3.6. Without loss of generality we may assume the existence of a one-dimensional foliation $\mathcal{F}^{\pitchfork}$ transverse to $\mathcal{F}$ defined by a flow $\Phi : M \times \mathbb{R} \to M$ (see [21] or [14] for more details). We use it for the construction of transverse cylinders and towers over pointed horizontal submanifolds.

1. So, let (Y, y) be a closed horizontal submanifold. For any compact transverse cylinder $C(Y)$ over Y there exists a compact neighborhood $[a, b]$ of 0 in $\mathbb{R}$ such that $C(Y) = \Phi(Y \times [a, b])$ with $Y = \Phi(Y \times \{0\})$; we denote it by $C_a^b(Y)$. For a tower $T(Y)$ we may assume, after a possible reparametrization of the flow Φ, that $\Phi : Y \times [a, b] \to T(Y)$ is a foliated homeomorphism; we identify $[a, b]$ with a compact arc in the transverse leaf T_y and denote the tower by $T_a^b(Y)$. Finally observe that cutting these cylinders and towers along Y, one gets *left and right half-cylinders and half-towers* over Y: they are parametrized by intervals of type $[a, y]$ and $[y, b]$, respectively.
2. In particular, if Y is contained in a leaf L of an exceptional minimal set $\mathcal{M}$, we will always choose the parametrized arc $[a, b]$ of a transverse cylinder or tower so that L_a and L_b are contained in $\mathcal{M}$, the first being semi-proper on the left and the second semi-proper on the right. We call the corresponding cylinders and towers *adapted* (to the embedding of the Cantor lamination). Finally if L has trivial intrinsic holonomy, the trace $\check{T}_a^b(Y)$ of $C_a^b(Y)$ on $\mathcal{M}$ will be a transverse tower over Y in the sense of (1) above; it is parametrized by the clopen Cantor set $J(a, b) = [a, b] \cap \mathcal{M}$. We call it a *restricted transverse tower* over Y. If the leaf L_y is semi-proper there exist such adapted towers parametrized by $[a, y]$ or $[y, b]$.

Now let us come to the description of vanishing separatrices Σ_u in the context of codimension one foliations. We assume Σ_u contained in a leaf of an exceptional minimal set $\mathcal{M}$ and vanishing in $\mathcal{M}$ but possibly not in M. According to Proposition 3.2, the minimal set $\mathcal{M}$ has trivial intrinsic holonomy so that $\mathcal{M}_* = \mathcal{M}$.

Observations and definitions 3.7. *Sporadically vanishing separatrices.*

Take a restricted adapted tower $\check{T}_a^b(\Sigma_u)$ as described in (2) of Notations 3.6 parametrized by the Cantor set $J(a, b)$. By definition of vanishing separatrices, there exists an open subset $S \subset J(a, b)$ such that $u \in \overline{S} \backslash S$ and Σ_t is null homologous for any $t \in S \cap J(a, b)$.

As S is the trace on $J(a, b)$ of some open set $\tilde{S} \subset Q$ we can restrict to a connected component $\tilde{S}'$ of $\tilde{S}$ and choosing $u \in \overline{S}' \backslash S'$, we obtain

(i) either a left half-tower $\check{T}_a^u(\Sigma_u)$ parametrized by a Cantor set $J(a, u)$ such that Σ_t is null homologous for any $t \in J(a, u), t \neq u$,
(ii) or a right half-tower $\check{T}_u^b(\Sigma_u)$ parametrized by a Cantor set $J(u, b)$ such that Σ_t is null homologous for any $t \in J(u, b), t \neq u$.

In the first case, we say that Σ_u is *vanishing to the left* in $\mathcal{M}$ and *vanishing to the right* in $\mathcal{M}$ in the second. In both cases it is *nontrivially vanishing* if Σ_u is not null homologous. In general, there does not exist any transverse tower over Σ_u in M. In other words Σ_u is a priori not vanishing in M that is why we say it is *sporadically vanishing in* M.

Finally note that in general, the half-towers $\check{T}_a^u(\Sigma_u)$ or $\check{T}_u^b(\Sigma_u)$ are not adapted towers in the sense of Notations 3.6 (2) unless L_u is semi-proper on the suitable side.

Next, we transfer the data given by the existence of a sporadically vanishing separatrix into the setting of singular homology. We consider a sporadically left-vanishing separatrix but of course a similar procedure would apply to a right-vanishing one.

Description 3.8. So, let $\Sigma_u \subset \mathcal{M}$ be a sporadically left-vanishing separatrix; let $C_a^u(\Sigma_u)$ and $\check{T}_a^u(\Sigma_u)$ be a corresponding pair of adapted left half-cylinder and restricted left half-tower over Σ_u as defined in Notations 3.6 (2). For any $t \in J(a,u), t \neq u$, Σ_t is null homologous, i.e., there exists a compact connected domain $K_t \subset L_t$ such that $\Sigma_t = \partial K_t$. Note that this domain is unique because if not we would produce a nontrivial n-cycle showing that the corresponding leaf L_t is compact contradicting the fact that it is exceptional.

As $\mathcal{L}$ has intrinsic trivial holonomy, we know by Notations 3.6 that there exists for each $t \in J(a,u)\backslash\{u\}$ an adapted pair $[C_{a_t}^{b_t}(K_t) \supset \check{T}_{a_t}^{b_t}(K_t)]$ such that the infinite sequence of intervals $[a_t, b_t]$ covers $J(a,u)\backslash\{u\}$. Now it is not difficult to show that we can select a countable increasing sequence $\{t_p\}_{p\in\mathbb{N}} \subset J(a,b)$ converging to u and for each p an arc $[a_p, b_p] \subset [a,b]$ such that

(i) $a_0 = a$ and we have an increasing sequence $\omega = \{a = a_0 < b_0 < \cdots < a_p < b_p < \ldots\}$ in $J(a,u)$ with upper bound u; in particular, all intervals $[a_p, b_p]$ are pairwise disjoint,
(ii) $\bigcup_p J(a_p, b_p) = J(a,u)\backslash\{u\}$.

Construction 3.9. *n-Cycles associated to a sporadically left-vanishing separatrix* Σ_u. We orient all manifolds under consideration: we choose an orientation for Σ_u, lift it to the foliation $\mathcal{C}(\Sigma_u)$ on $C_a^u(\Sigma_u)$, and endow the cylinder $C_a^u(\Sigma_u)$ itself with the product orientation by that of $\mathcal{F}^{\pitchfork}$. For any $t \in J(a,u)$ there exists a well-defined orientation for K_t such that $\Sigma_t = \partial K_t$ as oriented manifolds or as singular chains.

1. Then for any pair $x < y$ of elements in ω we define a n-cycle Δ_x^y of M by setting

$$\Delta_x^y = C_x^y(\Sigma_u) + K_x + K_y$$

with the orientations defined above. In particular, for any integer p, $\Delta_{a_p}^{b_p} = \partial C_{a_p}^{b_p}(K_{t_p})$ so that, up to a convenient subdivision of the chain $\Delta_a^{a_p}$, we get the relation

$$\Delta_a^{a_p} + \sum_{j=1}^{p-1} \partial C_{a_j}^{b_j}(K_{t_j}) = \sum_{j=1}^{p-1} \Delta_{b_j}^{a_{j+1}},$$

or at the level of homology classes in $H_n(M)$:

$$[\Delta_a^{a_p}] = \sum\nolimits_{j=1}^{p-1} [\Delta_{b_j}^{a_{j+1}}].$$

2. As M is compact, the subspace $E \subset H_n(M)$ generated by the infinite sequence of cycles $\{\Delta_{b_j}^{a_{j+1}}\}_{j\in\mathbb{N}}$ is finite dimensional and there exists an integer k such that E is generated by the finite subsequence $\{\Delta_{b_j}^{a_{j+1}}\}_{j=1}^{k}$. For any j, the intersection $\Delta_{b_j}^{a_{j+1}} \cap \mathcal{M}$ is reduced to the compact set $K_{a_{j+1}} \cup K_{b_j}$ contained in the union of two semi-proper leaves of $\mathcal{L}$. Then there exists a $(\mathcal{F}, \mathcal{F}^{\pitchfork})$-bidistinguished open cube U which meets $\mathcal{M}$ but not the compact set $[\bigcup_{j=1}^{k} \Delta_{b_j}^{a_{j+1}}]$, and for any totally exceptional leaf $L \subset \mathcal{M}$ there exist two different horizontal plaques in U with centers (z, w) in $U \cap Q \cap L$. We join these two points by a path $\sigma \subset L$ and a short transverse positive path τ in $U \cap Q$; the composition $\theta = \sigma * \tau$ is a loop in M such that

$$\theta \cap \Delta_{b_j}^{a_{j+1}} = \emptyset, \quad \text{thus} \quad \theta \wedge \Delta_{b_j}^{a_{j+1}} = 0$$

for any $1 \leq j \leq k$. Using (1) we obtain the final relation

$$\theta \wedge [\Delta_a^{a_p}] = 0 \quad \text{for any } p \in \mathbb{N}.$$

We reach to the central result of the paper (which is also the first part of Theorem D):

Theorem 3.10. *Any sporadically vanishing separatrix Σ_u of an exceptional minimal set $(\mathcal{M}, \mathcal{L})$ in a compact codimension one foliation $(M, \mathcal{F})$ is trivially vanishing.*

Proof. We suppose that Σ_u is left vanishing. According to Proposition 2.14, we may assume that $\mathcal{K} = \{K_{a_p}\}_{p\in\mathbb{N}}$ is a strong averaging sequence defining a transverse invariant measure μ for $\mathcal{F}$. As $\mathcal{K}$ is contained in $\mathcal{M}$, the support of μ equals $\mathcal{M}$.

Restricting to a subsequence if necessary, we may assume further that all domains K_{a_p} are positively oriented (with respect to the orientation of $\mathcal{F}$) so that the algebraic intersection of K_{a_p} with the transverse path τ introduced in Construction 3.9 (2) is given by

$$\tau \wedge K_{a_p} = \natural(\tau \cap K_{a_p}).$$

Moreover, note that the transverse loop $\theta = \sigma * \tau$ of Construction 3.9 satisfies

$$\sigma \cap K_a = \sigma \cap K_{a_p} = \tau \cap K_a = \tau \cap C_a^{a_p}(\Sigma_u) = \emptyset$$

so that finally

$$\theta \cap \Delta_a^{a_p} = [\sigma \cap C_a^{a_p}(\Sigma_u)] \cup [\tau \cap K_{a_p}].$$

Using again Construction 3.9 (2), we get for any p the relations

$$\begin{aligned} 0 = \theta \wedge \frac{1}{\natural(K_{a_p})}\Delta_a^{a_p} &= \frac{1}{\natural(K_{a_p})}(\sigma \wedge C_a^{a_p}(\Sigma_u)) + \frac{1}{\natural(K_{a_p})}(\tau \wedge K_{a_p}) \\ &= \frac{1}{\natural(K_{a_p})}(\sigma \wedge C_a^{a_p}(\Sigma_u)) + \frac{1}{\natural(K_{a_p})}\natural(\tau \cap K_{a_p}). \end{aligned}$$

But for any p, $\sigma \cap C_a^{a_p}(\Sigma_u)$ is a subset of the fixed finite set $\sigma \cap C_a^u(\Sigma_u)$; thus, $\lim_{p\to\infty} \frac{1}{\natural(K_{a_p})}(\sigma \wedge C_a^{a_p}(\Sigma_u)) = 0$ which implies immediately that

$$0 = \lim_{p\to\infty} \frac{1}{\natural(K_{a_p})}\natural(\tau \cap K_{a_p}) = \lim_{p\to\infty} \frac{1}{\natural(K_{a_p}^e)}\natural(\tau \cap K_{a_p}^e) = \bar{\mu}(\tau)$$

contradicting the fact that the measure μ supported by $\mathcal{M}$ is nontrivial. This achieves the proof for a left-vanishing separatrix but as the argument transposes readily to the case of right-vanishing separatrices, our proof is complete. □

Combining Theorem 2.4, Proposition 2.6, and Theorem 3.10, we get the following application:

Theorem 3.11. *An exceptional minimal set $(\mathcal{M}, \mathcal{L})$ of a codimension one, transversely orientable foliation $\mathcal{F}$ on a compact manifold M is end-rigid if one of the following conditions is satisfied:*

(i) $\mathcal{L}$ is tree-like and its leaves have generically one end,
(ii) the leaves of $\mathcal{L}$ have generically two or a Cantor set of ends.

Proof. Indeed, if the leaves of $\mathcal{L}$ have generically one end, we know by Theorem 3.10 that any sporadically vanishing separatrix is trivially vanishing, which means that $\mathcal{L}$ is end-rigid. □

Remark 3.12. It is worth noticing that the proof of Theorem 3.10 and consequently also that of Theorem 3.11 simplifies strongly for foliations of class C^2. Indeed, in this case, we know by a classical Theorem of Sacksteder that there exists a leaf in $\mathcal{M}$ with linear holonomy. This is an element of intrinsic holonomy and consequently $\mathcal{M}$ can support neither any transverse invariant measure nor any nontrivial vanishing separatrix.

This implies, in particular, the following.

Theorem 3.13. *For a foliation of class C^2, any exceptional minimal set is generically with one or a Cantor set of ends.*

4 Embeddability of Cantor Laminations

In this section, we discuss briefly the question of embeddability of Cantor laminations in codimension one. We also present more examples and state some related open problems.

Examples 4.1. Different types of exceptional minimal sets in C^0-foliations.

1. It is well known that for any minimal linear codimension one foliation $\mathcal{F}$ on the torus $\mathbb{T}^{n+1}$, all leaves are homeomorphic either to the n-plane or to a cylinder $\mathbb{T}^p \times \mathbb{R}^{n-p}$; these foliations are trivially tree-like and end-rigid. Performing a surgery along a closed transversal one implements infinitely many handles producing similar examples with all leaves non-tree-like.

   Now thickening one (or more) leaf of such a linear foliation in the same way as for the construction of the Denjoy homeomorphisms of $\mathbb{S}^1$ (compare [5]), one gets a foliation with a Cantor minimal set of the same type. This construction is possible in class C^1 but not in class C^2 according to Sacksteder's Theorem.
2. All leaves of an exceptional minimal set in an analytic foliation have a Cantor set of ends.

   Observe that any Ghys-Kenyon lamination is tree-like, without holonomy and not end-rigid. Using Theorem 3.11, we get the nonembeddability theorem (first part of Theorem E):

Theorem 4.2. *No Ghys-Kenyon lamination embeds as an exceptional minimal set in a codimension one foliation $(M, \mathcal{F})$ of class C^0.*

In a second step we extend the previous discussion to embeddings of n-dimensional Cantor laminations into $\mathbb{R}^{n+1}$.

Recall that Whitney's celebrated embedding theorem for compact manifolds extends `mutatis mutandis` to laminations: any compact Cantor lamination $(\mathcal{M}, \mathcal{L})$ of dimension n embeds into some Euclidean space $\mathbb{R}^p$ with $p > n$. Now one may ask whether it is possible to embed it into $\mathbb{R}^{n+1}$. To discuss this question we will use the following analogue of Theorem 3.10:

Theorem 4.3. *A Cantor lamination $(\mathcal{M}, \mathcal{L})$ of dimension n which embeds into $\mathbb{R}^{n+1}$ does not admit any transverse invariant measure. In particular, it is end-rigid if tree-like. Moreover, the same result holds true when replacing $\mathbb{R}^{n+1}$ by any simply connected $(n+1)$-manifold.*

Proof. (1) First given $(\mathcal{M}, \mathcal{L})$ embedded in $\mathbb{R}^{n+1}$, we construct a one-dimensional foliation $\mathcal{F}^{\pitchfork}$ pointwise transverse to $\mathcal{L}$ and defined on a simply connected neighborhood Ω of $\mathcal{M}$.

Indeed, dealing locally one constructs a germ of transverse foliation in a neighborhood W of $\mathcal{M}$ and extends it all over $\mathbb{R}^{n+1}$ with a closed set S of singularities. By the usual approximation trick, one reduces S to a countable family of isolated points and defines the wanted foliation $\mathcal{F}^{\pitchfork}$ by restricting the previous one to the complement Ω of S. As the singular points are isolated, Ω is simply connected and consequently $\mathcal{F}^{\pitchfork}$ is orientable.

Now any transverse invariant measure μ for $\mathcal{L}$ extends to the leaves of $\mathcal{F}^{\pitchfork}$, and dealing as in Sect. 3.1, one defines an associated cohomology class $\chi_\mu \in H^1(\Omega)$. It is of course trivial by simple connexity of Ω.

(2) To go on, we proceed as in Construction 3.9(2): we take a local coordinate chart V of $\mathbb{R}^{n+1}$ whose trace $U = V \cap \mathcal{M}$ is a distinguished local chart for $\mathcal{L}$. For any leaf $L \subset \mathcal{M}$ there exist two different horizontal plaques with centers (z, w) in $U \cap Q \cap L$. We join them by a path $\sigma \subset L$ and a short transverse positive path τ in $V \cap Q$ also contained in Ω; the composition $\theta = \sigma * \tau$ is a loop in Ω which verifies

$$\chi_\mu(\theta) = \mu(\theta) = \mu(\sigma) + \mu(\tau) = \mu(\tau) > 0,$$

contradicting the simple connexity of Ω. □

As a consequence of Theorem 4.3 and because all the laminations under consideration support a nontrivial transverse invariant measure, we get the following applications which complete the proofs of Theorems D and E:

Theorem 4.4. *(1) No Cantor lamination of dimension n whose leaves have generically two ends embeds into* $\mathbb{R}^{n+1}$.

(2) If $(\mathcal{M}, \mathcal{L})$ *is a Cantor lamination embedded in* $\mathbb{R}^{n+1}$ *which is generically tree-like with one end, then it is end-rigid.*

(3) No Ghys-Kenyon lamination embeds into $\mathbb{R}^{n+1}$.

Our study leaves open a number of natural questions about embeddability of Cantor laminations (in codimension 1); we provide here a nonexhaustive list:

Open questions 4.5.

1. Is it possible to extend Proposition 2.13 to general laminations with generically one end without assuming that the leaves are tree-like? Similarly, is Theorem 3.5 valid without the assumption that the foliations are tree-like?
2. One may also ask for an analogue of Blanc's Proposition 2.6. Is it true that a Cantor lamination, whose leaves have generically one end, has trivial holonomy when it is end-rigid ? Does it admit a transverse invariant measure and embed as an exceptional minimal set in a codimension one foliation?
3. We do not know any example of an end-rigid Cantor lamination with generically one end which embeds into $\mathbb{R}^{n+1}$. Do there exist such embeddable laminations?
4. It seems that all known examples of exceptional minimal sets of codimension one foliations which have generically a Cantor set of ends have nontrivial intrinsic holonomy. Might it be that this is a necessary condition for this sort of embeddings? A positive answer to this question would provide a kind of topological Sacksteder's Theorem!

5. In [18], Raymond constructs foliations on the three-sphere admitting an exceptional minimal set. Removing one point, one gets a Cantor lamination by surfaces with generically a Cantor set of ends embedded in $\mathbb{R}^3$; this lamination has non trivial intrinsic holonomy by Sacksteder's Theorem. Now one may ask whether there exist such examples with trivial intrinsic holonomy?

Remark 4.6 (Final remark). As observed in [16], any homeomorphism of the Cantor set embedded into the two-sphere $\mathbb{S}^2$ extends to the whole of $\mathbb{S}^2$. Thus, any group $\mathcal{G}$ of homeomorphisms of $\mathbb{K}$ extends to a group $\tilde{\mathcal{G}}$ of homeomorphisms of $\mathbb{S}^2$, in general, not isomorphic to $\mathcal{G}$. Suspending the action of $\tilde{\mathcal{G}}$, one defines a codimension two foliation which admits a minimal set defined by the suspension of $\mathcal{G}$. In other words, any minimal Cantor lamination defined by a group $\mathcal{G}$ embeds as an exceptional minimal set in a codimension two-foliation; this observation justifies our special interest in the codimension one case. Note that these foliations will bc only of class C^0 leaving open the corresponding question for differentiable foliations.

References

1. Alcalde-Cuesta F., Lojano-Rojo A. and Macho-Stadler M., *Dynamique transverse de la lamination de Ghys-Kenyon*, Astérisque, **323** (2009), 1–16.
2. Blanc E., *Propriétés génériques des laminations*, Thèse, Lyon (2001).
3. Blanc E., *Laminations minimales résiduellement à 2 bouts*, Comment. Math. Hlev., **78(4)** (2003), 845–864.
4. Cantwell J. and Conlon L., *Generic leaves*, Comment. Math. Helv., **73(2)** (1998), 306–336.
5. Denjoy A., *Sur les courbes définies par les équations différentielles à la surface du tore*, J. Math. Pures Appl., **11** (1932), 333–375.
6. Epstein D. B. A., Millet K. C. and Tischler D., *Leaves without holonomy*, J. London Math. Soc., **16** (1977), 548–552.
7. Ghys E., *Une variété qui n'est pas une feuille*, Topology, **24** (1985), 67–73.
8. Ghys E., *Topologie des feuilles génériques*, Ann. of Math., **141(2)** (1995), 387–422.
9. Ghys E., *Laminations par surfaces de Riemann*, Panor. Synthèses, **8** (1999), 49–95.
10. Goodman S. and Plante J., *Holonomy and averaging in foliated manifolds*, J. Differential Geom., **14** (1979), 401–407.
11. Hector G., *Feuilletages en cylindres*, Springer Lecture Notes **597** (1977), 252–270.
12. Hector G., *Topology of foliations and laminations*, ICMAT CSIC, Madrid, (2013).
13. Hector G., *Borel theory of foliations and laminations*, preprint, Lyon (2013).
14. Hector G. and Hirsch U., *Introduction to the geometry of Foliations*, Parts A and B, (1981) and (1983), Vieweg and Sohn, Braunschweig/Wiesbaden.
15. Levitt G., *Sur les mesures transverses invariantes d'un feuillletage de codimension* 1, C. R. Acad.Sci. Paris **290** (1980), 1139–1140.
16. Moise E., *Geometric Topology in dimensions 2 and 3*, GTM 47, Springer, Berlin (1977).
17. Novikov S. P., *Topology of foliations*, Trudy Mosk. Math. Obsch., **14** (1965), 248–278 and Trans. Moscow Math. Soc. 268–304.
18. Raymond B., *Ensembles de Cantor et feuilletages*, Publ. Math. d'Orsay, **191** (1976), 76–56.

19. Sacksteder R., *Foliations and pseudo-groups*, Amer. J. of Math., **87** (1965), 79–102.
20. Schweitzer P., *Surfaces not quasi-isometric to leaves of foliations of compact* 3-*manifolds*, Analysis and Geometry in foliated manifolds, Santiago de Compostela, World Sci. Publ., (1995), 223–238.
21. Siebenmann L. C., *Deformation of homeomorphisms on stratified sets*, Comment. Math. Helv. **47** (1972), 123–163.
22. Sullivan D., *Cycles for the dynamical study of foliated manifolds and complex manifolds*, Inv. Math., **36** (1976), 225–255.

Integral Formulas in Foliation Theory

Krzysztof Andrzejewski, Vladimir Rovenski, and Paweł Walczak

Abstract In this chapter we give an overview of integral formulas, and some of their consequences, appearing in the study of extrinsic geometry of foliations and distributions on Riemannian manifolds.

Keywords Riemannian manifold • Foliation • Integral formula • Curvature • Totally geodesic • Newton transformation

Mathematics Subject Classifications (2010): 53C12, 53C20

1 Introduction

Analyzing history of extrinsic geometry of foliations we see that from the origin it was related to some integral formulas containing the shape operator A (or the second fundamental form B) of leaves and its invariants (mean curvature h, higher order mean curvatures S_r, etc.) and some expressions corresponding to geometry (curvature) of M. These formulas are of some interest; in several geometric situations they provide obstructions to the existence of foliations with all the leaves

K. Andrzejewski (✉)
Department of Theoretical Physics and Computer Science, University of Łódź, Pomorska 149/153, 90-236 Łódź, Poland
e-mail: k-andrzejewski@uni.lodz.pl

V. Rovenski
Department of Mathematics, University of Haifa, Mount Carmel, Haifa 31905, Israel
e-mail: rovenski@math.haifa.ac.il

P. Walczak
Katedra Geometrii, Uniwersytet Lódzki, ul. Banacha 22, Lódź, Poland
e-mail: pawelwal@math.uni.lodz.pl

V. Rovenski and P. Walczak (eds.), *Geometry and its Applications*, Springer Proceedings in Mathematics & Statistics 72, DOI 10.1007/978-3-319-04675-4_4,

enjoying a given geometric property of foliations—totally geodesic (umbilical), minimal, constant mean curvature, etc. (see, [4, 8, 11, 19, 28, 30] and bibliographies therein). Such formulas have also applications in different areas of differential geometry and analysis on manifolds (see, for example [14, 16, 29]).

In this chapter we give an overview and summary of these formulas (certainly incomplete) and some of their consequences. Throughout the chapter everything (manifolds, submanifolds, foliations, etc.) is assumed to be C^∞-differentiable and oriented. For simplicity, we omit the volume form in integrals. Repeated indices denote summation over their range.

2 Integral Formulas

The first known integral formula (for codimension-one foliations) belongs to Reeb [22]. It says that the total mean curvature of the leaves of a codimension one foliation $\mathcal{F}$ on any closed Riemannian manifold equals zero, i.e.,

$$\int_M h = 0. \tag{1}$$

The proof of (1) is based on the divergence theorem and the identity $\operatorname{div} N = nh$ where N is a unit normal to $\mathcal{F}$ vector field and n the dimension of $\mathcal{F}$. One of the consequences of this formula (and its counterpart for foliated domains with boundary) provides the only obstruction for a function f on a closed foliated manifold to become the mean curvature with respect to some Riemannian metric. The conditions which are necessary and sufficient in this case read either $f = 0$ or f must change the sign (see [20]).

Formula (1) poses a generalization to the case of second-order mean curvature S_2 (see also [21, 31] for arbitrary codimension)

$$2\int_M S_2 = \int_M \operatorname{Ric}(N, N), \tag{2}$$

which is a direct consequences of Green's theorem applied to N. When $\dim \mathcal{F} = 1$ it reduces to Gauss theorem in the case Euler characteristic equals zero. Moreover, (2) posses a leaf-wise counterpart. Namely, for a closed leaf L we have

$$\int_L (\operatorname{Ric}(N, N) + \operatorname{tr}(A^2) + N(h) + |\nabla_N N|^2) = 0. \tag{3}$$

Both formulas have many applications. For example, (2) implies nonexistence of umbilical foliations on closed manifold of negative curvature, and in the case of constant mean curvature foliation $\mathcal{F}$ and nonnegative Ricci curvature, they imply that $\mathcal{F}$ is totally geodesic and N parallel [11]. Consequently, due to (2), we have the following assertion.

Theorem 1. *There is no codimension-one foliation of the Euclidean sphere whose leaves have constant mean curvature.*

Formulas (1) and (2) suggest the existence of similar ones for an arbitrary higher order mean curvatures (in general some functions of them) on a closed manifold. First step in this direction was done by Asimov, Brito et al. in [9, 13]. They showed the following theorem.

Theorem 2. *For a codimension one foliation of* $(n+1)$*-dimensional manifold* M *with constant sectional curvature* c *we have*

$$S_r^T := \int_M S_r = \begin{cases} c^{r/2}\binom{n/2}{r/2}\operatorname{vol}(M), & n, r \text{ even}, \\ 0, & n \text{ or } r \text{ odd}. \end{cases} \tag{4}$$

As a corollary we obtain that S_r^T depends only on geometry of M not $\mathcal{F}$. Proof of the above theorem is quite technical and it is based on some special differential forms (see (5) for $q = 1$).

Theorem 2 was generalized by Brito and Naveira [15] for a distribution $\mathcal{D}$ ($n = \dim \mathcal{D}$) of arbitrary codimension q. Namely, they introduce some differential forms Γ_r for even $r = 2s$ as follows

$$\Gamma_r = \sum\nolimits_{\sigma\in\Sigma_n} \varepsilon(\sigma)(\omega^{\sigma(1)\beta_1}\wedge\omega^{\sigma(2)\beta_1})\wedge\cdots\wedge(\omega^{\sigma(2s-1)\beta_s}\wedge\omega^{\sigma(2s)\beta_s})\wedge$$
$$\wedge\,\theta^{\sigma(2s+1)}\wedge\cdots\wedge\theta^{\sigma(n)}, \tag{5}$$

where $\omega^{i\alpha}(e_j) = \langle e_i, \nabla_{e_j} e_\alpha\rangle = -A^{\alpha i}{}_j$, θ^i orthonormal frames, for $i = 1,\ldots,n$; $\alpha = 1,\ldots,q$, Σ_n is the group of permutations of the set $\{1,\ldots,n\}$, $\varepsilon(\sigma)$ stands for the sign of the permutation σ. Furthermore, they define the total rth extrinsic mean curvature S_r^T of $\mathcal{D}$ on a compact manifold M as

$$S_r^T = \frac{1}{r!(n-r)!}\int_M \Gamma_r\wedge\nu,$$

where $\nu = \theta^{n+1}\wedge\cdots\wedge\theta^m$ ($m = \dim M$), and compute S_r^T for some distributions.

Theorem 3. *If* M *is a closed manifold of constant sectional curvature* $c \geq 0$ *and* $\mathcal{D}^\perp$ *is a totally geodesic distribution, then*

$$S_{2s}^T = \begin{cases} \binom{n/2}{s}\binom{q+2s-1}{2s}\binom{(q+2s-1)/2}{s}^{-1} c^s\operatorname{vol}(M) \\ \text{if } n \text{ is even and } q \text{ is odd}, \\ 2^{2s}(s!)^2((2s)!)^{-1}\binom{q/2+s-1}{s}\binom{n/2}{s} c^s\operatorname{vol}(M) \\ \text{if } n \text{ and } q \text{ are even}, \\ 0, \text{ otherwise}. \end{cases}$$

Let us note that, as in the case of codimension one, the integral does not depend on the distribution $\mathcal{D}$. However, the notion of higher order mean curvatures in arbitrary codimension, in contrast with codimension one, remained rather mysterious and more subtle.

Another method for the generalization of formula (4), and also more geometric definition of higher order mean curvatures in arbitrary codimension, has been proposed in [26, 27]. Using integration on the tangent sphere bundle ($S^\perp \subset \mathcal{D}^\perp$) authors define higher order mean curvatures in arbitrary codimension by the formula

$$\int_{N\in S^\perp(M)} S_r(C_N), \tag{6}$$

where $C_N(X) = -(\nabla_X N)^\top$ for $X \in \mathcal{D}$ is the *co-nullity tensor* of $\mathcal{D}$.

On the other hand, for a family of quadratic matrices $\mathbf{A} = (A_1, \dots, A_q)$ of order n and a multi-index $\boldsymbol{\lambda} = (\lambda_1, \dots, \lambda_q)$, we can define the *generalized elementary symmetric polynomials* $\sigma_{\boldsymbol{\lambda}}$ as the coefficients of the polynomial $\det(I + t_1 A_1 + \cdots, t_q A_q)$, i.e.,

$$\det(I + t_1 A_1 + \cdots + t_q A_q) = \sum\nolimits_{|\boldsymbol{\lambda}|\le n} \sigma_{\boldsymbol{\lambda}}\, t_1^{\lambda_1} \cdots t_q^{\lambda_q}, \tag{7}$$

where $|\boldsymbol{\lambda}| = \lambda_1 + \cdots + \lambda_q$. Observe that the coefficients $\sigma_{\boldsymbol{\lambda}} = \sigma_{\boldsymbol{\lambda}}(\mathbf{A})$ depend only on $\mathbf{A}$, and $\sigma_{(0,\dots,0)} = 1$. It is convenient to put $\sigma_{\boldsymbol{\lambda}} = 0$ for $|\boldsymbol{\lambda}| > n$.

Next, using "Jacobi tensor" $R_N^{\mathrm{mix}}(X) = R(X,N)N$ for $X \in \mathcal{D}$, authors obtain a series of integral formulas for foliated locally symmetric spaces.

Theorem 4. *Let $\mathcal{F}$ ($T\mathcal{F} = \mathcal{D}$) be a totally geodesic foliation and M locally symmetric then for any r we have*

$$\int_{S^\perp(M)} \Big(\sum\nolimits_{\|\boldsymbol{\lambda}\|=r} \sigma_{\boldsymbol{\lambda}}(B_{N,1}, \dots, B_{N,r})\Big) = 0, \tag{8}$$

where $\|\boldsymbol{\lambda}\| = \lambda_1 + 2\lambda_2 + \cdots + q\,\lambda_q$ and

$$B_{N,2k} = \frac{1}{(2k)!}(-R_N^{\mathrm{mix}})^k, \quad B_{N,2k+1} = \frac{1}{(2k+1)!}(-R_N^{\mathrm{mix}})^k C_N.$$

Since $\sigma_{(r,0,\dots,0)} = S_r$, one can find integral formula containing (6).

Next approach to higher order mean curvatures was proposed in [6]. The main idea is that we are looking for, naturally related to a geometry of a foliation, global vector fields on M and next we compute their divergence and use Stokes' theorem. In codimension-one, two canonical vector fields are hN (normal) and $\nabla_N N$ (tangent). Natural generalization of the first one is $S_r N$ and the generalization of the second one is based on the application of an operator T_r called Newton transformation (build of A) acting on $\nabla_N N$. More precisely, we define inductively operators

$$T_0 = I, \quad T_r = S_r I - A T_{r-1}, \quad 1 \leq r \leq n,$$

or, equivalently,

$$T_r = S_r I - S_{r-1} A + \cdots + (-1)^{r-1} S_1 A^{r-1} + (-1)^r A^r$$

which satisfy the following algebraic identities

$$\begin{aligned} \operatorname{tr}(T_r) &= (n-r) S_r, \\ \operatorname{tr}(A T_r) &= (r+1) S_{r+1}, \\ \operatorname{tr}(A^2 T_r) &= S_1 S_{r+1} - (r+2) S_{r+2}, \\ \partial_t (S_{r+1}) &= \operatorname{tr}(T_r \partial_t A). \end{aligned}$$

These operators arise naturally in the study of extrinsic geometry of hypersurfaces, see [18, 23, 24]. It is worth to notice that there is increasing number of applications of the Newton transformation in different areas of geometry in the last years (see, for example, [1–3, 10, 12, 23]).

Computing the divergence $\operatorname{div}_L(T_r \nabla_N N)$ along a leave L we obtain

$$\int_L (\langle \operatorname{div}_L T_r, \nabla_N N\rangle - N(S_{r+1}) + S_1 S_{r+1} - (r+2) S_{r+2} \\ + \operatorname{tr}(R_N T_r) + \langle \nabla_N N, T_r \nabla_N N\rangle) = 0, \quad (9)$$

where

$$\langle \operatorname{div}_L T_r, Y\rangle = \sum\nolimits_{j=1}^{r} \operatorname{tr}(R((-A)^{j-1} Y) T_{r-j}), \quad (10)$$

and the operator $R(Y) : \Gamma(\mathcal{F}) \to \Gamma(\mathcal{F})$ is given by

$$R(Y)(X) = R(X, Y)N, \quad X \in \Gamma(\mathcal{F}), \quad Y \in \Gamma(M).$$

Since we are interested in the full divergence we compute the divergence of the sum $S_{r+1} N + T_r(\nabla_N N)$; using Stokes' theorem and (9) one gets

Theorem 5. *Let $\mathcal{F}$ be a codimension-one foliation on a closed Riemannian manifold and N a unit normal of $\mathcal{F}$, then we have*

$$(r+2) \int_M S_{r+2} = \int_M \Big(\langle \operatorname{div} T_r, \nabla_N N\rangle + \operatorname{tr}(R_N T_r) \Big), \quad (11)$$

where $\operatorname{div} T_r$ *is given by* (10).

Example 1. Denote $Z = \nabla_N N$. For small r, $r = 1, 2$, (11) reads as

$$3\int_M S_3 = \int_M \Big(S_1 \mathrm{Ric}(N,N) - \mathrm{tr}(AR_N) + \mathrm{Ric}(N,Z)\Big), \tag{12}$$

$$4\int_M S_4 = \int_M \Big(\mathrm{tr}(T_2 R_N) - \langle T_2 Z, H^\perp\rangle - \langle T_2 Z, Z\rangle + \mathrm{tr}\left(T_2(R_{AZ,N} - R_{Z,N})\right)\Big), \tag{13}$$

where the linear operator $R_{X,Y} : \mathcal{D} \to \mathcal{D}$ is given by $R_{X,Y} : W \to R(W,X)Y^\top$.

In the case of constant sectional curvature we obtain (4). Moreover, the above result together with formula (9) implies the following theorem which generalizes Theorem 1 to the case of second mean curvature.

Theorem 6. *There is no codimension one foliation of the Euclidean sphere whose leaves have constant second-order mean curvature.*

Analyzing above applications of the Newton transformation in codimension one there arises a natural question about similar considerations in arbitrary codimension and their relations with the results obtained by Brito and Naveira. A formula for a general distribution $\mathcal{D}$ of codimension q on a closed Riemannian manifold (M, g) was obtained in [31] (see [21] for a foliation)

$$\int_M K^{\mathrm{mix}} = \int_M \left(\|H\|^2 + \|H^\perp\|^2 - \|B\|^2 - \|B^\perp\|^2 + \|T\|^2 + \|T^\perp\|^2\right) = 0, \tag{14}$$

where $K^{\mathrm{mix}} = \sum_{i\le n,\ \alpha\le q} g(R(e_i, e_\alpha)e_\alpha, e_i)$ is the *mixed scalar curvature* and e_i $(i \le n)$, e_α $(\alpha \le q)$ is a local orthonormal frame adapted to $\mathcal{D}$ and $\mathcal{D}^\perp$. Here, T and $T^\perp$ are the integrability tensors of the distributions, and H and $H^\perp$ are their mean curvature vectors. For $q = 1$, (14) reduces to (2).

One approach was proposed in [7] (and it is based on the suitable definitions for submanifolds [17]). Namely, let B_i^j be the matrix elements of the second fundamental form, then for even $r \in \{1, \dots, n\}$ we define rth mean curvature S_r of the distribution $\mathcal{D}$ by

$$S_r = \frac{1}{r!}\delta^{i_1\dots i_r}_{j_1\dots j_r}\langle B^{j_1}_{i_1}, B^{j_2}_{i_2}\rangle \cdots \langle B^{j_{r-1}}_{i_{r-1}}, B^{j_r}_{i_r}\rangle,$$

rth mean curvature vector field by

$$\mathbf{S}_{r+1} = \frac{1}{(r+1)!}\delta^{i_1\dots i_{r+1}}_{j_1\dots j_{r+1}}\langle B^{j_1}_{i_1}, B^{j_2}_{i_2}\rangle \cdots \langle B^{j_{r-1}}_{i_{r-1}}, B^{j_r}_{i_r}\rangle B^{j_{r+1}}_{i_{r+1}},$$

and finally Newton transformation by

$$T_r{}^i_j = \frac{1}{r!}\delta^{i_1\dots i_r i}_{j_1\dots j_r j}\langle B^{j_1}_{i_1}, B^{j_2}_{i_2}\rangle \cdots \langle B^{j_{r-1}}_{i_{r-1}}, B^{j_r}_{i_r}\rangle,$$

where the generalized Kronecker symbol $\delta^{i_1\ldots i_r}_{j_1\ldots j_r}$ is $+1$ or -1 according the i's are distinct, and the j's are either even or odd permutation of the i's; and is 0 in all other cases.

In spite of rather complicated definitions the main relations between S_r and T_r are similar to the case of codimension one (for instance, $\mathrm{tr}(T_r) = (n-r)S_r$) and we have the following relations between S_r and Γ_r:

$$\frac{1}{r!(n-r)!}\Gamma_r \wedge \nu = S_r \Omega.$$

Moreover, in the case of constant sectional curvature and a totally geodesic complementary distribution we can compute explicitly the divergence of $\boldsymbol{S}_{r+1}$, and using Stokes' theorem one obtains a recurrence

$$S^T_{r+2} = \int_M S_{r+2} = \int_M \frac{c(n-r)(q+r)}{(r+q)(r+2)} S_r, \tag{15}$$

which can be explicitly solved giving another proof of Theorem 3.

The second way of application of the Newton transformation in arbitrary codimension was proposed in [25]. In this approach author defines global $(1,1)$-tensor field C on $\mathcal{D}$ by taking the following integral at a point $x \in M$:

$$C(x) = \int_{N \in S^\perp_x(M)} T_r(C_N),$$

where $T_r(C_N)$ is the Newton transformation associated with the co-nullity tensor C_N. If we compute the divergence of $C(Z)$ for appropriate vector field $Z \in \Gamma(\mathcal{D})$, we obtain new series of integral formulas on M, as well as along leaves, containing higher order mean curvatures and some terms related to the curvature of M

$$\begin{aligned}(r+2)\int_{S^\perp(M)} S_{r+2}(N) = \int_{S^\perp(M)} \Big(&\underline{\langle \mathrm{div}_{\mathcal{D}}\, T^*_r(C_N), Z\rangle} - \langle T_r(C_N)Z, H^\perp\rangle \\ &+ \mathrm{tr}(T_r(C_N)\, R_N) + \sum\nolimits_{\alpha\le q} \langle T_r(C_N)(\nabla_{e_\alpha} N^\top), \nabla_N\, e_\alpha\rangle\Big),\end{aligned}$$

where $Z = \nabla_N N^\top$ and the underlined term is given by

$$\begin{aligned}\langle \mathrm{div}_{\mathcal{D}}\, T^*_r(C_N),\, Z\rangle = \sum\nolimits_{1\le j\le r} \Big(&\mathrm{tr}\left(T_{r-j}(C_N) R_{(-C_N)^{j-1}Z,\,N}\right) \\ &- \sum\nolimits_{\alpha\le q} \langle (-C^*_N)^{j-1}(C_{e_\alpha} - C^*_{e_\alpha}) T_{r-j}(C_N)\nabla_{e_\alpha} N^\top,\, Z\rangle\Big).\end{aligned}$$

For $r=0$ this yields (14). For $r=2$ and integrable $\mathcal{D}$ we find total S_4

$$
\begin{aligned}
4\int_{S^{\perp}(M)} S_4(N) = \int_{S^{\perp}}\Big(&\operatorname{tr}(T_2(A_N)R_N) - \langle T_2(A_N)Z, H^{\perp}\rangle \\
&+\operatorname{tr}\big(T_2(A_N)(R_{A_N Z,N} - R_{Z,N})\big) \\
&+\sum\nolimits_{\alpha\le q}\langle T_2(A_N)(\nabla_{e_\alpha} N^{\top}), \nabla_N\, e_\alpha\rangle\Big).
\end{aligned} \tag{16}
$$

For a codimension one foliation (tangent to $\mathcal{D}$), (16) reduces to (13).

The notion of extrinsic curvatures (for a distribution of arbitrary codimension) has been recently generalized in [5]. Namely, for any multi-index $\boldsymbol{\lambda}=(\lambda_1,\dots,\lambda_q)$ of length $|\boldsymbol{\lambda}|=\lambda_1+\dots+\lambda_q$, they introduce the transformation $T_{\boldsymbol{\lambda}}$ depending on a system of linear endomorphisms. More precisely, for a system of linear endomorphisms $\mathbf{A}=(A_1,\dots,A_q)$ they use invariants $\sigma_{\boldsymbol{\lambda}}$, see (7), to define

$$
T_{(0,\dots,0)} = 1, \tag{17}
$$

$$
\begin{aligned}
T_{\boldsymbol{\lambda}} &= \sigma_{\boldsymbol{\lambda}} 1 - \sum\nolimits_{\alpha} A_\alpha T_{\alpha_\flat(\boldsymbol{\lambda})} \\
&= \sigma_{\boldsymbol{\lambda}} 1 - \sum\nolimits_{\alpha} T_{\alpha_\flat(\boldsymbol{\lambda})} A_\alpha,
\end{aligned}
\qquad \text{if } |\boldsymbol{\lambda}|\ge 1 \tag{18}
$$

where $\alpha_\flat(i_1,\dots,i_q) = (i_1,\dots,i_{\alpha-1}, i_\alpha - 1, i_{\alpha+1},\dots,i_q)$. Since these transformations are similar to the classical Newton transformation they are called the *generalized Newton transformations*.

Let $\pi : P \to M$ be the principal bundle of orthonormal frames (oriented orthonormal frames, respectively) of $\mathcal{D}^{\perp}$ with the structure group G (which is always either the full orthogonal group or the special orthogonal group). Each element $(x,e)=(e_1,\dots,e_q)\in P_x$, $x\in M$, induces the system of endomorphisms

$$
\mathbf{A}(x,e) = (A_1(x,e),\dots,A_q(x,e))
$$

of $\mathcal{D}_x$, where $A_\alpha(x,e)$ is the shape operator corresponding to (x,e), i.e.,

$$
A_\alpha(x,e)(X) = -(\nabla_X e_\alpha)^{\top}, \quad X\in\mathcal{D}_x. \tag{19}
$$

Let $T_{\boldsymbol{\lambda}}(x,e)$ be the generalized Newton transformation associated with the operator $\mathbf{A}(x,e)$. Taking an average over a fiber we obtain a set of globally defined functions $\widehat{\sigma_{\boldsymbol{\lambda}}}$

$$
\widehat{\sigma_{\boldsymbol{\lambda}}}(x) = \int_{P_x}\sigma_{\boldsymbol{\lambda}}(x,e)\,de = \int_G \sigma_{\boldsymbol{\lambda}}(x,e_0 a)\,da, \tag{20}
$$

and we call them *extrinsic curvatures* of a distribution $\mathcal{D}$. Similarly to the codimension-one case (i.e., for S_r and T_r), we find total extrinsic curvatures in terms of generalized Newton transformations and second fundamental forms of

$\mathcal{D}$ and $\mathcal{D}^\perp$. In the special case of totally geodesic orthogonal distribution $\mathcal{D}^\perp$ and constant sectional curvature c of M, they reduce to the following recurrence formula:

$$\begin{aligned} |\boldsymbol{\lambda}|\sigma^M_{\boldsymbol{\lambda}} &= c\sum\nolimits_\alpha \int_P \operatorname{tr}(T_{\alpha^2_\flat(\boldsymbol{\lambda})}) = c\sum\nolimits_\alpha \int_P (n-|\boldsymbol{\lambda}|+2)\sigma_{\alpha^2_\flat(\boldsymbol{\lambda})} \\ &= c(n-|\boldsymbol{\lambda}|+2)\sum\nolimits_\alpha \sigma^M_{\alpha^2_\flat(\boldsymbol{\lambda})} \end{aligned} \tag{21}$$

with the initial conditions $\sigma^M_{(0,\ldots,0)} = \operatorname{vol}(P)$ and $\sigma^M_{\alpha^\sharp(0,\ldots,0)} = 0$, where $\alpha^\sharp(i_1,\ldots,i_q) = (i_1,\ldots,i_{\alpha-1},i_\alpha+1,i_{\alpha+1},\ldots,i_q)$. In this case, the total extrinsic curvatures do not depend on geometry of the distribution $\mathcal{D}$.

Summarizing our considerations we see that there has been the rise of integral formulas and their applications in foliation theory during the last years. Definitely, there are much more integral formulas and their consequences than we have indicated here.

Acknowledgements The first author and the third author (K. A. and P. W.) were supported by the Polish NSC grant N6065/B/H03/2011/40 and the second author (V. R.) was supported by the Marie-Curie actions grant EU-FP7-P-2010-RG, No. 276919.

References

1. L. J. Alias, A. G. Colares, Uniqueness of spacelike hypersurfaces with constant higher order mean curvature in generalized Robertson–Walker spacetimes, Math. Proc. Camb. Phil. Soc. 143, (2007) 703–729.
2. L. J. Alias, J. M. Malacarne, Constant scalar curvature hypersurfaces with spherical boundary in Euclidean space, Rev. Mat. Iberoamericana 18, (2002) 431–442.
3. L. J. Alias, S. de Lira, J. M. Malacarne, Constant higher-order mean curvature hypersurfaces in Riemannian spaces, J Inst. of Math. Jussieu 5(4), (2006) 527–562.
4. K. Andrzejewski, Jacobi fields and the stability of minimal foliations of arbitrary codimension, Tohoku Math. J. 62 (2010), 393–400.
5. K. Andrzejewski, W. Kozłowski, K. Niedziałomski, Generalized Newton transformation and its applications to extrinsic geometry, preprint (2012) arXiv:1211.4754.
6. K. Andrzejewski, P. Walczak, The Newton Transformations and New Integral Formulae for Foliated Manifolds, Ann. Glob. Anal. Geom. (2010), Vol. 37, 103–111.
7. K. Andrzejewski, P. Walczak, Extrinsic curvatures of distributions of arbitrary codimension, J. Geom. Phys. 60 (2010), no. 5, 708–713.
8. K. Andrzejewski, P. Walczak, Conformal fields and the stability of leaves with constant higher order mean curvature, Differential Geom. Appl. 29 (2011), no. 6, 723–729.
9. D. Asimov, Average Gaussian curvature of leaves of foliations. Bull. Amer. Math. Soc. 84, (1978), 131–133.
10. J. L. M. Barbosa, A. G. Colares, Stability of hypersurfaces with constant r–mean curvature, Ann. Global Anal. Geom. 15 (1997), 277–297.
11. J. L. M. Barbosa, K. Kenmotsu, G. Oshikiri, Foliations by hypersurfaces with constant mean curvature, Mat. Z. 207 (1991) 97–107.

12. A. Barros, P. Sousa, Compact graphs over a sphere of constant second order mean curvature. Proc. Am. Math. Soc. 137(9), (2009) 3105–3114.
13. F. Brito, R. Langevin, H. Rosenberg, Intégrales de courbure sur des variétés feuilletées, J. Diff. Geom. 16 (1980), 19–50.
14. F. Brito, P. Chac6n and A. M. Naveira, On the volume of unit vector fields on spaces of constant sectional curvature, Comment. Math. Helv. 79 (2004) 300–316.
15. F. Brito, A. M. Naveira, Total extrinsic curvature of certain distributions on closed spaces of constant curvature, Ann. Global Anal. Geom. 18, 371–383, (2000).
16. V. Brinzanescu, R. Slobodeanu, Holomorphicity and the Walczak formula on Sasakian manifolds. J. Geom. Phys. 57 (2006), no. 1, 193–207.
17. L. Cao, H. Li, r-Minimal submanifolds in space forms. Ann. Global Anal. Geom. 32, (2007) 311–341.
18. X. Cheng, H. Rosenberg, Embedded positive constant r–mean curvature hypersurfaces in $M^m \times \mathbb{R}$, An. Acad. Brasil. Cienc. 77 (2005), no. 2, 183–199.
19. M. Czarnecki, P. Walczak, Extrinsic geometry of foliations. In: Foliations 2005, ed. P. Walczak et all., pp 149–167. World Scientific, Singapore (2006).
20. G. Oshikiri, A characterization of the mean curvature functions of codimension-one foliations, Tohoku Math. J. 49 (1997), 557–563.
21. A. Ranjan, Structural equations and an integral formula for foliated manifolds, Geom. Dedicata 20 (1986), 85–91.
22. G. Reeb, Sur la courbure moyenne des varits intgrales dune quation de Pfaff $\omega = 0$, C. R. Acad. Sci. Paris 231 (1950), 101–102.
23. R. Reilly, Variational properties of functions of the mean curvatures for hypersurfaces in space forms, J. Differential Geom. 8 (1973), 465–477.
24. H. Rosenberg, Hypersurfaces of Constant Curvature in Space of Forms, Bull. Sci. Math. (1993), Vol. 117, 211–239.
25. V. Rovenski, Integral formulae for a Riemannian manifold with two orthogonal distributions. Cent. Eur. J. Math. 9 (2011), no. 3, 558–577.
26. V. Rovenski, P. Walczak, Integral formulae for foliations on Riemannian manifolds. In: Diff. Geom. and Appl., Proc. of Int. Conf., Olomouc, 2007 203–214. World Scientific, Singapore (2008)
27. V. Rovenski, P. Walczak, Integral formulae on foliated symmetric spaces. Math. Ann. 352 (2012), no. 1, 223–237.
28. V. Rovenski, P. Walczak, Topics in Extrinsic Geometry of Codimension-One Foliations, Springer Briefs in Mathematics, Springer (2011).
29. M. Svensson, Holomorphic foliations, harmonic morphisms and the Walczak formula. J. London Math. Soc. (2) 68 (2003), no. 3, 781–794.
30. P. Tondeur, Geometry of Foliations, Birkhäuser-Verlag (1997).
31. P. Walczak, An integral formula for a Riemannian manifold with two orthogonal complementary distributions. Colloq. Math. 58 (1990), no. 2, 243–252.

Prescribing the Mixed Scalar Curvature of a Foliation

Vladimir Rovenski and Leonid Zelenko

Abstract We introduce the flow of metrics on a foliated Riemannian manifold (M, g), whose velocity along the orthogonal (to the foliation $\mathcal{F}$) distribution $\mathcal{D}$ is proportional to the mixed scalar curvature, $\mathrm{Scal}_{\mathrm{mix}}$. The flow preserves harmonicity of foliations and is used to examine the question: When does a foliation admit a metric with a given property of $\mathrm{Scal}_{\mathrm{mix}}$ (e.g., positive/negative or constant)? If the mean curvature vector of $\mathcal{D}$ is leaf-wise conservative, then its potential function obeys the nonlinear heat equation $(1/n)\partial_t u = \Delta_{\mathcal{F}}\, u + (\beta_{\mathcal{D}} + \Phi/n)u + (\Psi_1^{\mathcal{F}}/n)u^{-1} - (\Psi_2^{\mathcal{F}}/n)u^{-3}$ with a leaf-wise constant Φ and known functions $\beta_{\mathcal{D}} \geq 0$ and $\Psi_i^{\mathcal{F}} \geq 0$. We study the asymptotic behavior of its solutions and prove that under certain conditions (in terms of spectral parameters of Schrödinger operator $\mathcal{H}_{\mathcal{F}} = -\Delta_{\mathcal{F}} - \beta_{\mathcal{D}}\,\mathrm{id}$) the flow of metrics admits a unique global solution, whose $\mathrm{Scal}_{\mathrm{mix}}$ converges exponentially to a leaf-wise constant. Hence, in certain cases, there exists a $\mathcal{D}$-conformal to g metric, whose $\mathrm{Scal}_{\mathrm{mix}}$ is negative, positive, or negative constant.

Keywords Foliation • Riemannian metric • Conformal • Mixed scalar curvature • Mean curvature vector • Parabolic PDE • Schrödinger operator • Twisted product

Mathematics Subject Classification (2010): Primary 53C12, Secondary 53C44

V. Rovenski (✉) • L. Zelenko
Mathematical Department, University of Haifa, Haifa, Israel
e-mail: rovenski@math.haifa.ac.il.

V. Rovenski and P. Walczak (eds.), *Geometry and its Applications*, Springer Proceedings in Mathematics & Statistics 72, DOI 10.1007/978-3-319-04675-4_5,

Introduction

In this section we discuss the question on prescribing the mixed scalar curvature of a foliation and define the flow of leaf-wise conformal metrics depending on this kind of curvature.

1. Geometry of Foliations. Let (M^{n+p}, g) be a connected closed (i.e., compact without boundary) Riemannian manifold, endowed with a p-dimensional foliation $\mathcal{F}$ (i.e., a partition into submanifolds (called leaves) of the same dimension p), and ∇ the *Levi-Civita connection* of g. The tangent bundle to M is decomposed orthogonally as $T(M) = \mathcal{D}_{\mathcal{F}} \oplus \mathcal{D}$, where the distribution $\mathcal{D}_{\mathcal{F}}$ is tangent to $\mathcal{F}$. Denote by $(\cdot)^{\mathcal{F}}$ and $(\cdot)^{\perp}$ projections onto $\mathcal{D}_{\mathcal{F}}$ and $\mathcal{D}$, respectively.

The second fundamental tensor and the mean curvature vector field of $\mathcal{F}$ are given by $h_{\mathcal{F}}(X,Y) = (\nabla_X Y)^{\perp}$ and $H_{\mathcal{F}} = \operatorname{Tr}_g h$, where $X, Y \in \mathcal{D}_{\mathcal{F}}$. A Riemannian manifold may admit many geometrically interesting foliations. Totally geodesic (i.e., $h_{\mathcal{F}} = 0$) and harmonic (i.e., $H_{\mathcal{F}} = 0$) foliations are among these kinds that enjoyed a lot of investigation of geometers (see survey in [8]). Simple examples are parallel circles or winding lines on a flat torus and a Hopf family of great circles on the sphere S^3. Similarly, we define the second fundamental tensor $h(X,Y) = \frac{1}{2}(\nabla_X Y + \nabla_Y X)^{\mathcal{F}}$ and the integrability tensor $T(X,Y) = \frac{1}{2}[X, Y]^{\mathcal{F}}$, where $X, Y \in \mathcal{D}$, of the distribution $\mathcal{D}$. The mean curvature vector of $\mathcal{D}$ is given by $H = \operatorname{Tr}_g h$. A foliation $\mathcal{F}$ is said to be *Riemannian*, or *transversely harmonic*, if, respectively, $h = 0$, or $H = 0$. *Conformal* foliations (i.e., $h = (1/n)\, H \cdot g_{|\mathcal{D}}$) were introduced by Vaisman [13] as foliations admitting a transversal conformal structure.

One of the principal problems of geometry of foliations is the following, see [10]:

Given a foliation $\mathcal{F}$ on a manifold M and a geometric property (P), does there exist a Riemannian metric g on M such that $\mathcal{F}$ enjoys (P) with respect to g?

Such problems of the existence and classification of metrics on foliations (first posed explicitly by H. Gluck for geodesic foliations) have been studied intensively by many geometers in the 1970s.

A foliation is *geometrically taut* if there is a Riemannian metric making $\mathcal{F}$ harmonic. Sullivan [12] provided a *topological tautness* condition for geometric tautness. By the Novikov Theorem (see [2]) and Sullivan's results, the sphere S^3 has no two-dimensional taut foliations. In the recent decades, several tools for proving results of this sort have been developed. Among them, one may find Sullivan's *foliated cycles* and new *integral formulae*, see [14] and a survey in [10].

2. The Mixed Scalar Curvature. There are three kinds of Riemannian curvature for a foliation: tangential, transversal, and mixed (a plane that contains a tangent vector to the foliation and a vector orthogonal to it is said to be mixed). The geometrical sense of the mixed curvature follows from the fact that for a totally geodesic foliation, certain components of the curvature tensor, see [8], regulate the deviation of leaves along the leaf geodesics. In general relativity, the *geodesic deviation equation* is an equation involving the Riemann curvature tensor, which measures the change in separation of neighboring geodesics or, equivalently, the tidal force experienced by a rigid body moving along a geodesic. In the language of mechanics it measures the rate of relative acceleration of two particles moving forward on neighboring geodesics.

Let $\{E_i, \mathcal{E}_a\}_{i\le n, a\le p}$ be a local orthonormal frame on $T(M)$ adapted to $\mathcal{D}$ and $\mathcal{D}_\mathcal{F}$. The *mixed scalar curvature* is the following function: $\mathrm{Scal}_{\mathrm{mix}} = \sum_{i=1}^n\sum_{a=1}^p R(\mathcal{E}_a, E_i, \mathcal{E}_a, E_i)$, see [8,10,14]. If either $\mathcal{D}$ or $\mathcal{D}_\mathcal{F}$ is one-dimensional and is tangent to a unit vector field N, then the mixed scalar curvature is simply the Ricci curvature $\mathrm{Ric}(N,N)$. On a foliated surface (M^2, g) this coincides with the Gaussian curvature K. Recall the formula, see [14]:

$$\mathrm{Scal}_{\mathrm{mix}}(g) = \mathrm{div}(H + H_\mathcal{F}) + \|H\|^2 + \|H_\mathcal{F}\|^2 + \|T\|^2 - \|h\|^2 - \|h_\mathcal{F}\|^2. \quad (1)$$

The norms of tensors are $\|h_\mathcal{F}\|^2 = \sum_{a,b}\|h_\mathcal{F}(\mathcal{E}_a, \mathcal{E}_b)\|^2$, $\|h\|^2 = \sum_{i,j}\|h(E_i,E_j)\|^2$, and $\|T\|^2 = \sum_{i,j}\|T(E_i,E_j)\|^2$. Integrating (1) over a closed manifold and using the Divergence Theorem, we obtain the integral formula with the total $\mathrm{Scal}_{\mathrm{mix}}(g)$. Thus, (1) yields decomposition criteria for foliated manifolds under constraints on the sign of $\mathrm{Scal}_{\mathrm{mix}}$ (see [14] and a survey in [8]).

The basic question that we address in the chapter is the following: *When a foliation admits a metric with a given property of* $\mathrm{Scal}_{\mathrm{mix}}$ (*e.g., constant, positive, or negative*)?

Example 1. For any $n \ge 2$ and $p \ge 1$ there exists a fiber bundle with a closed $(n+p)$-dimensional total space and compact p-dimensional totally geodesic fibers, having constant mixed scalar curvature. To show this, consider the Hopf fibration $\tilde\pi : S^3 \to S^2$ of a unit sphere (S^3, g_{can}) by great circles (closed geodesics). Let $(\tilde F, g_1)$ and $(\tilde B, g_2)$ be closed Riemannian manifolds with dimensions, respectively, $p-1$ and $n-2$. Let M be the product $\tilde F\times S^3\times\tilde B$ with the metric $g = g_1\times g_{\mathrm{can}}\times g_2$. Then $\pi : M \to S^2\times\tilde B$ is a fibration with a totally geodesic fiber $\tilde F\times S^1$. Certainly, $\mathrm{Scal}_{\mathrm{mix}} \equiv 2$.

We shall examine the question above using evolution equations. A *flow of metrics* on a manifold is a solution g_t of a differential equation $\partial_t g = S(g)$, where the geometric functional $S(g)$ is a symmetric $(0,2)$-tensor usually related to some kind of curvature. Rovenski and Walczak [10] (see also [11]) studied flows of metrics on a foliation that depend on the extrinsic geometry of the leaves and posed the following question:

Given a geometric property (P) *of a submanifold, can one find a flow* $\partial_t g = S(g)$ *on a foliation* $(M,\mathcal{F})$ *such that the solution metrics* g_t $(t \ge 0)$ *converge to a metric for which* $\mathcal{F}$ *enjoys* (P)?

The notion of the $\mathcal{D}$-*truncated* $(r,2)$-tensor $S^\perp$ (where $r = 0,1$) will be helpful: $S^\perp(X_1,X_2) = S(X_1^\perp, X_2^\perp)$. The $\mathcal{D}$-truncated metric tensor $g^\perp$ is given by $g^\perp(X_1,X_2) = g(X_1,X_2)$ and $g^\perp(Y,\cdot) = 0$ for all $X_i \in \mathcal{D}$, $Y \in \mathcal{D}_\mathcal{F}$. A flow of $\mathcal{D}$-*conformal* metrics is represented by $S^\perp(g) = s(g)\,g^\perp$, where $s(g)$ is a smooth function on the space of metrics on M. We study the flow of metrics g_t, see also [9]:

$$\partial_t g = -2\,(\mathrm{Scal}_{\mathrm{mix}}(g) - \Phi)\, g^{\perp}, \tag{2}$$

where $\Phi : M \to \mathbb{R}$ is a leaf-wise constant function; its value is clarified in what follows.

The flow (2) preserves harmonicity, total umbilicity, or total geodesy of foliations (see Sect. 1.3). We ask the following question (see [9] and also [4, Problem 15]):

Given a Riemannian manifold (M, g) with a harmonic foliation $\mathcal{F}$, when do solution metrics g_t of (2) *converge to a limit metric $\bar g$ with* $\mathrm{Scal}_{\mathrm{mix}}(\bar g)$ *positive, negative, or constant?*

In the case of a general foliation, the topology of the leaf through a point can change dramatically with the point; this gives many difficulties in studying truncated flows of metrics and leaf-wise parabolic PDEs. Therefore, we assume, at least at the first stage of study,

(a) the leaves to be compact, (b) the manifold M to be fibered (instead of being foliated). (3)

Example 2. (a) Let (M^2, g) be a two-dimensional Riemannian manifold (surface) of Gaussian curvature K, endowed with a unit geodesic vector field N. Certainly, (2) reduces to the following view:

$$\partial_t g = -2\,(K(g) - \Phi)\, g^{\perp} \tag{4}$$

which looks like the normalized Ricci flow on surfaces but uses the truncated metric $g^{\perp}$ instead of g. Let k be the geodesic curvature of curves orthogonal to N. From (4) we obtain the PDE $\partial_t k = K_{,x}$ (along a trajectory $\gamma(x)$ of N). The above yields the *Burgers equation* $\partial_t k + (k^2)_{,x} = k_{,xx}$, which is the prototype for advection–diffusion processes in gas and fluid dynamics, and acoustics. When k and K are known, the metrics may be recovered as $g_t^{\perp} = g_0^{\perp} \exp\big(-2\int_0^t (K(s,t) - \Phi)\,\mathrm{d}s\big)$.

(b) For the Hopf fibration $\pi : (S^{2m+1}, g_{\mathrm{can}}) \to \mathbb{C}P^m$ of a unit sphere with fiber S^1, by (1) we have $\mathrm{Scal}_{\mathrm{mix}} = 2m$. Thus, the metric g_{can} on S^{2m+1} is a fixed point of flow (2) with $\Phi = 2m$.

3. Structure of the Chapter. The solution strategy is based on deducing from (2) the forced Burgers-type equation

$$\partial_t H + \nabla^{\mathcal{F}} \|H\|^2 = n \nabla^{\mathcal{F}} (\mathrm{div}_{\mathcal{F}} H) + X,$$

for a certain vector field X, see Proposition 1. If H is leaf-wise conservative, i.e., $H = -n\nabla^{\mathcal{F}} \log u$ for a leaf-wise smooth function $u(x,t) > 0$, this and (1) yield the nonlinear heat equation

$$(1/n)\,\partial_t u = \Delta_{\mathcal{F}}\, u + (\beta_{\mathcal{D}} + \Phi/n)\, u + (\Psi_1^{\mathcal{F}}/n)\, u^{-1} - (\Psi_2^{\mathcal{F}}/n)\, u^{-3}, \quad u(\,\cdot\,,0) = u_0, \tag{5}$$

where functions $\beta_{\mathcal{D}}(x) \geq 0$ and $\Psi_i^{\mathcal{F}}(x) \geq 0$ are known, and $\Delta_{\mathcal{F}}$ is the leaf-wise Laplacian, see [2]. We study the asymptotic behavior of its solutions and prove that under certain conditions (in terms of spectral parameters of Schrödinger operator $\mathcal{H}_{\mathcal{F}} = -\Delta_{\mathcal{F}} - \beta_{\mathcal{D}}\,\mathrm{id}$) flow (4) has a unique global solution g_t, whose $\mathrm{Scal}_{\mathrm{mix}}$ converges exponentially to a leaf-wise constant. Thus, in certain cases, there exists a $\mathcal{D}$-conformal to g metric, whose $\mathrm{Scal}_{\mathrm{mix}}$ is negative, positive, or negative constant.

Section 1 contains main results (Theorems 1–4 and Corollaries 1–5), their proofs and examples for one-dimensional case and for twisted products. These are supported by results of Sect. 2 (Theorems 5–7) about nonlinear PDE (5) on a closed Riemannian manifold. In Sect. 2.2 we examine (5) for the modeling case when a leaf F is a circle S^1 and the coefficients $\beta_{\mathcal{D}}$ and $\Psi_i^{\mathcal{F}}$ are constants, and in Sects. 2.3–2.6 we study the general case.

1 Main Results, Proofs and Examples

1.1 Main Results

Define the operations with the leaf-wise derivatives: the divergence $\mathrm{div}_{\mathcal{F}} X := \sum_{\alpha=1}^{p} g(\nabla_\alpha X, \mathcal{E}_\alpha)$ of a vector field X and the Laplacian $\Delta_{\mathcal{F}}\, u := \mathrm{div}_{\mathcal{F}}(\nabla^{\mathcal{F}} u)$ of a leaf-wise smooth function u, where $\nabla^{\mathcal{F}} u := (\nabla u)^{\mathcal{F}}$. Notice that $\nabla^{\mathcal{F}}$, $\mathrm{div}_{\mathcal{F}}$, and $\Delta_{\mathcal{F}}$ are t-independent for the flow of metrics (4).

Based on the "linear algebra" inequality $n\,\|h\|^2 \geq \|H\|^2$ (with the equality when $\mathcal{D}$ is totally umbilical, i.e., $\mathcal{F}$ is conformal), we introduce the following measure of "nonumbilicity" of $\mathcal{D}$:

$$\beta_{\mathcal{D}} := n^{-2}\big(n\,\|h\|^2 - \|H\|^2\big) \geq 0. \tag{6}$$

For $p = 1$, we have $\beta_{\mathcal{D}} = n^{-2}\sum_{i<j}(k_i - k_j)^2$, where k_i are the principal curvatures of $\mathcal{D}$ see [6, Sect. 4.1].

The Schrödinger operator is central to all of quantum mechanics. By Lemma 3 (in Sect. 1.3), the flow of metrics (2) preserves the leaf-wise Schrödinger operator $\mathcal{H}_{\mathcal{F}}$ given by

$$\mathcal{H}_{\mathcal{F}}(u) = -\Delta_{\mathcal{F}}\, u - \beta_{\mathcal{D}}\, u. \tag{7}$$

The spectrum of $\mathcal{H}_{\mathcal{F}}$ on any compact leaf F is an infinite sequence of isolated real eigenvalues $\lambda_0^{\mathcal{F}} \leq \lambda_1^{\mathcal{F}} \leq \cdots \leq \lambda_j^{\mathcal{F}} \leq \cdots$ counting their multiplicities, and $\lim_{j\to\infty}\lambda_j^{\mathcal{F}} = \infty$. One may fix in $L_2(F)$ an orthonormal basis of corresponding eigenfunctions $\{e_j\}$, i.e., $\mathcal{H}_{\mathcal{F}}(e_j) = \lambda_j^{\mathcal{F}} e_j$. If all leaves are compact, then $\lambda_j^{\mathcal{F}}$ are leaf-wise constant functions and $\{e_j\}$ are leaf-wise smooth functions on M.

If the leaf $F(x)$ through $x \in M$ is compact, then $\lambda_0^{\mathcal{F}} \leq 0$ (since $\beta_{\mathcal{D}} \geq 0$) and the eigenfunction e_0 (called the *ground state*) may be chosen positive, see Proposition 3. The *fundamental gap* $\lambda_1^{\mathcal{F}} - \lambda_0^{\mathcal{F}} > 0$ of $\mathcal{H}_{\mathcal{F}}$ has mathematical and physical implications (e.g., in refinements of Poincaré inequality and a priori estimates); it is also used to control the rate of convergence in numerical methods of computation. Note that the least eigenvalue of operator $-\Delta_{\mathcal{F}}\, u - (\beta_{\mathcal{D}} + \frac{\Phi}{n})\, u$ is $\lambda_0^{\mathcal{F}} - \frac{\Phi}{n}$.

An important step in the study of evolutionary PDEs is to show short-time existence/uniqueness.

Theorem 1. *Let $\mathcal{F}$ be a harmonic foliation on a closed Riemannian manifold (M, g_0). Then the linearization of* (2) *at g_0 is a leaf-wise parabolic PDE; hence,* (2) *under assumptions* (3) *has a unique smooth solution g_t defined on a positive time interval* $[0, t_0)$.

We shall say that a smooth function $f(t,x)$ on $(0,\infty) \times F$ converges to $\bar f(x)$ as $t \to \infty$ in C^∞, if it converges in C^k-norm for any $k \ge 0$. It converges *exponentially fast* if there exists $\omega > 0$ (called the *exponential rate*) such that $\lim_{t\to\infty} e^{\omega t} \| f(t,\cdot) - \bar f\|_{C^k} = 0$ for any $k \ge 0$. Define $d_{u_0,e_0} := \min_F(u_0/e_0)/\max_F(u_0/e_0) > 0$.

The following theorems are central results of the work.

Theorem 2. *Let $\mathcal{F}$ be a harmonic foliation on a closed Riemannian manifold (M, g_0) with assumptions* (3)*, and $H_0 = -n\nabla^{\mathcal{F}} \log u_0$ for a smooth function $u_0 > 0$. If Φ obeys the inequality*

$$\Phi \ge n\lambda_0^{\mathcal{F}} + d_{u_0,e_0}^{-4} \max_F \|T\|_{g_0}^2, \tag{8}$$

then (2) *has a unique smooth global solution g_t $(t \ge 0)$, and for any $\alpha \in (0, \min\{\lambda_1^{\mathcal{F}} - \lambda_0^{\mathcal{F}}, 2\,(\Phi/n - \lambda_0^{\mathcal{F}})\})$ we have the leaf-wise convergence in C^∞, as $t \to \infty$, with the exponential rate $n\alpha$:*

$$\mathrm{Scal}_{\mathrm{mix}}(g_t) \to n\lambda_0^{\mathcal{F}} - \Phi \le 0, \qquad H_t \to -n\nabla^{\mathcal{F}} \log e_0, \qquad h_{\mathcal{F}}(g_t) \to 0.$$

For $T = 0$, condition (8) becomes $\Phi \ge n\lambda_0^{\mathcal{F}}$, and we have the following.

Corollary 1. *Let $\mathcal{F}$ be a harmonic foliation with integrable normal distribution on a closed Riemannian manifold (M, g_0) with assumptions* (3) *and $H_0 = -n\nabla^{\mathcal{F}} \log u_0$ for a smooth function $u_0 > 0$. If $\Phi \ge n\lambda_0^{\mathcal{F}}$, then the claim of Theorem 2 holds.*

Theorem 3. *Let $\mathcal{F}$ be a harmonic foliation on a closed Riemannian manifold (M, g_0) with assumptions* (3)*, and $H_0 = -n\nabla^{\mathcal{F}} \log u_0$ for a smooth function $u_0 > 0$. Suppose that $\sqrt{2}\,\max_M \|T\|_{g_0} < \min_M \|h_{\mathcal{F}}\|_{g_0}$. If $d_{u_0/e_0}^{\,2} > \sqrt{2}\,\max_M \|T\|_{g_0}/\min_M \|h_{\mathcal{F}}\|_{g_0}$ holds, then the interval*

$$I_0 = \Big(\max\big\{0,\ 3\, d_{u_0,e_0}^{-4} \max_M \|T\|_{g_0}^2 - \min_M \|h_{\mathcal{F}}\|_{g_0}^2\big\},\ \frac{1}{4}\, d_{u_0,e_0}^{\,4} \min_M \|h_{\mathcal{F}}\|_{g_0}^4 / \max_M \|T\|_{g_0}^2\Big) \tag{9}$$

is nonempty, and for any Φ satisfying the condition $n\lambda_0^{\mathcal{F}} - \Phi \in I_0$, flow (2) *admits a unique smooth solution g_t ($t \ge 0$), and it converges in C^∞ exponentially fast to a limit metric $\bar g = \lim_{t\to\infty} g_t$; moreover, we have the exponential convergence $\mathrm{Scal}_{\,\mathrm{mix}}(g_t) \to \Phi$, as $t \to \infty$, in C^∞ along the leaves.*

For $T = 0$ and $h_{\mathcal{F}} \ne 0$, the bounds of I_0 become simpler, and we have the following.

Corollary 2. *Let $\mathcal{F}$ be a harmonic foliation on a closed Riemannian manifold (M, g_0) with assumptions* (3). *Suppose that the normal distribution is integrable, $h_{\mathcal{F}} \ne 0$, and $H_0 = -n\nabla^{\mathcal{F}} \log u_0$ for a smooth function $u_0 > 0$. If $\Phi \le n\lambda_0^{\mathcal{F}}$, then the claim of Theorem 3 holds.*

The (co)dimension one versions of Corollaries 1 and 2 are discussed in Sect. 1.4. The above results are summarized (due to the basic question) in the following.

Corollary 3. *Let $\mathcal{F}$ be a harmonic foliation on a closed Riemannian manifold (M, g) with assumptions* (3) *and $H = -n\nabla^{\mathcal{F}} \log u_0$ for a smooth function $u_0 > 0$.*

(*i*) *Then for any $c > d_{u_0,e_0}^{-4} \max_F \|T\|_g^2$ there exists a $\mathcal{D}$-conformal to g metric $\bar g$ with $\mathrm{Scal}_{\mathrm{mix}}(\bar g) \le -c$.*
(*ii*) *If $\xi^2\|h_{\mathcal{F}}\|_g^2 < \xi^4\|T\|_g^2 + d_{u_0,e_0}^{-4} \max_F \|T\|_g^2$, where $\xi = u_0/(\tilde u_0^0 e_0)$ and $\tilde u_0^0$ is defined in Sect. 2.4, then there exists a $\mathcal{D}$-conformal to g metric $\bar g$ with $\mathrm{Scal}_{\mathrm{mix}}(\bar g) > 0$.*
(*iii*) *If $\sqrt{2}\max_F \|T\|_{g_0} < d_{u_0,e_0} \min_F \|h_{\mathcal{F}}\|_{g_0}$, then there exists a $\mathcal{D}$-conformal to g metric $\bar g$ with $\mathrm{Scal}_{\mathrm{mix}}(\bar g) = \mathrm{const} < 0$.*

Example 3. If $\beta_{\mathcal{D}} = 0$, then $\lambda_0^{\mathcal{F}} = 0$. This appears for twisted products $B \times_{\varphi_t} \bar M$ of (B, g) and $(\bar M, \bar g)$, i.e., $M = B \times \bar M$ with metrics $g_t = g + \varphi_t^2\,\bar g$ ($t \ge 0$), where $\varphi_t \in C^\infty(B \times \bar M)$ are positive functions, see [7]. The mean curvature vector $H = -n\,\nabla^{\mathcal{F}} \log\varphi$ is leaf-wise conservative.

The leaves $B \times \{y\}$ of a twisted product compose a totally geodesic foliation $\mathcal{F}$, while the fibers $\{x\} \times \bar M$ are totally umbilical with the leaf-wise conservative mean curvature vector $H = -n\,\nabla^{\mathcal{F}} \log\varphi$. If metrics g_t solve (2), then H obeys the Burgers-type equation $\partial_t H + \nabla^{\mathcal{F}}\|H\|^2 = n\nabla^{\mathcal{F}}(\mathrm{div}_{\mathcal{F}} H)$, see (19) with $T = 0$, $h_{\mathcal{F}} = 0$, and $\beta_{\mathcal{D}} = 0$. The function $\tilde\varphi := e^{-\Phi t}\varphi$ obeys the heat equation $\partial_t\tilde\varphi = n\,\Delta_{\mathcal{F}}\tilde\varphi$. Let B and $\bar M$ be closed and $g_t = g + \varphi_t^2\,\bar g$ solve (2) with $\Phi = 0$, then g_t converge as $t \to \infty$ in C^∞ with the exponential rate $\lambda_1^{\mathcal{F}}$ to the metric of the product $g_\infty = g + \varphi_\infty^2\,\bar g$, with $\varphi_\infty = \int_{\bar M} \varphi(0, \cdot, y)\,\mathrm{dy}_{\bar g}$; hence, $\mathrm{Scal}_{\mathrm{mix}}(g_\infty) = 0$.

Let us look at what happens when B has the boundary and $\varphi > 0$ in the interior of B. Simple example is a rotation surface in $\mathbb{R}^3$, the leaves are meridians. By the maximum principle [1, Sect. 3.73], the problem $\Delta_{\mathcal{F}} u = 0$, $u_{|\,\partial B} = 0$ has only zero solution; hence, $\lambda_0^{\mathcal{F}} = 0$ is not the eigenvalue. Let $\mu(t, x) = \varphi(t, x)_{|\,\partial B}$ be twice continuously differentiable in t, and there exist limits

$$\lim_{t\to\infty} \mu(t, x) = \tilde\mu(x), \quad \lim_{t\to\infty} \partial_t\mu(t, x) = 0, \quad \lim_{t\to\infty} \partial_t^2\mu(t, \cdot) = 0 \qquad (10)$$

for a smooth nonnegative function $\tilde{\mu} : \partial B \to \mathbb{R}$ uniformly for $x \in \partial B$. Define the functions $\delta(t,\cdot) := \mu(t,\cdot) - \tilde{\mu}$ and $\nu(t) := \max\{\|\delta(t,\cdot)\|_{C^0(\partial B)}, \|\partial_t \delta(t,\cdot)\|_{C^0(\partial B)}\}$.

Next theorem examines when for the twisted product initial metric on $B^p \times \bar{M}^n$ the global solution of (2) converges to a limit metric with leaf-wise constant mixed scalar curvature.

Theorem 4. *Let the metrics g_t on $B^p \times_\varphi \bar{M}^n$ solve* (2) *and any of conditions (i)–(iii) is satisfied:*

(i) $\Phi < 0$ and $(10)_{1,2}$, *(ii) $0 \le \Phi < n\lambda_1^{\mathcal{F}}$, $p < 4$, and* (10), *(iii) $\Phi = n\lambda_1^{\mathcal{F}}$, $p < 4$,* (10) *and*

$$\tilde{\mu} = 0, \qquad \int_0^\infty \nu(\tau)\, \mathrm{d}\tau < \infty. \tag{11}$$

Then g_t exist for all $t \ge 0$, and converge, as $t \to \infty$, uniformly on $B \times \bar{M}^n$ in C^0-norm to a limit metric $g_\infty = \mathrm{d}x^2 + \varphi^2_\infty(x)\, \bar{g}$ with $\mathrm{Scal}_{\mathrm{mix}}(g_\infty) = \Phi$. Moreover, (2) *has a single point global attractor for (i) and (ii), but for (iii) the metric g_∞ depends on initial and boundary conditions.*

1.2 Auxiliary Results

The *Levi-Civita connection* ∇ of a metric g on M is given by a well-known formula

$$\begin{aligned} 2\,g(\nabla_X Y, Z) = {} & X(g(Y,Z)) + Y(g(X,Z)) - Z(g(X,Y)) \\ & + g([X,Y],Z) - g([X,Z],Y) - g([Y,Z],X) \quad (X,Y,Z \in T(M)). \end{aligned} \tag{12}$$

Let g_t be a smooth family of metrics on $(M, \mathcal{F})$ and $S = \partial_t g$. Since the difference of two connections is a tensor, $\partial_t \nabla^t$ is a $(1,2)$-tensor on (M, g_t). Differentiating (12) with respect to t yields, see [10],

$$2\,g_t((\partial_t \nabla^t)(X,Y), Z) = (\nabla^t_X S)(Y,Z) + (\nabla^t_Y S)(X,Z) - (\nabla^t_Z S)(X,Y) \tag{13}$$

for all t-independent vector fields X, Y, Z on M. If $S = s(g)\, g^\perp$, for short we write

$$\partial_t g = s\, g^\perp \tag{14}$$

for a certain t-dependent function $s : M \to \mathbb{R}$.

Lemma 1. *For variations* (14) *of metrics we have*

$$\partial_t h_{\mathcal{F}} = -s\, h_{\mathcal{F}}, \qquad \partial_t H_{\mathcal{F}} = -s\, H_{\mathcal{F}}. \tag{15}$$

Variations of metrics (14) *preserve total umbilicity, total geodesy, and harmonicity of foliations.*

Proof. Let g_t $(t \geq 0)$ be a family of metrics on $(M, \mathcal{F})$ such that $\partial_t g_t = S(g)$, where the tensor $S(g)$ is $\mathcal{D}$-truncated. Using (13), we find for $X \in \mathcal{D}$ and $\xi, \eta \in \mathcal{D}_{\mathcal{F}}$,

$$
\begin{aligned}
2\, g_t(\partial_t h_{\mathcal{F}}(\xi,\eta), X) &= g_t(\partial_t(\nabla^t_\xi \eta) + \partial_t(\nabla^t_\eta \xi),\ X) \\
&= (\nabla^t_\xi S)(X,\eta) + (\nabla^t_\eta S)(X,\xi) - (\nabla^t_X S)(\xi,\eta) \\
&= -S(\nabla^t_\xi \eta, X) - S(\nabla^t_\eta \xi, X) = -2\, S(h_{\mathcal{F}}(\xi,\eta), X).
\end{aligned}
$$

Assuming $S(g) = s(g)\, g^{\perp}$, we have $(15)_1$. Tracing this yields $(15)_2$. By the theory of ODEs, if $H_{\mathcal{F}} = 0$ or $h_{\mathcal{F}} = 0$ at $t = 0$ then, respectively, $H_{\mathcal{F}} = 0$ or $h_{\mathcal{F}} = 0$ for all $t > 0$. By (15) we have

$$
\partial_t (h_{\mathcal{F}} - (1/p) H_{\mathcal{F}}\, g_{|\mathcal{F}}) = -s(h_{\mathcal{F}} - (1/p) H_{\mathcal{F}}\, g_{|\mathcal{F}}).
$$

By the theory of ODEs, if $h_{\mathcal{F}} = (1/p) H_{\mathcal{F}}\, g_{|\mathcal{F}}$ (i.e., $\mathcal{F}$ is totally umbilical) for $t = 0$, then $h_{\mathcal{F}} = (1/p) H_{\mathcal{F}}\, g_{|\mathcal{F}}$ for all $t > 0$. □

The *co-nullity operator* is defined by $C_N(X) = -(\nabla_X N)^{\perp}$ for $X \in T(M)$, $N \in \mathcal{D}_{\mathcal{F}}$. One may decompose C_N restricted to $\mathcal{D}$ into symmetric and antisymmetric parts as $C_N = A_N + T^{\sharp}_N$. The Weingarten operator A_N of $\mathcal{D}$ and the operator $T^{\sharp}_N$ are related with tensors h and T by

$$
g(A_N(X),\ Y) = g(h(X,Y),\ N), \quad g(T^{\sharp}_N(X),\ Y) = g(T(X,Y),\ N), \quad X, Y \in \mathcal{D}.
$$

The proof of the next lemma is based on (13) with $S = s\, g^{\perp}$.

Lemma 2 (See [9] and [11]). *For $\mathcal{D}$-conformal variations* (14) *of metrics we have*

$$
\begin{aligned}
\partial_t A_N &= -\frac{1}{2}\, N(s)\, \widehat{\mathrm{id}}, \quad \partial_t T^{\sharp}_N = -s\, T^{\sharp}_N \ (N \in \mathcal{D}_{\mathcal{F}}), \\
\partial_t H &= -\frac{n}{2}\, \nabla^{\mathcal{F}} s, \quad \partial_t(\mathrm{div}_{\mathcal{F}} H) = -\frac{n}{2}\, \Delta_{\mathcal{F}}\, s.
\end{aligned} \tag{16}
$$

By $(16)_{1,2}$, variations (14) preserve conformal foliations, i.e., the property $\beta_{\mathcal{D}} \equiv 0$.

Define the domain $U := \{x \in M :\ \Psi^{\mathcal{F}}_1 \Psi^{\mathcal{F}}_2 \neq 0\}$ and the functions

$$
\Psi^{\mathcal{F}}_1 := u_0^2\, \|h_{\mathcal{F}}\|^2_{g_0}, \quad \Psi^{\mathcal{F}}_2 := u_0^4\, \|T\|^2_{g_0}. \tag{17}
$$

Lemma 3 (Conservation laws). *Let g_t $(t \geq 0)$ be $\mathcal{D}$-conformal metrics* (14) *on a foliated manifold $(M, \mathcal{F}, \mathcal{D})$ such that $H_0 = -n\nabla^{\mathcal{F}} \log u_0$ for a positive function $u_0 \in C^{\infty}(M)$. Then the following two functions and two vector fields on U are t-independent*:

$$
\beta_{\mathcal{D}}, \quad \|h_{\mathcal{F}}\|^2/\|T\|, \quad H - (n/2)\nabla^{\mathcal{F}} \log \|T\|, \quad H - n\nabla^{\mathcal{F}} \log \|h_{\mathcal{F}}\|.
$$

Proof. Using Lemma 2 and $g^\perp(H,\cdot)=0$, we calculate

$$\begin{aligned}\partial_t \|h\|^2 &= \partial_t \sum\nolimits_\alpha \mathrm{Tr}\,(A^2_{\mathcal{E}_\alpha}) = 2\sum\nolimits_\alpha \mathrm{Tr}\,(A_{\mathcal{E}_\alpha}\partial_t A_{\mathcal{E}_\alpha})\\ &= -\sum\nolimits_\alpha \mathcal{E}_\alpha(s)\,\mathrm{Tr}\,A_{\mathcal{E}_\alpha} = -g(\nabla s, H),\\ \partial_t g(H,H) &= s\,g^\perp(H,H) + 2\,g(\partial_t H, H) = -n\,g(\nabla s, H).\end{aligned}$$

Hence, $n\,\partial_t\beta_{\mathcal{D}} = \partial_t \|h\|^2_g - \frac{1}{n}\,\partial_t g(H,H) = 0$, that is, the function $\beta_{\mathcal{D}}$ does not depend on t.

For any function $f\in C^1(M)$ and a vector $N\in\mathcal{D}_{\mathcal{F}}$, using $(\partial_t g)(\cdot\,,N)=0$, we find

$$g(\nabla^{\mathcal{F}}(\partial_t f), N) = N(\partial_t f) = \partial_t N(f) = \partial_t g(\nabla^{\mathcal{F}} f, N) = g(\partial_t(\nabla^{\mathcal{F}} f), N).$$

Hence $\nabla^{\mathcal{F}}(\partial_t f) = \partial_t(\nabla^{\mathcal{F}} f)$. By Lemma 2, we find

$$\begin{aligned}\partial_t \|T\|^2 &= -\partial_t \sum\nolimits_\alpha \mathrm{Tr}\,\big((T^\sharp_{\mathcal{E}_\alpha})^2\big) = -2\sum\nolimits_\alpha \mathrm{Tr}\,(T^\sharp_{\mathcal{E}_\alpha}\partial_t T^\sharp_{\mathcal{E}_\alpha})\\ &= 2s\sum\nolimits_\alpha \mathrm{Tr}\,\big((T^\sharp_{\mathcal{E}_\alpha})^2\big) = -2\,s\,\|T\|^2.\end{aligned}$$

Similarly, by Lemma 3 and using the proof of Lemma 1, we obtain $\partial_t\|h_{\mathcal{F}}\|^2 = -s\,\|h_{\mathcal{F}}\|^2$. By the above, $h_{\mathcal{F}}\neq 0\neq T$ on U, and we have $\partial_t \log\|T\|^2_{g_t} = -2\,s$ and $\partial_t\log\|h_{\mathcal{F}}\|^2_{g_t} = -s$. Using $\nabla^{\mathcal{F}}\partial_t = \partial_t\nabla^{\mathcal{F}}$, we obtain $\partial_t H_t = (n/2)\,\partial_t\nabla^{\mathcal{F}}\log\|T\|_{g_t}$ and $\partial_t H_t = n\,\partial_t\nabla^{\mathcal{F}}\log\|h_{\mathcal{F}}\|_{g_t}$; moreover, $\partial_t(\|h_{\mathcal{F}}\|^2/\|T\|) = 0$. From the above the claim follows. □

Next lemma allows us to reduce (2) to the leaf-wise PDE (with space derivatives along $\mathcal{F}$ only).

Lemma 4 (See also [9]). *Let $\mathcal{F}$ be a harmonic foliation on (M,g). Then* (1) *reads*

$$\mathrm{Scal}_{\mathrm{mix}} = \mathrm{div}_{\mathcal{F}} H - \|H\|^2/n + \|T\|^2 - \|h_{\mathcal{F}}\|^2 - n\,\beta_{\mathcal{D}}. \qquad (18)$$

Proof. From (1), using $H_{\mathcal{F}}=0$ and identity $\mathrm{div}\,H = \mathrm{div}_{\mathcal{F}} H - \|H\|^2_g$, we obtain $\mathrm{Scal}_{\mathrm{mix}} = \mathrm{div}_{\mathcal{F}} H - \|h\|^2 + \|T\|^2 - \|h_{\mathcal{F}}\|^2$. Substituting $\|h\|^2 = n\,\beta_{\mathcal{D}} + \|H\|^2/n$ due to (6), we get (18). □

Proposition 1. *Let $\mathcal{F}$ be a harmonic foliation on a Riemannian manifold (M,g_0) and a family of metrics g_t $(0\le t<t_0)$ solves* (2)*. Then*

$$\partial_t H + \nabla^{\mathcal{F}}\|H\|^2 = n\nabla^{\mathcal{F}}(\mathrm{div}_{\mathcal{F}} H) + n\nabla^{\mathcal{F}}(\|T\|^2_{g_t} - \|h_{\mathcal{F}}\|^2_{g_t} - n\,\beta_{\mathcal{D}}). \qquad (19)$$

Suppose that $H_0 = -n\nabla^{\mathcal{F}}\log u_0$ for a leaf-wise smooth function $u_0>0$, then $H_t = -n\nabla^{\mathcal{F}}\log u$ for some positive function $u: M\times[0,t_0)$; moreover,

(*i*) *if* $\Psi_2^{\mathcal{F}} \neq 0$, *then* $u = (\Psi_2^{\mathcal{F}})^{1/4} \|T\|_{g_t}^{-1/2}$, *and the nonlinear PDE* (5) *is satisfied.*
(*ii*) *if* $\Psi_1^{\mathcal{F}} \equiv 0 \equiv \Psi_2^{\mathcal{F}}$, *then the potential function u may be chosen as a solution of the linear PDE*

$$(1/n)\,\partial_t u = \Delta_{\mathcal{F}}\, u + \beta_{\mathcal{D}}\, u, \qquad u(\,\cdot\,,0) = u_0. \tag{20}$$

Proof. By Theorem 1, (2) admits a unit local leaf-wise smooth solution g_t $(0 \le t < t_0)$. The functions $\mathrm{Scal}_{\mathrm{mix}}(g_t)$, H_t, $\|T\|_{g_t}$, and $\|h_{\mathcal{F}}\|_{g_t}$ are then uniquely determined for $0 \le t < t_0$. From $(16)_3$ with $s = -2\,(\mathrm{Scal}_{\mathrm{mix}}(g) - \Phi)$ and using (18) we obtain (19).

(i) By Lemma 3(ii), $H_t - (n/4)\nabla^{\mathcal{F}} \log \|T\|_{g_t}^2 = X$ for some vector field X on M. Since H_0 is conservative, $X = -(n/4)\nabla^{\mathcal{F}} \log \psi$ for some leaf-wise smooth function $\psi > 0$ on M. Hence, $H_t = -n\nabla^{\mathcal{F}} \log\big(\psi^{1/4}\|T\|_{g_t}^{-1/2}\big)$ and, by condition $H_0 = -n\nabla^{\mathcal{F}} \log u_0$, one may take $\psi = u_0^4\, \|T\|_{g_0}^2$. Define a leaf-wise smooth function $u := (\Psi_2^{\mathcal{F}})^{1/4}\|T\|_{g_t}^{-1/2}$ on $U \times [0, t_0)$ and calculate

$$\partial_t(\log \|T\|_{g_t}^2) = -4\,\partial_t \log(\|T\|_{g_t}^{-1/2}) = -4\,\partial_t \log((\Psi_2^{\mathcal{F}})^{-1/4}u) = -4\,\partial_t \log u.$$

By Lemma 3 and (17) $\|h_{\mathcal{F}}\|_{g_t}^2/\|T\|_{g_t} = \Psi_1^{\mathcal{F}}(\Psi_2^{\mathcal{F}})^{-1/2}$; thus, $u = (\Psi_1^{\mathcal{F}})^{1/2}\|h_{\mathcal{F}}\|_{g_t}^{-1}$ on $U \times [0, t_0)$ and

$$\partial_t(\log \|h_{\mathcal{F}}\|_{g_t}^2) = -2\,\partial_t \log(\|h_{\mathcal{F}}\|_{g_t}^{-1}) = -2\,\partial_t \log((\Psi_1^{\mathcal{F}})^{-1/2}u) = -2\,\partial_t \log u.$$

From the above and (18) we then obtain

$$\begin{aligned}\partial_t \log u &= -(1/4)\,\partial_t(\log \|T\|_{g_t}^2) = s/2 = -\mathrm{Scal}_{\mathrm{mix}}(g_t) + \Phi\\ &= n\,\Delta_{\mathcal{F}} \log u + n\, g(\nabla^{\mathcal{F}} \log u, \nabla^{\mathcal{F}} \log u) + n\,\beta_{\mathcal{D}}\\ &\quad +\Phi + \Psi_1^{\mathcal{F}}\, u^{-2} - \Psi_2^{\mathcal{F}}\, u^{-4}.\end{aligned}$$

Substituting $\partial_t \log u = u^{-1}\partial_t u$, $\nabla^{\mathcal{F}} \log u = u^{-1}\nabla^{\mathcal{F}} u$, and $\Delta_{\mathcal{F}} \log u = u^{-1}\Delta_{\mathcal{F}}\, u - u^{-2} g(\nabla^{\mathcal{F}} u, \nabla^{\mathcal{F}} u)$, we find that u solves the nonlinear heat equation (5).

(ii) Note that H obeys a forced leaf-wise Burgers equation [a consequence of (19)]

$$\partial_t H + \nabla^{\mathcal{F}} \|H\|_{g_t}^2 = n\nabla^{\mathcal{F}}(\mathrm{div}_{\mathcal{F}}\, H) - n^2\nabla^{\mathcal{F}} \beta_{\mathcal{D}}.$$

The rest of the proof can be seen in [9, Proposition 2]. □

Under certain conditions, (19) and (20) have single-point exponential attractors. The authors of [5] proved the polynomial convergence of a solution to the forced Burgers PDE on $\mathbb{R}^n$.

1.3 Proofs of Main Results

Proof of Theorem 1. Let $g_t = g_0 + s\, g_0^{\perp}$ $(0 \le t < \varepsilon)$ be $\mathcal{D}$-conformal metrics on a foliated manifold $(M, \mathcal{F})$, where $s : M \times [0, \varepsilon) \to \mathbb{R}$ is a smooth function. By Lemma 1, $\mathcal{F}$ is harmonic with respect to all g_t. We differentiate (18) by t, and apply Lemmas 2 and 3 to obtain

$$\partial_t \mathrm{Scal}_{\mathrm{mix}}(g_t) = -(n/2)\, \Delta_{\mathcal{F}}\, s + g(\nabla s, H) + s\big(\|h_{\mathcal{F}}\|^2_{g_t} - 2\, \|T\|^2_{g_t}\big).$$

Hence, the linearization of (2) at g_0 is the following linear PDE for s on the leaves:

$$\partial_t s = n\, \Delta_{\mathcal{F}}\, s - 2\, g_0(\nabla\, s,\ H_0) - 2\, (\mathrm{Scal}_{\mathrm{mix}}(g_0) + \|h_{\mathcal{F}}\|^2_{g_0} - 2\, \|T\|^2_{g_0})s.$$

The result follows from the theory of parabolic PDEs [1] and assumption (3), see also Sect. 2.6. □

Proof of Theorem 2. By Theorem 1, there exists a unique local solution g_t on $M \times [0, t_0)$. By Proposition 1(ii), H obeys (19), and $H = -n\nabla^{\mathcal{F}} \log u$ for some positive function u satisfying (5) with $u(\cdot\,, 0) = u_0$. Note that conditions (8) yield $(u_0^-)^4 \ge (\Psi_2^{\mathcal{F}})^+/(\Phi - n\lambda_0^{\mathcal{F}})$, see (50) with $\beta = \beta_{\mathcal{D}} + \Phi/n$ and $\lambda_0^{\mathcal{F}} - \Phi/n < 0$ and definitions (17) and (44). By Theorem 5, one may leaf-wise smoothly extend a solution of (5) on $M \times [0, \infty)$; hence, $H_t(x)$ is defined for $t \ge 0$ and is smooth on the leaves. By Theorem 6(i), $u \to \infty$ as $t \to \infty$ with exponential rate $n\alpha$ for $\alpha \in (0, \min\{\lambda_1^{\mathcal{F}} - \lambda_0^{\mathcal{F}}, 2(\Phi/n - \lambda_0^{\mathcal{F}}\})$. Hence, $\Psi_2^{\mathcal{F}} u^{-4}$ is leaf-wise smooth; moreover, $\|T\|_{g_t} \to 0$ and $h_{\mathcal{F}}(g_t) \to 0$ as $t \to \infty$. By Theorem 6(ii), $H_t = -n\nabla^{\mathcal{F}} \log u$ approaches in C^∞, as $t \to \infty$, to the vector field $\bar{H} = -n\nabla^{\mathcal{F}} \log e_0$; hence, $\mathrm{div}_{\mathcal{F}}\, H_t$ approaches to the leaf-wise smooth function $-n\, \Delta_{\mathcal{F}} \log e_0$. Since $-\Delta_{\mathcal{F}}\, e_0 - (\beta_{\mathcal{D}} + \Phi/n)\, e_0 = \lambda_0^{\mathcal{F}}\, e_0$, we have, as $t \to \infty$,

$$\mathrm{div}_{\mathcal{F}}\, H_t - \|H\|^2_{g_t}/n \to -n(\Delta_{\mathcal{F}}\, e_0)/e_0 = n(\lambda_0^{\mathcal{F}} + \beta_{\mathcal{D}}) - \Phi.$$

By (18), $\mathrm{Scal}_{\mathrm{mix}}(\cdot\,, t)$ approaches exponentially to $n\lambda_0^{\mathcal{F}} - \Phi$ as $t \to \infty$. Then a smooth solution to (2) is $g_t = g_0 \exp(-2 \int_0^t (\mathrm{Scal}_{\mathrm{mix}}(\cdot\,, \tau) - \Phi)\, d\tau)$, where $t \ge 0$, see also Sect. 2.6. □

Proof of Theorem 3. By Theorem 1, there exists a unique local solution g_t on $M \times [0, t_0)$. By Proposition 1(ii), H obeys (19), and $H = -n\nabla^{\mathcal{F}} \log u$ for some positive function u satisfying (5). Note that the inequality $\sqrt{2} \max\limits_M \|T\|_{g_0} < d^2_{u_0,e_0} \min\limits_M \|h_{\mathcal{F}}\|_{g_0}$ yields

$$3\, d^{-4}_{u_0,e_0} \max_M \|T\|^2_{g_0} - \min_M \|h_{\mathcal{F}}\|^2_{g_0} < \frac{1}{4}\, d^{\,4}_{u_0,e_0} \min_M \|h_{\mathcal{F}}\|^4_{g_0} / \max_M \|T\|^2_{g_0};$$

hence, I_0 is nonempty. Next, we find that the inequalities $0 < n\lambda_0^{\mathcal{F}} - \varPhi < \frac{1}{4}\, d^{\,4}_{u_0,e_0} \min\limits_M \|h_{\mathcal{F}}\|^4_{g_0} / \max\limits_M \|T\|^2_{g_0}$ yield $0 < n\lambda_0^{\mathcal{F}} - \varPhi < \frac{1}{4} \min\limits_M (\|h_{\mathcal{F}}\|^4_{g_0} u_0^4/e_0^4)/$

$\max\limits_M(\|T\|^2_{g_0} u_0^4/e_0^4)$, which represent conditions (72). Finally, the condition $n\lambda_0^{\mathcal F} - \Phi \ge 3\, d_{u_0,e_0}^{-4} \max\limits_M \|T\|^2_{g_0} - \min\limits_M \|h_{\mathcal F}\|^2_{g_0}\}$, see the definition of I_0, yield

$$n\lambda_0^{\mathcal F} - \Phi > 3\, \max_M(\|T\|^2_{g_0} u_0^4/e_0^4)/(u_0/e_0)^4 - \min_M(\|h_{\mathcal F}\|^2_{g_0} u_0^2/e_0^2)/(u_0/e_0)^2,$$

which means $u_0 \in \mathcal U_1$. By Theorem 7, there exists u_*—a unique solution in $\mathcal U_1$ of the stationary PDE for (5),

$$n\, \Delta_{\mathcal F}\, u + (n\, \beta_{\mathcal D} + \Phi)\, u + \Psi_1^{\mathcal F} u^{-1} - \Psi_2^{\mathcal F} u^{-3} = 0,$$

and $H_t \to -n\nabla^{\mathcal F} \log u_*$ and $\mathrm{Scal}_{\mathrm{mix}}(g_t) \to \Phi$ as $t \to \infty$ with the exponential rate. □

Proof of Corollary 3. Claim (i) follows from Theorem 2. The metrics g_t ($g_0 = g$) of Theorem 2 diverge as $t \to \infty$ with the exponential rate $\mu = \Phi - n\lambda_0^{\mathcal F}$:

$$\exists C > 1,\ \forall X \in \mathcal D,\ \forall t \ge 0:\ C^{-1} e^{2\mu t} g(X,X) \le g_t(X,X) \le C e^{2\mu t} g(X,X).$$

Consider $\mathcal D$-conformal metrics $\bar g_t = g_{\mathcal F} + e^{-2\mu t}(g_t)^\perp$. By $(16)_3$, $\bar H_t = H_t$. Let $(\cdot\,,\cdot)_0$ be the inner product and the norm in $L_2(F)$ for any leaf F. The function $v = e^{-\mu t} u$ converges as $t \to \infty$ to $\tilde u_0^0 e_0$, where $\tilde u_0^0 = (\tilde u, e_0)_0 = u_0^0 + \int_0^\infty q_0(\tau)\, d\tau$, see Theorem 6 and (66). For $t \to \infty$ we have

$$\|h_{\mathcal F}\|^2_{\bar g_t} = e^{2\mu t} \|h_{\mathcal F}\|^2_{g_t} = \Psi_1^{\mathcal F}/v^2 \to \xi^2 \|h_{\mathcal F}\|^2_g,$$

$$\|T\|^2_{\bar g_t} = e^{4\mu t} \|T\|^2_{g_t} = \Psi_2^{\mathcal F}/v^4 \to \xi^4 \|T\|^2_g,$$

and the metrics $\bar g_t$ converge as $t \to \infty$ to the metric $\bar g_\infty = g_{\mathcal F} + \xi^{-2} g^\perp$. By (18), we find

$$\mathrm{Scal}_{\mathrm{mix}}(\bar g_\infty) = n\lambda_0^{\mathcal F} - \Phi + \xi^4 \|T\|^2_g - \xi^2 \|h_{\mathcal F}\|^2_g\,.$$

Comparing with (8) completes the proof of (ii). Claim (iii) follows from Theorem 3. □

1.4 One-Dimensional Case

Let (M, g) be a Riemannian manifold with a unit vector field N, i.e., $p = 1$ or $n = 1$. In this case, $\mathrm{Scal}_{\mathrm{mix}}$ is the Ricci curvature $\mathrm{Ric}(N, N)$.

Case $p = 1$. Let N be tangent to a geodesic foliation $\mathcal F$, h the scalar second fundamental form, and $H = \mathrm{Tr}_g\, h$ the mean curvature of $\mathcal D = N^\perp$. We have $h_{\mathcal F} = 0$, and (18) reads $\mathrm{Ric}(N, N) = \|T\|^2 + N(H) - H^2/n - n\,\beta_{\mathcal D}$. Let the metric evolve as, see (2),

$$\partial_t g = -2\,(\mathrm{Ric}_g(N, N) - \Phi)\, g^\perp, \tag{21}$$

then H obeys the PDE along N-curves, see (19), $\partial_t H + N(H^2) = n\,N(N(H)) + n\,\big(N(\|T\|^2) - n\,\beta_{\mathcal{D}}\big)$. Suppose that $H = -n\,N(\log u_0)$ for a leaf-wise smooth function $u_0 > 0$ on M, then we assume $H = -n\,N(\log u)$ for a positive function $u : M \times [0, t_0)$, see Proposition 1.

If $\mathcal{D}$ is integrable, then the function $u(\,\cdot\,, t) > 0$ may be chosen as a solution of the following linear heat equation, see (20), $\partial_t u = n\,N(N(u)) + n\,\beta_{\mathcal{D}}\,u$, where $u(\,\cdot\,, 0) = u_0$. By Theorem 2, flow (21) admits a unique global solution g_t $(t \geq 0)$. If $\lambda_0^{\mathcal{F}} - \Phi/n < 0$, then we have exponential convergence as $t \to \infty$ of $g_t \to \bar{g}$, $H \to -n\,N(\log e_0)$, and $\operatorname{Ric}_{g_t}(N, N) \to n\lambda_0^{\mathcal{F}} - \Phi$.

If $\mathcal{D}$ is nowhere integrable, then $u = (\Psi_2^{\mathcal{F}})^{1/4}\|T\|_{g_t}^{-1/2}$ (with $\Psi_2^{\mathcal{F}} := u_0^4\,\|T\|_{g_0}^2 > 0$); moreover, the potential function $u > 0$ solves the nonlinear heat equation, see (5),

$$(1/n)\,\partial_t u = N(N(u)) + (\beta_{\mathcal{D}} + \Phi/n)\,u - (\Psi_2^{\mathcal{F}}/n)\,u^{-3}, \qquad u(\,\cdot\,, 0) = u_0.$$

If (8) are satisfied, then (21) admits a unique solution g_t $(t \geq 0)$. We have exponential convergence as $t \to \infty$ of functions $H \to -n\,N(\log e_0)$ and $\operatorname{Ric}_{g_t}(N, N) \to n\lambda_0^{\mathcal{F}} - \Phi$. By Theorem 2 we have

Corollary 4. *Let N be a unit vector field tangent to a geodesic foliation $\mathcal{F}$ on (M, g) and* (3) *hold.*

(i) Then for any $c > d_{u_0,e_0}^{-4}\max_F \|T\|_g^2$ there is a $\mathcal{D}$-conformal to g metric $\bar{g}$ with the property $\operatorname{Ric}_{\bar{g}}(N, N) \leq -c < 0$.
(ii) If $\xi^2\|h_{\mathcal{F}}\|_g^2 < \xi^4\|T\|_g^2 + d_{u_0,e_0}^{-4}\max_F \|T\|_g^2$, where $\xi = u_0/(\tilde{u}_0^0 e_0)$ and $\tilde{u}_0^0$ is defined in Sect. 2.4, then there exists a $\mathcal{D}$-conformal to g metric $\bar{g}$ such that $\operatorname{Ric}_{\bar{g}}(N, N) > 0$.

Case $n = 1$. Let N be orthogonal to a compact harmonic foliation $\mathcal{F}$ of codimension one. Then $T = \beta_{\mathcal{D}} = H_{\mathcal{F}} = 0$, $H = \nabla_N N$, $\Psi_1^{\mathcal{F}} = u_0^2\|h_{\mathcal{F}}\|_{g_0}^2$, operator (7) coincides with $-\Delta_{\mathcal{F}}$ (hence, $\lambda_0^{\mathcal{F}} = 0$ and $e_0 = \text{const}$), and (18) reads $\operatorname{Ric}(N, N) = \operatorname{div}_{\mathcal{F}} H - \|H\|^2 + \|h_{\mathcal{F}}\|^2$. By (19), we have

$$\partial_t H + \nabla^{\mathcal{F}}\|H\|_{g_t}^2 = \nabla^{\mathcal{F}}(\operatorname{div}_{\mathcal{F}} H) - \nabla^{\mathcal{F}}(\|h_{\mathcal{F}}\|_{g_t}^2).$$

Suppose the condition $H_0 = -\nabla^{\mathcal{F}}\log u_0$ for a leaf-wise smooth function $u_0 > 0$ on M. Then $H = -\nabla^{\mathcal{F}}\log u$, where, see (5),

$$\partial_t u = \Delta_{\mathcal{F}}\,u + \Phi\,u + \Psi_1^{\mathcal{F}}\,u^{-1}, \qquad u(\,\cdot\,, 0) = u_0.$$

If $\Phi > 0$, then (8) holds and, by Theorem 2, flow (21) admits a unique global solution g_t $(t \geq 0)$. As $t \to \infty$, we have convergence $H \to 0$, $\operatorname{Ric}_{g_t}(N, N) \to -\Phi$, $h_{\mathcal{F}}(g_t) \to 0$ with the exponential rate α for any $\alpha \in (0, \min\{\lambda_1^{\mathcal{F}}, 2\,\Phi\})$. By Theorems 2 and 3, we have the following.

Corollary 5. *Let $\mathcal{F}$ be a codimension one harmonic foliation with a unit normal vector field N and assumptions* (3).

(i) *Then for any $c > d_{u_0,e_0}^{-4} \max_F \|T\|_g^2$ there is a $\mathcal{D}$-conformal to g metric $\bar g$ with* $\mathrm{Ric}_{\bar g}(N,N) \le -c$.

(ii) *If $h_{\mathcal{F}} \ne 0$, then there exists a $\mathcal{D}$-conformal to g metric $\bar g$ with* $\mathrm{Ric}_{\bar g}(N,N) = \mathrm{const} < 0$.

For a totally geodesic foliation $\mathcal{F}$, i.e., $h_{\mathcal{F}} \equiv 0$, (18) reads $\mathrm{Ric}(N,N) = \mathrm{div}_{\mathcal{F}} H - \|H\|^2 = \mathrm{div}\, H$.

Let the metric evolve by (21). By Proposition 1, H obeys the homogeneous Burgers equation $\partial_t H + \nabla^{\mathcal{F}} \|H\|^2 = \nabla^{\mathcal{F}}(\mathrm{div}_{\mathcal{F}} H)$. Suppose that the curvature vector H of N-curves is leaf-wise conservative: $H = -\nabla^{\mathcal{F}} \log u$ for a function $u > 0$. This yields the heat equation $\partial_t u = \Delta_{\mathcal{F}} u$. Solution of the above PDE satisfies on the leaves $\bar u := \lim_{t\to\infty} u(t,x) = \int_{F_x} u_0(x)\,\mathrm{d}x / \mathrm{Vol}(F_x, g)$. Since $\nabla^{\mathcal{F}} e_0 = 0$, we have $\bar H = \lim_{t\to\infty} H(t,\cdot) = 0$. Then $\mathrm{Ric}_{\bar g}(N,N) = 0$, where $\bar g = \lim_{t\to\infty} g_t$.

1.5 Twisted Products

Definition 1 (See [7]). Let $(B^p, \mathrm{d}x^2)$ and $(\bar M^n, \bar g)$ be Riemannian manifolds, and $\varphi \in C^\infty(B \times \bar M)$ a positive function. The *twisted product* $B \times_\varphi \bar M$ is the manifold $M = B \times \bar M$ with the metric $g = \mathrm{d}x^2 + \varphi^2\, \bar g$. When the warping function φ depends on B only, the twisted product becomes a *warped product*. (When $\varphi = 1$, $B \times_\varphi \bar M$ is a direct product.) The *rotational symmetric metrics*, i.e., $\bar M$ is a unit n-sphere, are the particular case of a warped product; such metrics appear on rotation surfaces in space forms.

The *leaves* $B \times \{y\}$ of a twisted product compose a totally geodesic foliation $\mathcal{F}$ on M, while the *fibers* $\{x\} \times \bar M$ are totally umbilical with the mean curvature vector $H = -n\, \nabla^{\mathcal{F}} \log \varphi$.

One may apply the existence/uniqueness Theorem 1 to conclude that (2) preserves total umbilicity of foliations with integrable orthogonal distribution. Thus we have the following.

Corollary 6. *Flow* (2) *preserves twisted* (*warped*) *product metrics.*

For a twisted product we have $h_{\mathcal{F}} = 0$, $T = 0$, and

$$h = -\nabla^{\mathcal{F}}(\log \varphi)\, g^\perp, \quad H = -n\, \nabla^{\mathcal{F}} \log \varphi \quad (\text{when} \quad \varphi \ne 0).$$

Since $R(X,N,Y,N) = -\frac{1}{\varphi} N(N(\varphi))\, g^\perp(X,Y)$ $(N \in \mathcal{D}_{\mathcal{F}})$ when $\varphi \ne 0$, we conclude that

$$\mathrm{Scal}_{\mathrm{mix}} = -n\, \Delta_{\mathcal{F}} \varphi / \varphi \quad (\text{when } \varphi \ne 0). \tag{22}$$

Let $0 < \lambda_1^{\mathcal{F}} \le \cdots \le \lambda_i^{\mathcal{F}} \cdots$ be eigenvalues (leaf-wise constant functions on M) counting their multiplicities, $\lim_{j\to\infty} \lambda_j^{\mathcal{F}} = \infty$, and $\{e_i\}_{1\le i<\infty}$ the corresponding L_2-orthonormal basis of eigenfunctions with $e_i = 0$ on ∂B of the eigenvalue problem in B: $-\Delta_{\mathcal{F}}\, e_i = \lambda_i^{\mathcal{F}} e_i$. As in Proposition 3, $\lambda_1^{\mathcal{F}}$ has multiplicity 1 (hence $\lambda_1^{\mathcal{F}} < \lambda_2^{\mathcal{F}}$) and $e_1 > 0$ on the interior of B. Denote by $(\cdot\,,\,\cdot)_0$ and $\|\cdot\|_0$ the inner product and the norm in $L_2(B)$.

Proof of Theorem 4. If a family of twisted product metrics $g_t = \mathrm{dx}^2 + \varphi^2(t,x,y)\bar g$ solves (2) on $M = B\times\bar M$, then $\partial_t(\varphi^2)\,\bar g = 2(n\Delta_{\mathcal{F}}\,\varphi/\varphi)\,\bar g + 2\,\Phi\,\varphi^2\,\bar g$. This yields the leaf-wise parabolic Cauchy's problem with Dirichlet boundary conditions for the warping function φ (we omit the parameter y)

$$\partial_t\varphi = n\,\Delta_{\mathcal{F}}\,\varphi + \Phi\,\varphi,\quad \varphi(0,\,\cdot) = \varphi_0,\quad \varphi(t,\,\cdot)_{|\,\partial B} = \mu(t,\,\cdot). \tag{23}$$

Linear problem (23) has a unique classical solution $\varphi : [0,\infty)\times B \to \mathbb{R}$. We shall study convergence of φ as $t\to\infty$ to a stationary state, i.e., to a solution $\tilde\varphi : B \to \mathbb{R}$ of the problem

$$-\,\Delta_{\mathcal{F}}\,\tilde\varphi = (\Phi/n)\,\tilde\varphi,\qquad \tilde\varphi_{|\,\partial B} = \tilde\mu. \tag{24}$$

Similar problems can be studied for Neumann boundary conditions. One may assume $n = 1$.

Let $U : [0,\infty)\times B \to \mathbb{R}$ solves the Dirichlet problem on B, where t plays the role of a parameter,

$$\Delta_{\mathcal{F}}\,U = 0,\qquad U_{|\,\partial B} = \delta(t,\,\cdot).$$

Since $U(t,\,\cdot)$ is harmonic on B, by the maximum principle, see [1, Sect. 3.73], we have $\|U(t,\,\cdot)\|_{C^0} = \|\delta(t,\,\cdot)\|_{C^0(\partial B)}$ for any $t > 0$. It is easy to check that the function

$$v(t,x) = \varphi(t,x) - \tilde\varphi(x) - U(t,x),\quad (t,x)\in[0,\infty)\times B, \tag{25}$$

solves the Cauchy's problem

$$\partial_t v = \Delta_{\mathcal{F}}\,v + \Phi\,v + f,\quad v(0,\,\cdot) = v_0,\quad v(t,\,\cdot)_{|\,\partial B} = 0, \tag{26}$$

where

$$v_0 := \varphi_0 - \tilde\varphi - U(0,\,\cdot),\qquad f := \Phi\,U - \partial_t U. \tag{27}$$

Since $\mu(t,\,\cdot)$ is twice differentiable in t, the functions $\partial_t U(t,\,\cdot)$ and $\partial_t^2 U(t,\,\cdot)$ are also harmonic on B, their boundary values are $\partial_t U_{|\,\partial B} = \partial_t\delta(t,\,\cdot)$ and $\partial_t^2 U_{|\,\partial B} = \partial_t^2\delta(t,\,\cdot)$, and we have $\|\partial_t U\|_{C^0} = \|\partial_t\delta(t,\,\cdot)\|_{C^0(\partial B)}$ and $\|\partial_t^2 U\|_{C^0} = \|\partial_t^2\delta(t,\,\cdot)\|_{C^0(\partial B)}$, respectively, for $t > 0$. Hence,

$$|f(t,\,\cdot)|_{C^0} \le (|\Phi| + 1)\nu(t),\qquad |\partial_t f(t,\,\cdot)|_{C^0} \le (|\Phi| + 1)\tilde\nu(t), \tag{28}$$

where $\tilde{\nu}(t) := \max\{\|\delta(t,\cdot)\|_{C^0(\partial B)}, \|\partial_t\delta(t,\cdot)\|_{C^0(\partial B)}, \|\partial_t^2\delta(t,\cdot)\|_{C^0(\partial B)}\}$. Consider Fourier series

$$v(t,x) = \sum\nolimits_{j=1}^{\infty} v_j(t)\, e_j(x), \quad f(t,x) = \sum\nolimits_{j=1}^{\infty} f_j(t)\, e_j(x),$$
$$v_0(x) = \sum\nolimits_{j=1}^{\infty} v_j^0\, e_j(x), \tag{29}$$

where $\int_B e_i(s)e_j(s)\,\mathrm{ds} = \delta_{ij}$ and

$$v_j = \int_B v(\cdot,s)\, e_j(s)\,\mathrm{ds}, \quad f_j = \int_B f(\cdot,s)\, e_j(s)\,\mathrm{ds}, \quad v_j^0 = \int_B v_0(s)\, e_j(s)\,\mathrm{ds}\,.$$

For cases (i) and (ii) we will obtain $\varphi_\infty = \tilde{\varphi}$ and for $(\boldsymbol{iii})$ we will get $\varphi_\infty := (v_1^0 + \int_0^\infty f_1(\tau)\,\mathrm{d}\tau)e_1$.

(i) By (25), we obtain

$$|\varphi(t,x) - \tilde{\varphi}(x)| = |v(t,x) + U(t,x)| \leq |v(t,x)| + |\delta(t,x)|. \tag{30}$$

Then from $(26)_1$ and $(28)_1$ we get the estimate

$$\begin{aligned}\partial_t v - \Delta_{\mathcal{F}}\, v + \Phi\, v - (|\Phi| + 1)\nu(t) &\leq \partial_t v - \Delta_{\mathcal{F}}\, v + \Phi\, v - f \\ &\leq \partial_t v - \Delta_{\mathcal{F}}\, v + \Phi\, v + (|\Phi| + 1)\nu(t).\end{aligned}$$

By the maximum principle for parabolic equations with Dirichlet's boundary conditions, see [1, Sect. 4.46], we have $|v(t,\cdot)|_{C^0} \leq \bar{v}(t)$, where $\bar{v}(t)$ is a solution of the Cauchy's problem for the ODE,

$$\frac{\mathrm{d}}{\mathrm{dt}}\,\bar{v} = \Phi\,\bar{v} + (|\Phi| + 1)\nu(t), \qquad \bar{v}(0) = |v_0|_{C^0}.$$

Using (30) and Lemma 5 below with $a = \Phi < 0$ and $s(t) = (|\Phi| + 1)\nu(t)$, we prove case (i).

(ii) Substituting (29) into $(26)_{1,2}$ and comparing the coefficients yield the Cauchy's problem

$$v_j' = (\Phi - \lambda_j^{\mathcal{F}})v_j + f_j(t), \qquad v_j(0) = v_j^0, \tag{31}$$

for $v_j(t)$, whose solution is

$$v_j(t) = v_j^0 e^{(\Phi - \lambda_j^{\mathcal{F}})t} + \textstyle\int_0^t e^{(\Phi - \lambda_j^{\mathcal{F}})(t-\tau)} f_j(\tau)\,\mathrm{d}\tau. \tag{32}$$

Denote by $w(t,\cdot) = (-\Delta_{\mathcal{F}} - \Phi\,\mathrm{id})\,v(t,\cdot)$ and $w_j(t) = (w(t,\cdot), e_j)_0$. By Elliptic Regularity Theorem the operator $(-\Delta_{\mathcal{F}} - \Phi\,\mathrm{id})^{-1}$ maps L_2 into H^2

and, since $p < 4$, by Sobolev Embedding Theorem the space H^2 is embedded continuously into C^0 (see Sect. 2.1). Therefore, the operator $(-\Delta_{\mathcal{F}} - \Phi\ \mathrm{id}\,)^{-1}$ acts continuously from L_2 into C^0; hence,

$$|v(t,\cdot)|_{C^0} \le |(-\Delta_{\mathcal{F}}-\Phi\ \mathrm{id}\,)^{-1} w(t,\cdot)|_{C^0} \le |(-\Delta_{\mathcal{F}}-\Phi\ \mathrm{id}\,)^{-1}|_{\mathcal{B}(L_2,C^0)} \cdot |w(t,\cdot)|_0. \tag{33}$$

We denote by $\mathcal{B}(L_2, C^0)$ the Banach space of all bounded linear operators $A : L_2 \to C^0$ with the norm $\|A\|_{\mathcal{B}(L_2,C^0)} = \sup_{v\in L_2\setminus 0} \|A(v)\|_C/\|v\|_0$, see Sect. 2. Furthermore, we obtain

$$v_j(t)=(v(t,\cdot), e_j)_0=(\lambda_j^{\mathcal{F}}-\Phi)^{-1}(v(t,\cdot), (\lambda_j^{\mathcal{F}}-\Phi)e_j)_0=(\lambda_j^{\mathcal{F}}-\Phi)^{-1}(v(t,\cdot),$$
$$(-\Delta_{\mathcal{F}}-\Phi\ \mathrm{id}\,)e_j)_0=(\lambda_j^{\mathcal{F}}-\Phi)^{-1}\big((-\Delta_{\mathcal{F}}-\Phi\ \mathrm{id}\,)\,v(t,\cdot), e_j\big)_0=(\lambda_j^{\mathcal{F}}-\Phi)^{-1}\,w_j(t).$$

In particular, $v_j^0 = (\lambda_j^{\mathcal{F}} - \Phi)^{-1} w_j(0)$. From (32), using integration by parts, we obtain

$$w_j(t)=w_j(0)\,e^{(\Phi-\lambda_j^{\mathcal{F}})t}+f_j(t) - e^{(\Phi-\lambda_j^{\mathcal{F}})t} f_j(0) - \int_0^t e^{(\Phi-\lambda_j^{\mathcal{F}})(t-\tau)} f_j'(\tau)\,\mathrm{d}\tau.$$

Applying Schwarz's inequality to the integral, we get the following:

$$\begin{aligned}\Big(\int_0^t e^{(\Phi-\lambda_j^{\mathcal{F}})(t-\tau)} f_j'(\tau)\,\mathrm{d}\tau\Big)^2 &\le \int_0^t e^{(\Phi-\lambda_j^{\mathcal{F}})(t-\tau)}\,\mathrm{d}\tau \times \int_0^t e^{(\Phi-\lambda_j^{\mathcal{F}})(t-\tau)} (f_j'(\tau))^2\,\mathrm{d}\tau\\ &= \frac{1-e^{(\Phi-\lambda_j^{\mathcal{F}})t}}{\lambda_j^{\mathcal{F}}-\Phi} \int_0^t e^{(\Phi-\lambda_j^{\mathcal{F}})(t-\tau)} (f_j'(\tau))^2\,\mathrm{d}\tau\\ &\le \frac{1}{\lambda_j^{\mathcal{F}}-\Phi} \int_0^t e^{(\Phi-\lambda_j^{\mathcal{F}})(t-\tau)} (f_j'(\tau))^2\,\mathrm{d}\tau.\end{aligned}$$

Taking into account these circumstances and using Parseval's equality, we have

$$\begin{aligned}|w(t,\cdot)|_0 &\le |w(0,\cdot)|_0\,e^{(\Phi-\lambda_1^{\mathcal{F}})t} + |f(t,\cdot)|_0 + |f(0,\cdot)|_0\,e^{(\Phi-\lambda_1^{\mathcal{F}})t}\\ &\quad+(\lambda_1^{\mathcal{F}}-\Phi)^{-1/2}\Big(\int_0^t e^{(\Phi-\lambda_j^{\mathcal{F}})(t-\tau)} |\partial_t f(\tau,\cdot)|_0^2\,\mathrm{d}\tau\Big)^{1/2}.\end{aligned} \tag{34}$$

Then using estimate $(28)_2$ and the arguments from the proof of Lemma 5, we get

$$\begin{aligned}|w(t,\cdot)|_0 &\le |w(0,\cdot)|_0\,e^{(\Phi-\lambda_1^{\mathcal{F}})t}+\sqrt{\mathrm{vol}\,B}\,(|\Phi|+1)\big(\nu(t)+\nu(0)\,e^{(\Phi-\lambda_1^{\mathcal{F}})t}\big)\\ &\quad+\frac{\sqrt{\mathrm{vol}\,B}}{\lambda_1^{\mathcal{F}}-\Phi}\Big(e^{(1-\theta)(\Phi-\lambda_1^{\mathcal{F}})t}\sup_{\tau\in[0,\theta t]}\tilde{\nu}^2(\tau)+\sup_{\tau\in[\theta t,t]}\tilde{\nu}^2(\tau)\Big)^{1/2},\\ &\quad\theta\in(0,1).\end{aligned} \tag{35}$$

Using (30), (33), and equality (22), we prove case (ii).

(iii) Since $\tilde{\mu} = 0$, we can choose $\tilde{\varphi} = 0$ as a solution of (24). Hence $v(t,x) = \varphi(t,x) - U(t,x)$, see (25), and $v_0(x) = \varphi_0(x) - U(0,x)$, see (27). By (25), we have

$$|\varphi(t,x) - \varphi_\infty(x)| = |v(t,x) - \varphi_\infty(x)| + \|\delta(t,\cdot)\|_{C^0(B)}, \tag{36}$$

where $\varphi_\infty := (v_1^0 + \int_0^\infty f_1(\tau)\,d\tau)e_1$. For $j = 1$, we obtain from (31):

$$v_1' = f_1(t), \quad v_1(0) = v_1^0 \quad \Rightarrow \quad v_1(t) = v_1^0 + \int_0^\infty f_1(\tau)\,d\tau - \int_t^\infty f_1(\tau)\,d\tau,$$

where the improper integrals converge in view of condition (11) and definition (27). Hence

$$\Big|v_1(t) - v_1^0 - \int_0^\infty f_1(\tau)\,d\tau\Big| = \Big|\int_t^\infty f_1(\tau)\,d\tau\Big| \le (\mathrm{vol}\,B)^{1/2}\int_t^\infty \nu(\tau)\,d\tau, \tag{37}$$

which converges to 0 as $t \to \infty$. Define $\tilde{w}(t,\cdot) = (-\Delta_{\mathcal{F}} - \gamma\ \mathrm{id})(v(t,\cdot) - \varphi_\infty)$ with fixed $\gamma < \lambda_1^{\mathcal{F}}$. Since $p < 4$, we get as in the proof of (ii):

$$|v(t,\cdot) - \varphi_\infty|_{C^0} \le |(-\Delta_{\mathcal{F}} - \Phi\ \mathrm{id})^{-1}|_{\mathcal{B}(L_2, C^0)} \cdot |\tilde{w}(t,\cdot)|_0,$$

where $\tilde{w}_j(t) = (\tilde{w}(t,\cdot), e_j)$. Clearly, $\tilde{w}_j(t)$ for $j > 1$ coincides with $w_j(t)$ defined in the proof of claim (ii). Then as in this proof of (ii), we obtain for $j > 1$:

$$\tilde{w}_j(t) = \tilde{w}_j(0)\,e^{(\lambda_1^{\mathcal{F}} - \lambda_j^{\mathcal{F}})t} + f_j(t) - e^{(\lambda_1^{\mathcal{F}} - \lambda_j^{\mathcal{F}})t} f_j(0) - \int_0^t e^{(\lambda_1^{\mathcal{F}} - \lambda_j^{\mathcal{F}})(t-\tau)} f_j'(\tau)\,d\tau.$$

Using the above and estimates (36) and (37), we complete the proof of case (iii) similarly as the proof of (ii) (see (34), (35), and all further arguments).

□

Lemma 5. *Let $y(t)$ solve the Cauchy's problem (for the ODE) $y' = \alpha(t)y + s(t)$, $y(0) = y_0$, where the functions $\alpha, \nu \in C[0,\infty)$, $\alpha(t) \le a < 0$, and the function $s(t)$ is bounded. Then*

$$|y(t)| \le |y_0|e^{at} + |a|^{-1}e^{(1-\theta)at}\sup\nolimits_{\tau\in[0,\,\theta t]}|s(\tau)| + |a|^{-1}\sup\nolimits_{\tau\in[\theta t,\,t]}|s(\tau)| \tag{38}$$

for any $\theta \in (0,1)$. In particular, if $\lim_{t\to\infty} s(t) = 0$, then $\lim_{t\to\infty} y(t) = 0$.

Proof. As is known, $y(t) = y_0 e^{\int_0^t \alpha(\xi)\,d\xi} + \int_0^t e^{\int_\tau^t \alpha(\xi)\,d\xi} s(\tau)\,\mathrm{d}\tau$. Hence, we have the estimate

$$
\begin{aligned}
|y(t)| &= |y_0| e^{\,at} + \int_0^{\theta t} e^{\,a(t-\tau)} |s(\tau)|\,\mathrm{d}\tau + \int_{\theta t}^{t} e^{\,a(t-\tau)} |s(\tau)|\,\mathrm{d}\tau \\
&\le |y_0| e^{\,at} + \sup_{\tau\in[0,\,\theta t]} |s(\tau)| \int_0^{\theta t} e^{\,a(t-\tau)}\,\mathrm{d}\tau + \sup_{\tau\in[\theta t,\,t]} |s(\tau)| \int_{\theta t}^{t} e^{\,a(t-\tau)}\,\mathrm{d}\tau.
\end{aligned}
$$

The above and $\int_0^{\theta t} e^{\,a(t-\tau)}\,\mathrm{d}\tau = (e^{\,at} - e^{(1-\theta)\,at})/a$, $\int_{\theta t}^{t} e^{\,a(t-\tau)}\,\mathrm{d}\tau = (e^{(1-\theta)\,at} - 1)/a$ yield (38). □

Example 4 (Rotation surfaces). The metric on a rotation surface in $\mathbb{R}^3$ belongs to warped products, see Example 3. Let $M_t^2 \subset \mathbb{R}^3 : [\varphi(t,x)\cos\theta,\ \varphi(t,x)\sin\theta,\ \psi(t,x)]$, where $0 \le x \le l$, $|\theta| \le \pi$, $\varphi \ge 0$ be a one-parameter family of rotation surfaces such that $(\partial_x\varphi)^2 + (\partial_x\psi)^2 = 1$. The profile curves $\theta = \text{const}$ are unit speed geodesics tangent to the vector field N. The θ-curves are circles in $\mathbb{R}^3$; their geodesic curvature is $k = -(\log\varphi)_{,x}$. The metric $g_t = \mathrm{d}x^2 + \varphi^2(t,x)\,\mathrm{d}\theta^2$ is rotational symmetric and its Gaussian curvature is $K = -\varphi_{,xx}/\varphi$. Let g_t obeys (4), then φ solves the Cauchy's problem

$$
\partial_t\varphi = \varphi_{,xx} + \Phi\,\varphi, \quad \varphi(0,x) = \varphi_0(x), \quad \varphi(t,0) = \mu_0(t) \ge 0, \quad \varphi(t,l) = \mu_1(t) \ge 0, \tag{39}
$$

where $\varphi(x) > 0$ for $x \in (0,l)$, $\mu_0, \mu_1 \in C^1[0,\infty)$ and there exist limits $\lim_{t\to\infty} \mu_j(t) = \tilde\mu_j \in [0,\infty)$.

The solution of stationary problem (e.g., (24) with $B = [0,l]$ and $\lambda_j^{\mathcal{F}} = (\pi\, j/l)^2$) has the view

$$
\tilde\varphi(x) = \begin{cases}
\dfrac{\tilde\mu_1 \sin(\sqrt{\Phi}\,x) + \tilde\mu_0 \sin(\sqrt{\Phi}\,(l-x))}{\sin(\sqrt{\Phi}\,l)} & \text{if } 0 < \Phi < \lambda_1^{\mathcal{F}}, \\
\tilde\mu_0 + (\tilde\mu_1 - \tilde\mu_0)(x/l) & \text{if } \Phi = 0, \\
\dfrac{\tilde\mu_1 \sinh(\sqrt{-\Phi}\,x) + \tilde\mu_0 \sinh(\sqrt{-\Phi}\,(l-x))}{\sinh(\sqrt{-\Phi}\,l)} & \text{if } \Phi < 0.
\end{cases}
$$

For the resonance case, $\Phi = \lambda_1^{\mathcal{F}} = (\pi/l)^2$, the stationary problem is solvable if and only if $\tilde\mu_0 = \tilde\mu_1 = 0$, and in this case the solutions are $\tilde\varphi(x) = C\sin(\pi x/l)$, where $C > 0$ is constant.

By Theorem 4, if $\Phi > (\pi/l)^2$, then g_t diverge as $t \to \infty$; otherwise, g_t converge to a limit metric $g_\infty = \mathrm{d}x^2 + \varphi_\infty^2(x)\mathrm{d}\theta^2$ with $K(g_\infty) = \Phi$. Certainly, if $\Phi = (\pi/l)^2$ and additional assumptions hold

$$
\tilde\mu_j = 0, \quad \int_0^\infty |\mu_j(\tau) - \tilde\mu_j|\,\mathrm{d}\tau < \infty, \quad \int_0^\infty |\mu_j'(\tau)|\,\mathrm{d}\tau < \infty \qquad (j = 0,1), \tag{40}
$$

see (11), then $\varphi_\infty = (v_1^0 + \int_0^\infty f_1(\tau)\,d\tau)\,\sin(\pi x/l)$, and if $\Phi < (\pi/l)^2$, then $\varphi_\infty = \tilde\varphi$. When a solution $\varphi(x,t)$ $(t \ge 0)$ of (39) is known and $|\varphi_{,x}| \le 1$, we find $\psi(t,x) = \psi(t,0) + \int_0^x \sqrt{1-(\varphi_{,x})^2}\,dx$. Note that rotation surfaces in $\mathbb{R}^3$ of constant Gaussian curvature are locally classified.

Assume for simplicity that $\mu_j(t) \equiv \tilde\mu_j$. Then $\delta(t) \equiv 0$ and $U \equiv 0$. Hence, $f \equiv 0$, $v_0 = \varphi_0 - \tilde\varphi_0$, and (40) is satisfied. For $\Phi = (\pi/l)^2$, we get $\varphi_\infty = C\,\sin(\pi x/l)$, where $C = v_1^0 = \int_0^l v_0(s)\,ds$, and then $\psi_\infty = \frac{l}{\pi}\,\mathrm{EllipticE}\,(\cos(\pi x/l), C\pi/l)$. Here $\mathrm{EllipticE}(z,k) = \int_0^z \sqrt{(1-k^2s^2)/(1-s^2)}\,ds$ is the incomplete elliptic integral. To provide a numerical example for $\Phi = (\pi/l)^2$, let $l = \pi$, $C = 1$ and $\mu_j = 0$. In this case, the limit profile curve is a semicircle $[\sin x, \cos x]$ $(0 \le x \le \pi)$, and the limit rotation surface is a round sphere of radius 1.

2 Results for PDEs

Let (F, g) be a closed p-dimensional Riemannian manifold, e.g., a leaf of a compact foliation $\mathcal{F}$. Functional spaces over F will be denoted without writing (F), for example, L_2 instead of $L_2(F)$.

Let H^l be the Hilbert space of differentiable by Sobolev real functions on F, with the inner product $(\cdot,\cdot)_l$ and the norm $\|\cdot\|_l$. In particular, $H^0 = L_2$ with the product $(\cdot,\cdot)_0$ and the norm $\|\cdot\|_0$.

If E is a Banach space, we denote by $\|\cdot\|_E$ the norm of vectors in this space. If B and C are real Banach spaces, we denote by $\mathcal{B}^r(B,C)$ the Banach space of all bounded r-linear operators $A : \prod_{i=1}^r B \to C$ with the norm $\|A\|_{\mathcal{B}^r(B,C)} = \sup_{v_1,\dots,v_r \in B\setminus 0} \frac{\|A(v_1,\dots,v_r)\|_C}{\|v_1\|_B\cdot\ldots\cdot\|v_r\|_B}$. If $r = 1$, we shall write $\mathcal{B}(B,C)$ and $A(\cdot)$, and if $B = C$ we shall write $\mathcal{B}^r(B)$ and $\mathcal{B}(B)$, respectively.

If M is a k-regular manifold or an open neighborhood of the origin in a real Banach space and N is a real Banach space, we denote by $C^k(M,N)$ $(k \ge 1)$ the Banach space of all C^k-regular functions $f : M \to N$, for which the following norm is finite:

$$\|f\|_{C^k(M,N)} = \sup_{x\in M} \max\{\|f(x)\|_N,\ \max_{1\le|r|\le k} \|d^r f(x)\|_{\mathcal{B}^r(T_xM,N)}\}.$$

Denote by $\|\cdot\|_{C^k}$, where $0 \le k < \infty$, the norm in the Banach space C^k; certainly, $\|\cdot\|_C$ when $k = 0$. In coordinates $(x_1,\dots,x_p)$ on F, we have $\|f\|_{C^k} = \max_{x\in F} \max_{|\alpha|\le k} |d^\alpha f(x)|$, where $\alpha \ge 0$ is the multi-index of order $|\alpha| = \sum_{i=1}^p \alpha_i$ and d^α is the partial derivative.

Sobolev embedding Theorem (See [1]). *If a nonnegative $k \in \mathbb{Z}$ and $l \in \mathbb{N}$ are such that $2l > p + 2k$, then H^l is continuously embedded into C^k.*

We shall also use the following scalar maximum principle [3, Theorem 4.4].

Proposition 2. *Suppose that $X(t)$ is a smooth family of vector fields on a closed Riemannian manifold (F, g), and $f \in C^\infty(\mathbb{R} \times [0, T))$. Let $u : F \times [0, T) \to \mathbb{R}$ be a C^∞ supersolution to*

$$\partial_t u \geq \Delta_g u + \langle X(t), \nabla u\rangle + f(u, t).$$

Let $\varphi : [0, T] \to \mathbb{R}$ solve the Cauchy's problem for ODEs $\frac{\mathrm{d}}{\mathrm{dt}}\varphi = f(\phi(t), t)$, $\varphi(0) = C$. If $u(\cdot, 0) \geq C$, then $u(\cdot, t) \geq \varphi(t)$ for $t \in [0, T)$. (Claim also holds with the sense of all three inequalities reversed).

2.1 The Schrödinger Operator

For a smooth (non-constant in general) function $\beta : F \to \mathbb{R}$, the Schrödinger operator, see (7),

$$\mathcal{H}(u) := -\Delta u - \beta\, u \tag{41}$$

is self-adjoint and bounded from below (but it is unbounded). The domain of definition of $\mathcal{H}$ is H^2.

Elliptic Regularity Theorem (See [1]). *If $\mathcal{H}$ is given by* (41) *and $0 \notin \sigma(\mathcal{H})$, then for any nonnegative $k \in \mathbb{Z}$ we have $\mathcal{H}^{-1} : H^k \to H^{k+2}$.*

The spectrum $\sigma(\mathcal{H})$ consists of an infinite sequence of isolated eigenvalues $\lambda_0 \leq \lambda_1 \leq \cdots \leq \lambda_j \leq \cdots$ of $\mathcal{H}$ counting their multiplicities, and $\lim_{j\to\infty} \lambda_j = \infty$. If we fix in L_2 an orthonormal basis of corresponding eigenfunctions $\{e_j\}$ (i.e., $\mathcal{H}(e_j) = \lambda_j e_j$), then any function $u \in L_2$ is expanded into the series (converging to u in the L_2-norm) $u(x) = \sum_{j=0}^{\infty} c_j\, e_j(x)$, where $c_j = (u, e_j)_0 = \int_F u(x)\, e_j(x)\, \mathrm{d}x$. The proof is based on the following facts. One can add a constant to β such that $\mathcal{H}$ becomes invertible in L_2 (e.g., $\lambda_0 > 0$) and $\mathcal{H}^{-1}$ is bounded in L_2. Since by the Elliptic Regularity Theorem with $k = 0$, we have $\mathcal{H}^{-1} : L_2 \to H^2$, and the embedding of H^2 into L_2 is continuous and compact, see [1], then the operator $\mathcal{H}^{-1} : L_2 \to L_2$ is compact. This means that the spectrum $\sigma(\mathcal{H})$ is discrete; hence, by the spectral expansion theorem for compact self-adjoint operators, $\{e_j\}_{j\geq 0}$ form an orthonormal basis in L_2.

Proposition 3 (See [9]). *Let β be a smooth function on a closed Riemannian manifold (F, g). Then the eigenspace of operator* (41)*, corresponding to the least eigenvalue, λ_0, is one-dimensional, and it contains a positive smooth eigenfunction, e_0.*

Suppose that $\beta(x) \geq \beta^-$ on F; hence, $(\beta(x)u,\, u)_0 \geq \beta^-(u,\, u)_0$. Thus,

$$\begin{aligned}(\mathcal{H}u,\, u)_0 &= \int_F (|\nabla u(x)|^2 - \beta(x)|u(x)|^2)\, \mathrm{d}x \leq \int_F (|\nabla u(x)|^2 - \beta^-|u(x)|^2)\, \mathrm{d}x\\ &= (-\Delta u - \beta^- u,\, u)_0\end{aligned}$$

for any $u \in \mathrm{Dom}(\mathcal{H})$. Since β^- is the maximal eigenvalue of the linear operator $\Delta + \beta^- \,\mathrm{id}$ (id is the identity operator), by the variational principle for eigenvalues, we obtain $\lambda_0 \le -\beta^- < 0$. Similarly, one may show that the condition $\beta(x) \le \beta^+$ on F provides $\lambda_0 \ge \beta^+$.

2.2 *The Nonlinear Heat Equation*

The Cauchy's problem for the *heat equation with a linear reaction term*, see (5), has the form

$$\partial_t u = \Delta\, u + \beta\, u, \qquad u(x,0) = u_0(x). \tag{42}$$

After scaling the time and replacement of functions

$$t/n \to t, \quad \Psi_i^{\mathcal{F}}/n \to \Psi_i, \quad \beta_{\mathcal{D}} + \Phi/n \to \beta, \quad \lambda_0^{\mathcal{F}} - \Phi/n \to \lambda_0,$$

problem (5) reads as the following Cauchy's problem for the *nonlinear heat equation* on (F, g):

$$\partial_t u = \Delta\, u + \beta\, u + \Psi_1(x)\, u^{-1} - \Psi_2(x)\, u^{-3}, \qquad u(x,0) = u_0(x). \tag{43}$$

By [1, Theorem 4.51], the parabolic PDE (43) has a unique smooth solution $u(\cdot\,,t)$ for $t \in [0, t_0)$. Denote by $\mathcal{C}_t = F \times [0,t)$ the cylinder with the base F. Define the quantities

$$\begin{aligned} &\Psi_i^+ = \max{}_F\,(\Psi_i/e_0^{2i}), \quad \Psi_i^- = \min{}_F\,(\Psi_i/e_0^{2i}), \quad i = 1,2,\\ &u_0^+ = \max{}_F\,(u_0/e_0), \quad u_0^- = \min{}_F\,(u_0/e_0), \quad \beta^- = \min{}_F\,|\beta|. \end{aligned} \tag{44}$$

The following examples show that (43) may have

- solutions on closed manifolds F (i.e., periodic solutions when $p = 1$);
- attractors (which are not global) when $\beta < 0$, and no attractors when $\beta > 0$.

Example 5. First, we shall examine (43) for modeling case when β and $\Psi_i \ge 0$ are real constants. Denote $f(u) := \beta u + \Psi_1\, u^{-1} - \Psi_2\, u^{-3}$.

1. The corresponding Cauchy's problem for ODE with $y(t)$ is

$$y\,' = f(y), \qquad y(0) = y_0 > 0. \tag{45}$$

(1a) Assume $\beta < 0$. The stationary positive solutions of (45) are the roots of equation $f(y) = 0$, which is biquadratic: $|\beta|\, y^4 - \Psi_1 y^2 + \Psi_2 = 0$. If $4\,|\beta|\,\Psi_2 < \Psi_1^2$, then we have two positive solutions $y_{1,2} = \big(\frac{\Psi_1 \pm (\Psi_1^2 - 4\,|\beta|\,\Psi_2)^{1/2}}{2|\beta|}\big)^{1/2}$ and $y_1 > y_2$.

The linearization of (45) at the point y_k is

$$v' = f'(y_k)v \qquad (k = 1, 2),$$

We have

$$f'(y_k) = -|\beta|\frac{\mathrm{d}}{\mathrm{dy}}(y^{-3}(y^2 - y_1^2)(y^2 - y_2^2))|_{y=y_k}.$$

Hence $f'(y_1) < 0$ and $f'(y_2) > 0$, and the stationary solution y_1 of (45) is asymptotically stable but y_2 is unstable. The solution $y(t)$ of (45) satisfying $y(0) = y_0$ is given in implicit form by

$$\frac{(y^2-y_1^2)^A}{(y^2-y_2^2)^B}=\frac{(y_0^2-y_1^2)^A}{(y_0^2-y_2^2)^B}\,e^{-2|\beta|t}, \quad \text{where} \quad A=\frac{y_1^2}{y_1^2-y_2^2}, \qquad B=\frac{y_2^2}{y_1^2-y_2^2}.$$

(1b) Next, assume $\beta > 0$, then $f(y) = 0$ is the biquadratic equation $\beta\, y^4 + \Psi_1 y^2 - \Psi_2 = 0$, which has only one positive root $y_1 = \big(\frac{-\Psi_1+(\Psi_1^2+4\,\beta\,\Psi_2)^{1/2}}{2\,\beta}\big)^{1/2}$. We calculate

$$f'(y_1) = \beta\,\frac{\mathrm{d}}{\mathrm{dy}}\Big(y^{-3}(y^2 - y_1^2)\big(y^2 + \frac{\Psi_2}{\beta\, y_1^2}\big)\Big)|_{y=y_1} > 0;$$

hence, a unique positive stationary solution y_1 of (45) is unstable. One may also show that in the case $\beta = 0$, (45) has a unique positive stationary solution, which is unstable.

(1c) Let $\Psi_2 = 0$ and $\Psi_1 > 0$, then $\partial_t u = \Delta u + f(u)$, where $f(u) = \beta\, u + \Psi_1 u^{-1}$.
For $\beta < 0$ the zero-mode approximation $\partial_t u = f(u)$ of the equation above has a unique positive stationary (also called equilibrium) solution $u_* = (\Psi_1/|\beta|)^{1/2}$ (root of f). The solution u_* is stable (attractor) since

$$f'(u_*) = -|\beta|\,\frac{\mathrm{d}}{\mathrm{du}}(u^{-1}(u - u_*)(u + u_*))|_{u=u_*} < 0.$$

If $\beta \geq 0$, then $\partial_t u = f(u)$ has no positive stationary solutions, and (48) has no cycles (since it has no fixed points), hence (47) has no solutions.

2. Let F be a circle S^1. Then (43) corresponds to the Cauchy's problem

$$u_{,t} = u_{,xx} + f(u), \quad u(x,0) = u_0(x) > 0 \qquad (x \in S^1,\ t \geq 0). \qquad (46)$$

The stationary equation with $u(x)$ for (46) with periodic boundary conditions has the form

$$u'' + f(u) = 0, \quad u(0) = u(l), \quad u'(0) = u'(l), \quad l > 0 \qquad (47)$$

(i.e., S^1 is a circle of length l). Denote $v = \frac{\mathrm{d}}{\mathrm{dx}} u$. Then (47) is reduced to the dynamical system

$$u' = v, \quad v' = -f(u) \quad (x \geq 0,\ u > 0). \tag{48}$$

The existence of a periodic solution to (47) is equivalent to the existence of a solution with the same period of (48). The system (48) is Hamiltonian, since $\partial_u v - \partial_v f(u) = 0$, its *Hamiltonian* $\mathrm{H}(u, v)$ (the first integral) is a solution of the system $\partial_u \mathrm{H}(u, v) = f(u),\ \partial_v \mathrm{H}(u, v) = v$. Then

$$\mathrm{H}(u, v) = \frac{1}{2}\,(v^2 + \beta u^2) + \Psi_1 \ln u + \frac{1}{2}\,\Psi_2\, u^{-2}.$$

The trajectories of (48) lie in the level lines of $\mathrm{H}(u, v)$. Consider two cases.

(2a) Assume $\beta < 0$. By results above, system (48) has two fixed points: $(y_i, 0)$ $(i = 1, 2)$ with $y_1 > y_2$. To clear up the character of these points, we linearize (48) at $(y_i, 0)$,

$$\vec{\eta}\,' = A_i\, \vec{\eta}, \qquad A_i = \begin{pmatrix} 0 & 1 \\ -f'(y_i) & 0 \end{pmatrix}.$$

As we have shown in the previous section, $f'(y_1) < 0$ and $f'(y_2) > 0$. Hence the point $(y_1,\ 0)$ has the "saddle" type and $(y_2,\ 0)$ is the "center." The separatrix (the level line of $\mathrm{H}(u, v)$ passing through the saddle point $(y_1,\ 0)$) is given by $\mathrm{H}(u, v) = \mathrm{H}(y_1, 0)$, i.e.,

$$v^2 = |\beta|(u^2 - y_1^2) - 2\,\Psi_1 \ln(u/y_1) - \Psi_2(u^{-2} - y_1^{-2}).$$

The separatrix divides the half-plane $u > 0$ into three simply connected areas. Then $(y_2,\ 0)$ is a unique minimum point of $\mathrm{H}(u, v)$ in the area $D = \{(u,\ v) : \mathrm{H}(u, v) < \mathrm{H}(y_1, 0),\ 0 < u < y_1\}$. The phase portrait of (48) in D consists of the fixed point $(y_2,\ 0)$ and the cycles surrounding this point all correspond to non-constant solutions of (47) with various l. Other two areas do not contain cycles of the system, since they have no fixed points.

Assume $\beta \geq 0$. By results above, system (48) has one fixed point: $(y_1, 0)$ and $f'(y_1) > 0$. Hence, $(y_1, 0)$ is the "center." Since $(y_1,\ 0)$ is a unique minimum point of $\mathrm{H}(u, v)$ in the semiplane $u > 0$, the phase portrait of (48) consists of the fixed point $(y_1,\ 0)$ and the cycles surrounding this point all correspond to non-constant solutions of (47) with various l.

(2b) For $\Psi_2 = 0$ and $\Psi_1 > 0$, the Hamiltonian of (48) is $H(u, v) = \frac{1}{2}(v^2 + \beta\, u^2) + \Psi_1 \ln u$. Solving $H(u, v) = C$ with respect to v and substituting to the first equation of the system, we get $\frac{\mathrm{du}}{\mathrm{dx}} = \sqrt{-\beta\, u^2 - 2\,\Psi_1 \ln u + 2\,C}$. In the case $\beta < 0$, the separatrix is $H(u, v) = H(u_*, 0)$, i.e.,

$$v^2 = |\beta|(u^2 - u_*^2) - 2\,\Psi_1 \ln(u/u_*).$$

The separatrix divides the half-plane $u > 0$ into four simply connected areas. Since in each of these areas there are no fixed points of (48), this system has no cycles. Hence, (47) has no solutions.

(2c) Consider (47) for $\Psi_1 = 0$ and $l = 2\pi$. Define $p = u'$ and search for $p = p(u)$ as a function of u. Then $u'' = \frac{\mathrm{d}}{\mathrm{du}}\, p$, and we obtain

$$(p^2)_u{}' = -2\,\beta\,u + 2\,\Psi_2\,u^{-3} \implies |u'|^2 = C_1 - \beta\,u^2 - \Psi_2\,u^{-2}.$$

After separation of variables and integration we get

$$u = \begin{cases} \sqrt{\frac{C_1}{2\beta} + \frac{1}{2\beta}\sqrt{C_1^2 - 4\beta\,\Psi_2}\,\sin(2\sqrt{\beta}(x + C_2))}, & (C_1^2 - 4\beta\Psi_2 \geq 0)\ \beta > 0, \\ \sqrt{-\frac{C_1}{2|\beta|} + \frac{1}{2|\beta|}\sqrt{C_1^2 + 4|\beta|\Psi_2}\,\cosh(2\sqrt{|\beta|}(x + C_2))} & \beta < 0, \\ \sqrt{\Psi_2/C_1 + C_1(x + C_2)^2} & \beta = 0. \end{cases} \tag{49}$$

By $(49)_{2,3}$, for $\beta \leq 0$ and $\Psi_1 = 0$, (47) has no positive solutions. By $(49)_1$, for $\beta > 0$ and $\Psi_2 > 0$ the solution $u(x)$ is 2π-periodic and positive only in two cases:

- $\beta \neq \frac{n^2}{4}$ $(n \in \mathbb{N})$ and $C_1 = 2\,(\beta\,\Psi_2)^{1/2}$; such a solution $u_* = (\Psi_2/\beta)^{1/4}$ is unique.
- $\beta = \frac{n^2}{4}$ $(n \in \mathbb{N})$; such solutions form a two-dimensional manifold:

$$u_0(C_1, C_2) = \frac{1}{n}\big(2\,C_1 + 2(C_1^2 - n^2\Psi_2)^{1/2}\sin(n(x + C_2))\big)^{1/2}.$$

2.3 Long-Time Solution to (43) with $\lambda_0 < 0$

Lemma 6. *Let $\lambda_0 < 0$ for (F, g) and $u(x,t) > 0$ be a solution in $\mathcal{C}_{t_0}$ of (43) with the condition*

$$(u_0^-)^4 \geq \Psi_2^+/|\lambda_0|, \tag{50}$$

see (8). Then the following a priori estimates are valid:

$$w_-(t) \leq u(x,t)/e_0(x) \leq w_+(t), \qquad (x,t) \in \mathcal{C}_{t_0}, \tag{51}$$

where $\lambda_0 < 0$ and

$$w_-(t) = e^{-\lambda_0 t}\Big((u_0^-)^4 + \frac{\Psi_2^+}{\lambda_0} - e^{4\lambda_0 t}\frac{\Psi_2^+}{\lambda_0}\Big)^{\frac{1}{4}}, \quad w_+(t) = e^{-\lambda_0 t}\Big((u_0^+)^2 - \frac{\Psi_1^+}{\lambda_0} + e^{2\lambda_0 t}\frac{\Psi_1^+}{\lambda_0}\Big)^{\frac{1}{2}}. \tag{52}$$

Proof. Since $e_0(x) > 0$ on F, we can change the unknown function in (43):

$$u(x,t) = e_0(x)\, w(x,t).$$

Substituting into (43) and using $\Delta e_0 + \beta e_0 = -\lambda_0 e_0$, we obtain the Cauchy's problem for $w(x,t)$:

$$\partial_t w = \Delta w - \lambda_0 w + 2\, g(\nabla \log e_0,\, \nabla w) \\ +e_0^{-2}(x)\Psi_1(x)\, w^{-1} - e_0^{-4}(x)\Psi_2(x)\, w^{-3},\ w(\cdot\,,0) = u_0/e_0. \quad (53)$$

Then using $(44)_1$, we obtain the differential inequalities

$$\Delta w - \lambda_0 w + 2\, g(\nabla \log e_0, \nabla w) - \Psi_2^+ w^{-3} \le \partial_t w \le \Delta w - \lambda_0 w + 2\, g(\nabla \log e_0, \nabla w) + \Psi_1^+ w^{-1}.$$

By the scalar maximum principle of Proposition 2 and $(44)_{2,3}$, we conclude that (51) holds, where $w_-(t)$ and $w_+(t)$ are solutions of the following Cauchy's problems for ODEs

$$\frac{\mathrm{d}}{\mathrm{dt}}\, w_- = -\lambda_0\, w_- - \Psi_2^+ w_-^{-3},\ w_-(0) = u_0^-;\ \frac{\mathrm{d}}{\mathrm{dt}}\, w_+ = -\lambda_0\, w_+ + \Psi_1^+ w_+^{-1},\ w_+(0) = u_0^+.$$

One may check that these solutions are expressed by (52) and $w_-(t) < w_+(t)$ for all $t \ge 0$. □

Note that if $\Psi_i^+ = 0$ (i.e., $\Psi_i \equiv 0$), then (51) reads $u_0^- e^{-\lambda_0 t} \le u(\cdot,t)/e_0 \le u_0^+ e^{-\lambda_0 t}$. Define

$$v(x,t) = e^{\lambda_0 t} u(x,t),$$

see (43), and obtain the Cauchy's problem

$$\partial_t v = \Delta\, v + (\beta + \lambda_0)\, v + Q, \quad v(x,0) = u_0(x), \quad (54)$$

where $Q := \sum_{i=1}^{2} (-1)^{i+1} \Psi_i(x)\, v^{1-2i}(x,t)\, e^{\,2i\lambda_0 t}$. Certainly, $Q = Q_1 - Q_2$, where

$$Q_1(x,t) = \Psi_1(x) v^{-1}(x,t) e^{2\lambda_0 t}, \quad Q_2(x,t) = \Psi_2(x) v^{-3}(x,t) e^{4\lambda_0 t}.$$

Lemma 7. *Let $v(x,t)$ be a positive solution of* (54) *in $C_{t_0} = F \times [0,\, t_0)$, where $\lambda_0 < 0$, the functions $u_0 > 0$ and $\Psi_i \ge 0$ belong to C^∞, and* (50) *is satisfied. Then*

(i) for any multi-index $\alpha = (\alpha_1, \alpha_2, \ldots, \alpha_p)$ there exists a real $C_\alpha \ge 0$ such that

$$|\partial_x^\alpha v(x,t)| \le C_\alpha (1+t)^{|\alpha|}, \quad (x,t) \in C_{t_0}.$$

(ii) for any multi-index α there exist real $\bar{Q}_{i\alpha} \ge 0$ $(i = 1, 2)$ such that

$$|\partial_x^\alpha Q_i(x,t)| \le \bar{Q}_{i\alpha}(1+t)^{|\alpha|} e^{2i\lambda_0 t}, \quad (x,t) \in \mathcal{C}_{t_0}, \quad i = 1,2. \tag{55}$$

Proof. Using (51) and (52), we estimate the solution $v(x,t)$ of (54) when (50) holds:

$$v_- \le v(x,t)/e_0(x) \le v_+, \qquad (x,t) \in F \times [0,\infty), \tag{56}$$

where the constants are given by $v_- = \big((u_0^-)^4 - \Psi_2^+/|\lambda_0|\big)^{\frac{1}{4}}$ and $v_+ = \big((u_0^+)^2 + \Psi_1^+/|\lambda_0|\big)^{\frac{1}{2}}$.

(i) Denote for brevity $D_j = \partial_{x_j}$ $(j = 1, 2, \dots, p)$ and $D_\alpha = \partial_x^\alpha = D_{\alpha_1} D_{\alpha_2} \cdots D_{\alpha_p}$. Differentiating (54) by $x_1, \dots, x_p$, we obtain the following PDEs for the functions $p_\alpha(x,t) := \partial_x^\alpha v(x,t)$:

$$\begin{aligned}
\partial_t p_j &= (\Delta + (\lambda_0 + \beta)\,\mathrm{id})\, p_j + D_j(\beta) v + D_j(Q),\\
\partial_t p_{jk} &= (\Delta + (\lambda_0 + \beta)\,\mathrm{id})\, p_{jk} + D_j(\beta) p_k + D_k(\beta) p_j + D_{jk}(\beta) v + D_{jk}(Q),
\end{aligned} \tag{57}$$

and so on, where $1 \le j,k \le p$ and

$$\begin{aligned}
D_j(Q) &= \sum\nolimits_i (-1)^{i+1} e^{2i\lambda_0 t} v^{-2i} \big(D_j(\Psi_i) v + (1-2i)\Psi_i p_j\big),\\
D_{jk}(Q) &= \sum\nolimits_i (-1)^{i+1} e^{2i\lambda_0 t} v^{-2i} \big(D_{jk}(\Psi_i) v + (1-2i) D_j(\Psi_i) p_k\\
&\quad + (1-2i) D_k(\Psi_i) p_j - 2i(1-2i) v^{-1} \Psi_i p_j p_k + (1-2i)\Psi_i p_{jk}\big),
\end{aligned} \tag{58}$$

and so on. Let us change unknown functions in (57) and so on:

$$p_j = \tilde{p}_j\, e_0, \quad p_{jk} = \tilde{p}_{jk}\, e_0, \ \dots \tag{59}$$

Then in the same manner, as (53) have been obtained from (43), we get for $j,k = 1, 2, \dots, p$

$$\begin{aligned}
\partial_t \tilde{p}_j &= \Delta \tilde{p}_j + 2\, g(\nabla \log e_0, \nabla \tilde{p}_j) + a\tilde{p}_j + b_j/e_0 + D_j(\beta) v/e_0,\\
\partial_t \tilde{p}_{jk} &= \Delta \tilde{p}_{jk} + 2\, g(\nabla \log e_0, \nabla \tilde{p}_{jk}) + a\tilde{p}_{jk} + b_{jk}/e_0 + D_{jk}(\beta) v/e_0,
\end{aligned} \tag{60}$$

and so on, where

$$a = \sum_i (-1)^{i+1}(1-2i)\Psi_i\, v^{-2i} e^{2i\lambda_0 t}, \quad b_j = \sum_i (-1)^{i+1} D_j(\Psi_i)\, v^{1-2i} e^{2i\lambda_0 t},$$

$$b_{jk} = \sum_i (-1)^{i+1} e^{2i\lambda_0 t} v^{-2i} \big(D_{jk}(\Psi_i)\frac{v}{e_0} + (1-2i)$$

$$\big(D_j(\Psi_i)\tilde p_k + D_k(\Psi_i)\tilde p_j - 2i\,\Psi_i \frac{e_0}{v}\,\tilde p_j\,\tilde p_k\big)\big).$$

From (56) and (58)–(60) we get the differential inequalities

$$-a^+(t)|\tilde p_j| - b_j^+ - \beta_j^+ v_+ \le \partial_t \tilde p_j - \Delta \tilde p_j - 2\,g(\nabla \log e_0, \nabla \tilde p_j)$$

$$\le a^+(t)|\tilde p_j| + b_j^+ + \beta_j^+ v_+\,,$$

$$-a^+(t)|\tilde p_{jk}| - b_{jk}^+ - \beta_{jk}^+ v_+ \le \partial_t \tilde p_{jk} - \Delta \tilde p_{jk} - 2\,g(\nabla \log e_0, \nabla \tilde p_{jk})$$

$$\le a^+(t)|\tilde p_{jk}| + b_{jk}^+ + \beta_{jk}^+ v_+$$

for $j = 1, 2, \dots, p$, where

$$a^+(t) = \sum_i (2i-1)\Psi_i^+ (v_-)^{-2i} e^{2i\lambda_0 t}, \quad b_j^+ = \sum_i \big((v_-)^{1-2i} \max_F |D_j(\Psi_i)|\big),$$

$$\beta_j^+ = \max_F |D_j(\beta)|,$$

$$b_{jk}^+ = \sum_i e^{2i\lambda_0 t}(v_-)^{-2i}\big(\max_F |D_{jk}(\Psi_i)/e_0^{2i}|\, v_+ + (2i-1)\big(\max_F |D_j(\Psi_i)/e_0^{2i}|\,\tilde p_k$$

$$+ \max_F |D_k(\Psi_i)/e_0^{2i}|\,\tilde p_j + 2i\,(v_-)^{-1}\Psi_i^+\,\tilde p_j\,\tilde p_k\big)\big),$$

$$\beta_{jk}^+ = \max_F |D_{jk}(\beta)|. \tag{61}$$

By the maximum principle of Proposition 2, the estimate $|\tilde p_j(x,t)| \le \tilde p_j^+(t)$ is valid for any $(x,t) \in \mathcal{C}_\infty = F \times [0,\infty)$, where $p_j^+(t)$ solves the Cauchy's problem for the ODE:

$$\frac{\mathrm{d}}{\mathrm{dt}}\, p_j^+ = a^+(t)|p_j^+| + b_j^+ + \beta_j^+ v_+, \quad p_j^+(0) = \bar p_j^0 := \max{}_F\, |\tilde p_j(\cdot\,,0)|\,.$$

As is known,

$$p_j^+(t) = \bar p_j^0 \exp\Big(\int_0^t a^+(\tau)\,\mathrm{d}\tau\Big) + \int_0^t (b_j^+ + \beta_j^+ v_+) \exp\Big(\int_s^t a^+(\tau)\,\mathrm{d}\tau\Big)\,\mathrm{ds}\,.$$

In view of $(61)_1$, we have

$$\int_0^\infty a^+(\tau)\,\mathrm{d}\tau < \infty.$$

Above yield that for any $j \in \{1, 2, \dots, p\}$ there exists a real $\tilde{C}_j > 0$ such that

$$|\tilde{p}_j(x,t)| \le \tilde{C}_j(1+t), \qquad (x,t) \in \mathcal{C}_\infty .$$

In view of (59), this completes the proof of (*i*) for $|\alpha| = 1$.

Similarly we obtain that for any $j, k \in \{1, 2, \dots, p\}$ there exists a real $\tilde{C}_{jk} \ge 0$ such that $|\tilde{p}_{jk}(x,t)| \le \tilde{C}_{jk}(1+t)^2$ for $(x,t) \in \mathcal{C}_\infty$. By (59), we obtain claim (*i*) for $|\alpha| = 2$. By induction with respect to $|\alpha|$ we prove (*i*) for any α.

(ii) Estimates (55) for $|\alpha| = 0$, $|Q_i(x,t)| \le (\max_F \Psi_i)\,(v_-)^{1-2i} e^{2i\lambda_0 t}$, follow immediately from (56). Estimates (55) for $|\alpha| = 1, 2$ follow from claim (*i*), estimates (56), and equalities (58). By induction with respect to $|\alpha|$ we prove (ii) for any α.

□

Theorem 5. *Cauchy's problem* (43) *on* F*, with* $\lambda_0 < 0$ *and the initial value* $u_0(x)$ *satisfying* (50)*, admits a unique smooth solution* $u(x,t) > 0$ *in the cylinder* $\mathcal{C}_\infty = F \times [0, \infty)$.

Proof. The positive solution $u(x,t)$ of (43) satisfies a priori estimates (51) on any cylinder $\mathcal{C}_{t_\star}$ where it exists. By standard arguments, using the local theorem of the existence and uniqueness for semiflows, we obtain that this solution can be uniquely prolonged on the cylinder $\mathcal{C}_\infty$. Then by Lemma 7, all partial derivatives by x of $u(x,t)$ exist in $\mathcal{C}_\infty$. Hence, u is smooth on $\mathcal{C}_\infty$. □

2.4 *Asymptotic Behavior of Solutions to* (43) *with* $\lambda_0 < 0$

Recall that λ_0 and $e_0 > 0$ are the least eigenvalue and the ground state of operator (41).

Theorem 6. *Let* $u > 0$ *be a smooth solution on* $\mathcal{C}_\infty$ *of* (43) *with* $\lambda_0 < 0$ *and the initial value* $u_0(x)$ *satisfying* (50) (*see Theorem 5). Then there exists a solution* $\tilde{u}$ *on* $\mathcal{C}_\infty$ *of the linear PDE*

$$\partial_t \tilde{u} = \Delta \tilde{u} + (\beta(x) + \lambda_0)\,\tilde{u} \tag{62}$$

such that for any $\alpha \in \big(0, \min\{\lambda_1 - \lambda_0,\, 2\,|\lambda_0|\}\big)$ *and any* $k \in \mathbb{N}$

$$(i)\; u = e^{-\lambda_0 t}(\tilde{u} + \theta(x,t)), \qquad (ii)\; \nabla \log u = \nabla \log e_0 + \theta_1(x,t),$$

where $\|\theta(\cdot\,,t)\|_{C^k} = O(e^{-\alpha t})$ *and* $\|\theta_1(\cdot\,,t)\|_{C^k} = O(e^{-\alpha t})$ *as* $t \to \infty$.

Proof. (i) Let $G_0(t,x,y)$ be the fundamental solution of (62), called the *heat kernel*. As is known, $G_0(t,x,y) = \sum_j e^{(\lambda_0-\lambda_j)t} e_j(x)\, e_j(y)$. Due to the Duhamel's principle, the solution $v = e^{\lambda_0 t} u$ of Cauchy's problem (54) satisfies the nonlinear integral equation

$$v(x,t) = \int_F G_0(t,x,y)\, u_0(y)\,\mathrm{d}y + \int_0^t \Big(\int_F G_0(t-\tau,x,y) Q(y,\tau)\,\mathrm{d}y\Big)\,\mathrm{d}\tau\,. \tag{63}$$

Expand v, u_0, and Q into Fourier series by eigensystem $\{e_j\}$:

$$\begin{aligned} v(x,t) &= \textstyle\sum_{j=0}^\infty v_j(t)\, e_j(x), \quad v_j(t) = (v(\cdot\,,t), e_j)_0 = \int_F v(y,t)\, e_j(y)\,\mathrm{d}y,\\ u_0(x) &= \textstyle\sum_{j=0}^\infty u_j^0\, e_j(x), \qquad\qquad u_j^0 = (u_0, e_j)_0 = \int_F u_0(y)\, e_j(y)\,\mathrm{d}y,\\ Q(x,t) &= \textstyle\sum_{j=0}^\infty q_j(t)\, e_j(x),\ q_j(t) = (Q(\cdot\,,t), e_j)_0 = \int_F Q(y,t)\, e_j(y)\,\mathrm{d}y\,. \end{aligned} \tag{64}$$

Then we obtain from (63)

$$v_j(t) = u_j^0\, e^{(\lambda_0-\lambda_j)t} + \int_0^t e^{(\lambda_0-\lambda_j)(t-\tau)} q_j(\tau)\,\mathrm{d}\tau \qquad (j=0,1,\ldots). \tag{65}$$

Substituting $v_j(t)$ of (65) into (64), we represent v in the form $v = \tilde u + \theta$, where

$$\tilde u = \tilde u_0^0\, e_0 + \sum\nolimits_{j=1}^\infty u_j^0\, e^{(\lambda_0-\lambda_j)t} e_j, \quad \tilde u_0^0 = u_0^0 + \int_0^\infty q_0(\tau)\,\mathrm{d}\tau, \tag{66}$$

$$\theta = -\Big(\int_t^\infty q_0(\tau)\,\mathrm{d}\tau\Big) e_0 + \sum\nolimits_{j=1}^\infty \tilde v_j\, e_j,\ \tilde v_j = \int_0^t e^{(\lambda_0-\lambda_j)(t-\tau)} q_j(\tau)\,\mathrm{d}\tau. \tag{67}$$

Observe that $\tilde u$ solves (62) with the initial condition $\tilde u(\cdot\,,0) = u_0 + \big(\int_0^\infty q_0(\tau)\,\mathrm{d}\tau\big) e_0$.

Let us take $k\in\mathbb{N}$, $l = \big[p/4 + k/2\big] + 1$, and $\gamma < \lambda_0$. Using assumption $u_0 \in C^\infty(F)$ and the fact that $Q(\cdot\,,t)\in C^\infty(F)$ for any $t\ge 0$, we may consider the functions $w_0 := (\mathcal{H}-\gamma\,\mathrm{id})^l u_0$ and $P(\cdot\,,t) := (\mathcal{H}-\gamma\,\mathrm{id})^l Q(\cdot\,,t)$, which have the same properties: $w_0\in C^\infty(F)$ and $P(\cdot\,,t)\in C^\infty(F)$ for any $t\ge 0$. Let us represent

$$\begin{aligned} (u_0,\, e_j)_0\, e_j &= ((\mathcal{H}-\gamma\,\mathrm{id})^{-l} w_0,\, e_j)_0\, e_j = (w_0,\, (\mathcal{H}-\gamma\,\mathrm{id})^{-l} e_j)_0\, e_j\\ &= (w_0,\, e_j)_0 \frac{e_j}{\lambda_j-\gamma} = (\mathcal{H}-\gamma\,\mathrm{id})^{-l}\big((w_0,\, e_j)_0\, e_j\big). \end{aligned}$$

Similarly, we obtain

$$(Q(\cdot\,,t), e_j)_0\, e_j = (\mathcal{H}-\gamma\,\mathrm{id})^{-l}\big((P(\cdot\,,t), e_j)_0\, e_j\big).$$

Using (67) and taking into account that the operator $(\mathcal{H} - \gamma\,\mathrm{id})^{-l}$ acts continuously in L_2 and that the series in (66) and (67) converge in L_2, we obtain the representations

$$\sum\nolimits_{j=1}^{\infty} u_j^0 e^{(\lambda_0-\lambda_j)t} e_j = (\mathcal{H}-\gamma\,\mathrm{id})^{-l} \sum\nolimits_{j=1}^{\infty} e^{(\lambda_0-\lambda_j)t}(w_0, e_j)_0\, e_j,$$

$$\sum\nolimits_{j=1}^{\infty} \tilde{v}_j(t) e_j = (\mathcal{H}-\gamma\,\mathrm{id})^{-l} \int_0^t \Big(\sum\nolimits_{j=1}^{\infty} e^{(\lambda_0-\lambda_j)(t-\tau)}(P(\cdot,t), e_j)_0\, e_j\Big)\,\mathrm{d}\tau.$$

By the Elliptic Regularity Theorem and the Sobolev Embedding Theorem (see Sect. 2.1), the operator $(\mathcal{H} - \gamma\,\mathrm{id})^{-l}$ acts continuously from L_2 into C^k. Then we have

$$\begin{aligned}
\Big\|\sum\nolimits_{j=1}^{\infty} u_j^0 e^{(\lambda_0-\lambda_j)t} e_j\Big\|_{C^k} &\le \|(\mathcal{H}-\gamma\,\mathrm{id})^{-l}\|_{\mathcal{B}(L_2,C^k)} \cdot \Big\|\sum\nolimits_{j=1}^{\infty} e^{(\lambda_0-\lambda_j)t}(w_0, e_j)_0\, e_j\Big\|_0 \\
&= \|(\mathcal{H}-\gamma\,\mathrm{id})^{-l}\|_{\mathcal{B}(L_2,C^k)} \Big(\sum\nolimits_{j=1}^{\infty} e^{2(\lambda_0-\lambda_j)t}(w_0, e_j)_0^2\Big)^{1/2} \\
&\le \|(\mathcal{H}-\gamma\,\mathrm{id})^{-l}\|_{\mathcal{B}(L_2,C^k)}\, e^{(\lambda_0-\lambda_1)t}\|w_0\|_0,
\end{aligned} \tag{68}$$

$$\begin{aligned}
\Big\|\sum\nolimits_{j=1}^{\infty} \tilde{v}_j(t) e_j\Big\|_{C^k} &\le \|(\mathcal{H}-\gamma\,\mathrm{id})^{-l}\|_{\mathcal{B}(L_2,C^k)} \\
&\quad \cdot \Big\|\int_0^t \sum\nolimits_{j=1}^{\infty} e^{(\lambda_0-\lambda_j)(t-\tau)}(P(\cdot,t), e_j)_0\, e_j\,\mathrm{d}\tau\Big\|_0 \\
&\le \|(\mathcal{H}-\gamma\,\mathrm{id})^{-l}\|_{\mathcal{B}(L_2,C^k)} \int_0^t \Big\|\sum\nolimits_{j=1}^{\infty} e^{(\lambda_0-\lambda_j)(t-\tau)}(P(\cdot,t), e_j)_0\, e_j\Big\|_0 \mathrm{d}\tau \\
&\le \|(\mathcal{H}-\gamma\,\mathrm{id})^{-l}\|_{\mathcal{B}(L_2,C^k)} \int_0^t \Big(\sum\nolimits_{j=1}^{\infty} e^{2(\lambda_0-\lambda_j)(t-\tau)}(P(\cdot,t), e_j)_0^2\Big)^{1/2} \mathrm{d}\tau \\
&\le \|(\mathcal{H}-\gamma\,\mathrm{id})^{-l}\|_{\mathcal{B}(L_2,C^k)} \int_0^t e^{(\lambda_0-\lambda_1)(t-\tau)}\|P(\cdot,t)\|_0\,\mathrm{d}\tau.
\end{aligned} \tag{69}$$

On the other hand, by Lemma 7(ii),

$$\|P(\cdot,t)\|_0 \le \sqrt{\mathrm{Vol}(F,g)}\,\|(\mathcal{H}-\gamma\,\mathrm{id})^l(Q(\cdot,t))\|_{C^0} \le \bar{Q}(1+t)^{2l} e^{2\lambda_0 t}$$

for some $\bar{Q} \ge 0$. Then continuing (69), we find

$$\begin{aligned}
\int_0^t e^{(\lambda_0-\lambda_1)(t-\tau)}\|P(\cdot,t)\|_0\,\mathrm{d}\tau &\le \bar{Q}\int_0^t e^{(\lambda_0-\lambda_1)(t-\tau)}(1+\tau)^{2l} e^{2\lambda_0\tau}\,\mathrm{d}\tau \\
&< \bar{Q} e^{(\lambda_0-\lambda_1)t}(1+t)^{2l}\int_0^t e^{(\lambda_1-\lambda_0+2\lambda_0)\,\tau}\,\mathrm{d}\tau \\
&= \bar{Q}(1+t)^{2l}\begin{cases} \frac{e^{2\lambda_0 t}-e^{(\lambda_0-\lambda_1)t}}{\lambda_1-\lambda_0+2\lambda_0} & \text{if } \ 2\lambda_0 \ne \lambda_0-\lambda_1, \\ e^{(\lambda_0-\lambda_1)t}\,t & \text{if } \ 2\lambda_0 = \lambda_0-\lambda_1. \end{cases}
\end{aligned} \tag{70}$$

From (66)–(70) we get claim (i).

(ii) From (66) and (67) we obtain

$$u = e^{-\lambda_0 t}\big(\tilde u_0^0\, e_0 + \bar\theta(\cdot\,,t)\big),\quad \nabla u = e^{-\lambda_0 t}\big(\tilde u_0^0 \nabla e_0 + \nabla\bar\theta(\cdot\,,t)\big),$$

where $\bar\theta(\cdot\,,t) = \theta(\cdot\,,t) + \sum_{j=1}^{\infty} u_j^0 e^{(\lambda_0-\lambda_j)t} e_j$. In view of (68), $\|\bar\theta(\cdot\,,t)\|_{C^k} = O(e^{-\alpha t})$ for any $k \in \mathbb{N}$. Furthermore, since $\tilde u(\cdot,0) > 0$ on F, then $\tilde u_0^0 = (\tilde u(\cdot,\,0),\ e_0)_0 > 0$. Using

$$w(\cdot,t,\tau) := \tau u(\cdot,t) + (1-\tau)\,\tilde u_0^0\, e^{-\lambda_0 t} e_0 = e^{-\lambda_0 t}\,(\tilde u_0^0\, e_0 + \tau\,\bar\theta(\cdot,t)),$$

we have

$$\begin{aligned}\theta_1(\cdot,t) &= \nabla \log u(\cdot,t) - \nabla \log e_0 = \int_0^1 \frac{\partial}{\partial\tau}\Big(\nabla \log w(\cdot,t,\tau)\Big)\,\mathrm{d}\tau\\ &= \int_0^1 \Big(\frac{\nabla\bar\theta(\cdot,t)}{\tilde u_0^0 e_0 + \tau\,\bar\theta(\cdot,t)} - \frac{\bar\theta(\cdot,t)(\tilde u_0^0\nabla e_0 + \tau\nabla\bar\theta(\cdot,t))}{(\tilde u_0^0 e_0 + \tau\,\bar\theta(\cdot,t))^2}\Big)\,\mathrm{d}\tau\,.\end{aligned}$$

By the above, and the fact that $\inf\{\,|\tilde u_0^0 e_0 + \tau\,\bar\theta(\cdot,t)| : x \in F,\ t \in [t_0,\infty),\ \tau \in [0,1]\} > 0$ holds for $t_0 > 0$ large enough, follows claim (*ii*).

□

2.5 *Attractor of* (43) *with* $\boldsymbol{\lambda_0 > 0}$

In this section we assume that $\lambda_0 > 0$ for (F,g).

Along with Cauchy's problem (43) consider the Cauchy's problem for the ODE:

$$y' = \phi(y),\quad y(0) = y_0, \tag{71}$$

where $\phi(y) = -\lambda_0 y + \Psi_1^- y^{-1} - \Psi_2^+ y^{-3}$ $(y > 0)$. Assume that

$$0 < \lambda_0 < (\Psi_1^-)^2/(4\,\Psi_2^+), \tag{72}$$

hence, $\Psi_1^- > 0$. Then the equation $\phi(y) = 0$ has two distinct positive solutions $y_2 < y_1$,

$$y_1 = \Big(\frac{\Psi_1^- + \sqrt{(\Psi_1^-)^2 - 4\,\Psi_2^+\lambda_0}}{2\lambda_0}\Big)^{1/2},\quad y_2 = \Big(\frac{\Psi_1^- - \sqrt{(\Psi_1^-)^2 - 4\,\Psi_2^+\lambda_0}}{2\lambda_0}\Big)^{1/2},$$

which are stationary solutions of $(71)_1$. Denote by $y_3 = \big(\frac{-\Psi_1^- + \sqrt{(\Psi_1^-)^2 + 12\,\Psi_2^+\lambda_0}}{2\lambda_0}\big)^{1/2}$ a unique positive solution of equation $\phi'(y) = 0$. Notice that $\phi(y) > 0$ for $y \in (y_2,\ y_1)$ and $\phi(y) < 0$ for $y \in (0,\ \infty) \setminus [y_2,\ y_1]$; $\phi(y)$ increases in $(0, y_3)$ and decreases in $(y_3,\ \infty)$. It is clear that $y_3 \in (y_2,\ y_1)$. The line $z = -\lambda_0 y$ is the asymptote for the graph of $\phi(y)$ for $y \to \infty$. We have $\lim_{y\downarrow 0}\phi(y) = -\infty$, $\phi'(y)$ decreases in $(0, y_4)$ and increases in $(y_4,\ \infty)$, where $y_4 = \sqrt{6\,\Psi_2^+/\Psi_1^-} > y_3$, and $\lim\limits_{y\to\infty}\phi'(y) = -\lambda_0$. Hence,

$$\mu(\varepsilon) := \inf{}_{y\in[y_1-\varepsilon,\,\infty)}(-\phi'(y)) = \min\{|\phi'(y_1-\varepsilon)|,\ \lambda_0\}$$

for $\varepsilon \in (0,\ y_1 - y_3)$. Define the closed in C sets $\mathcal{U}_2^\varepsilon \subset \mathcal{U}_1^\varepsilon$ by

$$\begin{aligned}\mathcal{U}_1^\varepsilon &:= \{u_0 \in C\ :\ u_0/e_0 \ge y_1 - \varepsilon\},\\ \mathcal{U}_2^\varepsilon &:= \{u_0 \in C\ :\ y_1 - \varepsilon \le u_0/e_0 \le (\Psi_1^+/\lambda_0)^{1/2}\}\end{aligned}$$

with $\varepsilon \in (0,\ y_1 - y_3)$ under assumption (72). Define the set $\mathcal{U}_1 := \{u_0 \in C\ :\ u_0/e_0 > y_3\}$. Notice that $\mathcal{U}_1^\varepsilon \subset \mathcal{U}_1$ for all $\varepsilon \in (0,\ y_1 - y_3)$.

Let $\mathcal{S}_t\ :\ u_0 \to u(\cdot\,, t)$ $(t \ge 0)$ be the one-parameter semigroup for (43) with (72) in $F \times [0, \infty)$.

Proposition 4. *If $\varepsilon \in (0,\ \bar{y}_1 - y_3)$, then* (43) *with* (72) *and $u_0 \in \mathcal{U}_1^\varepsilon$ admits a unique global solution. Furthermore, the sets $\mathcal{U}_1^\varepsilon$ and $\mathcal{U}_2^\varepsilon$ are invariant for the operators $\mathcal{S}_t$ $(t \ge 0)$.*

Proof. Let $u(\cdot\,, t) = e_0\, w(\cdot\,, t)$ and $w_0 = w(\cdot\,, 0)$. From (53) we obtain the differential inequalities

$$\begin{aligned}\Delta w + 2\,\langle\nabla\log e_0,\ \nabla w\rangle + \phi(w) \le \partial_t w \le \Delta w + 2\,\langle\nabla\log e_0,\ \nabla w\rangle\\ -\lambda_0 w + \Psi_1^+ w^{-1}.\end{aligned} \tag{73}$$

Suppose that $u_0 \in \mathcal{U}_1^\varepsilon$; hence, $w_0 \ge y_1 - \varepsilon$. By the maximum principle of Proposition 2 and Lemma 8, in the maximal domain D_M of the existence of the solution $w(x,t)$ of (53), the estimate

$$w(\cdot\,, t) \ge y_1 - \varepsilon e^{-\mu(\varepsilon)\,t}$$

is valid, which, in particular, implies that this solution cannot "blowdown" to zero. Applying the maximum principle to the right inequality of (73), we get that in D_M

$$w(\cdot\,, t) \le \big(((u_0^+)^2 - \Psi_1^+/\lambda_0)e^{-2\lambda_0 t} + \Psi_1^+/\lambda_0\big)^{1/2}. \tag{74}$$

From the last estimate we conclude that the solution $u(x,t)$ of (43) exists for all $(x,t) \in F \times [0,\infty)$. Furthermore, we have proved above that if $u_0 \in \mathcal{U}_1^\varepsilon$, then $u(\cdot\,,t) \in \mathcal{U}_1^\varepsilon$ for any $t > 0$. This means that the set $\mathcal{U}_1^\varepsilon$ is invariant for the operators $\mathcal{S}_t$ $(t \geq 0)$. This fact and the following from (74) estimate $w(\cdot\,,t) \leq \max\{u_0^+,\ (\Psi_1^+/\lambda_0)^{1/2}\}$ imply that also the set $\mathcal{U}_2^\varepsilon$ is invariant for all $\mathcal{S}_t$. □

Lemma 8. *(i) If $y_0 > y_2$, then the solution $y(t)$ of Cauchy's problem* (71) *obeys $\lim_{t\to\infty} y(t) = y_1$. Furthermore, if $y_0 \in (y_2,\ y_1)$, then $y(t)$ is increasing and if $y_0 > y_1$, then $y(t)$ is decreasing.*

(ii) If $y_0 \geq y_1 - \varepsilon$, where $\varepsilon \in (0,\ y_1 - y_3)$, then the estimate is valid:

$$|y(t) - y_1| \leq |y_0 - y_1| e^{-\mu(\varepsilon)\,t}. \tag{75}$$

Proof. (i) Assume that $y_0 \in (y_2,\ y_1)$. Since $\phi(y)$ is positive in $(y_2,\ y_1,\ y(t)$ is increasing. The graph of $y(t)$ cannot intersect the graph of the stationary solution y_1; hence, the solution $y(t)$ exists and is continuous on the whole $[0,\infty)$, and it is bounded there. There exists $\lim_{t\to\infty} y(t)$, which coincides with y_1, since y_1 is a unique solution of $\phi(y) = 0$ in (y_2,∞). The case $y_0 > y_1$ is treated similarly.

(ii) For $y_0 \geq y_1 - \varepsilon$, where $\varepsilon \in (0,\ y_1 - y_3)$, denote $z(t) = y_1 - y(t)$. We obtain from $(71)_1$, using definition of $\mu(\varepsilon)$ and the fact that $\phi(y_1) = 0$,

$$(z^2)' = 2\,z z' = 2z^2 \int_0^1 \phi'(y + \tau z)\,\mathrm{d}\tau \leq -2\,\mu(\varepsilon)\,z^2.$$

This differential inequality implies (75). The case $y_0 > y_1$ is treated similarly.

Define $d(e_0) := e_0^{\max}/e_0^{\min} \geq 1$, where $e_0^{\max} = \max_F e_0$ and $e_0^{\min} = \min_F e_0$.

Lemma 9. *If* (72) *holds and $\varepsilon \in (0,\ y_1 - y_3)$, then the operators $\mathcal{S}_t$ $(t \geq 0)$ satisfy in $\mathcal{U}_1^\varepsilon$ the Lipschitz condition with respect to C-norm with the Lipschitz constant $d(e_0)e^{-\mu(\varepsilon)\,t}$.*

Proof. Let $u(\cdot\,,t)$ be a solution of problem (43). Recall that the function $w(\cdot\,,t) = u(\cdot\,,t)/e_0$ is the solution of Cauchy's problem (53), which we shall write in the form

$$\partial_t w = \Delta w + 2\,\langle\nabla \log e_0,\ \nabla w\rangle + f(w,\ \cdot),\quad w(\cdot\,,0) = u_0/e_0, \tag{76}$$

where $f(w,\ \cdot) = -\lambda_0 w + \Psi_1\, w^{-1}(e_0)^{-2} - \Psi_2\, w^{-3}(e_0)^{-4}$. By Proposition 4, the set $\mathcal{U}_1^\varepsilon$ is invariant for (43), i.e., $\mathcal{S}_t\big(\mathcal{U}_1^\varepsilon\big) \subseteq \mathcal{U}_1^\varepsilon$ for any $t \geq 0$. Take $u_i^0 \in \mathcal{U}_1^\varepsilon$ $(i = 1,2)$ and denote by $u_i(\cdot\,,t) = \mathcal{S}_t(u_i^0)$, $w_i(\cdot,t) = u_i(\cdot,t)/e_0$ and $w_i^0 = u_i^0/e_0$. Using (76) and the equalities

$$2\,\psi\Delta\,\psi = \Delta(\psi^2) - 2\,\|\nabla\,\psi\|^2,\quad \nabla\,(\psi^2) = 2\psi\nabla\,\psi$$

with $\psi = w_2 - w_1$, we obtain

$$\partial_t\big((w_2 - w_1)^2\big) = 2\,(w_2 - w_1)\,\partial_t(w_2 - w_1) \leq \Delta\big((w_2 - w_1)^2\big) + \\ + < \nabla \log e_0,\ \nabla(w_2 - w_1)^2 > +2(f(w_2, \cdot) - f(w_1, \cdot))(w_2 - w_1).$$

Since $w_i \geq y_1 - \varepsilon > y_3$ $(i = 1, 2)$, we get

$$(f(w_2, \cdot) - f(w_1, \cdot))(w_2 - w_1) = (w_2 - w_1)^2 \int_0^1 \partial_w f(w_1 + \tau(w_2 - w_1), \cdot)\,\mathrm{d}\tau$$

$$\leq (w_2 - w_1)^2 \int_0^1 \phi'(w_1 + \tau(w_2 - w_1))\,\mathrm{d}\tau \leq -\mu(\varepsilon)(w_2 - w_1)^2.$$

Thus, the function $v(\cdot\,, t) = (w_1(\cdot\,, t) - w_1(\cdot\,, t))^2$ satisfies the differential inequality

$$\partial_t v \leq \Delta v - 2\,\mu(\varepsilon)\,v + \langle \nabla \log e_0,\ \nabla v \rangle.$$

By the maximum principle of Proposition 2, we have $v(\cdot\,, t) \leq v_+(t)$, where $v_+(t)$ solves the ODE

$$v'_+ = -2\,\mu(\varepsilon)\,v_+(t), \qquad v(0) = \|w_2^0 - w_1^0\|_C^2.$$

Thus, we have the estimate

$$\|\mathcal{S}_t(u_2^0) - \mathcal{S}_t(u_1^0)\|_C \leq e_0^{\max}\|w_2(\cdot\,, t) - w_1(\cdot\,, t)\|_C \\ \leq e_0^{\max} e^{-\mu(\varepsilon)\,t}\|w_2^0 - w_1^0\|_C \leq d(e_0) e^{-\mu(\varepsilon)\,t}\|u_2^0 - u_1^0\|_C \tag{77}$$

which implies the desired claim. □

Theorem 7. *If* (72) *is satisfied, then the stationary equation of* (43) *on* (F, g)

$$\Delta\,u + \beta\,u + \Psi_1 u^{-1} - \Psi_2 u^{-3} = 0$$

has in the set $\mathcal{U}_1$ *a unique solution* $u_\star$, *and* $y_1 \leq u_\star/e_0 \leq (\Psi_1^+/\lambda_0)^{1/2}$ *holds. Furthermore, for* $\varepsilon \in (0,\ y_1 - y_3)$,

(i) $\mathcal{U}_1^\varepsilon$ *is attracted by* (43) *to the point* $u_\star$ *in the sense of an exponential* C*-convergence, i.e.,*

$$\big\|u(\cdot\,, t) - u_\star\big\|_C \leq d(e_0)\,e^{-\mu(\varepsilon)\,t}\,\big\|u_0 - u_\star\big\|_C \qquad (t > 0,\ u_0 \in \mathcal{U}_1); \tag{78}$$

(ii) $\mathcal{U}_1^\varepsilon$ *is attracted by* (43) *to the point* $u_\star$ *also in the sense of exponential* C^∞*-convergence: for any* $u_0 \in \mathcal{U}_1 \cap C^\infty$, *multi-index* α *with* $|\alpha| \geq 1$ *and* $\delta \in (0,\ \mu(\varepsilon))$ *there is* $C(\alpha, \delta) > 0$ *such that*

$$\left\| D^{\alpha}\big(u(\cdot,t)-u_{\star}\big)\right\|_{C} \le C(\alpha,\delta)\, e^{-(\mu(\varepsilon)-\delta)\,t}\left\|u_0-u_{\star}\right\|_{C^{|\alpha|}} \qquad (t>0). \tag{79}$$

Proof. By Proposition 4 and Lemma 9, for any $t\ge 0$ the operator $\mathcal{S}_t$ maps the set $\mathcal{U}_1^{\varepsilon}$, which is closed in C, into itself, and for $t>\frac{1}{\mu(\varepsilon)}\ln d(e_0)$ it is a contraction there. Since all operators $\mathcal{S}_t$ commute with one another, they have a unique common fixed point $u_\star$ in $\mathcal{U}_1^{\varepsilon}$, which is a stationary solution of (43). By Proposition 4, $\mathcal{U}_2^{\varepsilon}\subset\mathcal{U}_1^{\varepsilon}$ is also $\mathcal{S}_t$-invariant; hence, $u_\star\in\mathcal{U}_2^{\varepsilon}$. Since $\varepsilon\in(0,\ y_1-y_3)$ is arbitrary, we obtain the desired bounds for $u_\star$.

(i) Estimate (78) follows directly from (77).

(ii) First, consider the case when $|\alpha|=1$. Denote by

$$u(\cdot,t)=\mathcal{S}_t(u_0),\quad v(\cdot,t)=u(\cdot,t)-u_\star,\quad y_i(\cdot,t)=\partial_{x_i}v(\cdot,t)$$
$$(i\in\{1,2,\dots,p\}).$$

Since $u_\star$ is a stationary solution of (43), we have the PDE

$$\partial_t v=\Delta v+\beta v+b(u,\cdot)-b(u_\star,\cdot),\quad \text{where}\quad b(u,\cdot)=\Psi_1 u^{-1}-\Psi_2 u^{-3}.$$

Differentiating the equality above, and denoting $w_i(\cdot,t)=y_i(\cdot,t)/e_0$ we get

$$\partial_t w_i=\Delta w_i+2\,g\,(\nabla\log e_0,\nabla w_i)+h\,w_i+\theta_i\,v,$$

where $h=-\lambda_0+\int_0^1\partial_u b(u_\star+\tau(u-u_\star))\,\mathrm{d}\tau$ and $\theta_i=(\partial_{x_i}\beta+\partial_{x_i}h)/e_0$. Since $u_\star$ and $u(\cdot,t)$ $(t\ge 0)$ belong to the convex set $\mathcal{U}_1^{\varepsilon}$, we obtain in the same manner as in Lemma 9 that $h\le\int_0^1\phi'(u_\star+\tau(u-u_\star))\,\mathrm{d}\tau\le-\mu(\varepsilon)$. Furthermore, since $\mathcal{U}_1^{\varepsilon}$ is separated from $u_0=0$, the functions $\theta_i(\cdot,t)$ and all their derivatives are bounded in $F\times[0,\infty)$. Applying (78) and

$$|\theta_i\,v\,w_i|\le(2\,\delta)^{-1}\,(\bar\theta_i)^2v^2+2\,\delta\,w_i^2,\quad \delta\in(0,\ \mu(\varepsilon))$$

with $\bar\theta_i=\sup_{F\times[0,\infty)}|\theta_i(\cdot,t)|$, and using the same arguments as in the proof of Lemma 9, we conclude that the function $z_i(\cdot,t)=w_i^2(\cdot,t)$ satisfies the differential inequality

$$\partial_t z\le\Delta z+<\nabla\log e_0,\nabla z>-2\,(\mu(\varepsilon)-\delta)z+\bar\phi_i e^{-2\mu(\varepsilon)\,t},$$

where $\bar\phi_i=d(e_0)(2\delta)^{-1}(\bar\theta_i)^2\|v(\cdot,0)\|_C^2$. By the maximum principle of Proposition 2, we obtain the estimate $z(\cdot,t)\le z_i^+(t)$, where $z_i^+(t)$ solves the ODE

$$(z_i^+)'=-2\,(\mu(\varepsilon)-\delta)z_i^++\bar\phi_i e^{-2\mu(\varepsilon)\,t},\quad z_i^+(0)=\|w_i(\cdot,0)\|_C^2.$$

Hence,

$$z_i^+(t) = \big(\|w_i(\cdot\,,0)\|_C^2 + \bar{\phi}_i/(2\,\delta)\big)e^{-2\,\mu(\varepsilon)\,t} - \bar{\phi}_i e^{-2(\mu(\varepsilon)-\delta)\,t}/(2\,\delta).$$

Coming back from $w_i(\cdot\,,t)$ to $y_i(\cdot\,,t) = \partial_{x_i}(u(\cdot\,,t) - u_\star)$, we obtain (79) with $|\alpha| = 1$. Considering the second partial derivatives $y_{,ij} = \partial^2_{x_i x_j} v$ $(i, j \in \{1, 2, \dots, p\})$ and denoting $w_{ij}(\cdot\,,t) = y_{,ij}(\cdot\,,t)/e_0$, we obtain the equations

$$\partial_t w_{,ij} = \Delta w_{,ij} + 2\,\langle \nabla \log e_0, \nabla w_{,ij}\rangle + h\, w_{,ij} + \theta_{ij}(\cdot\,,t),$$

where the term $\theta_{ij}(\cdot\,,t)$ contains only $v(\cdot\,,t)$ and $y_i(\cdot\,,t)$, which have been estimated above. Then we get (79) for $|\alpha| = 2$ in a similar manner as above. By induction we prove (79) for any $|\alpha|$. □

Remark 1. As in the proof of Theorem 6(ii), we may show in Theorem 7 that $\nabla \log u \to \nabla \log u_*$ as $t \to \infty$ in C^∞ with the exponential rate $\mu(\varepsilon) - \delta$ for small $\delta > 0$.

2.6 Nonlinear Heat Equation with Parameter

Let the metric g, the connection ∇, and the Laplacian Δ smoothly depend on q, which belongs to an open subset Q of $\mathbb{R}^n$. Consider the Cauchy's problem on a closed Riemannian manifold (F, g)

$$\partial_t u = \Delta u + f(x, u, q), \quad u(x, 0, q) = u_0(x, q). \tag{80}$$

Here, f is defined in the domain $D = F \times I \times Q$, where $I \subseteq \mathbb{R}$ is an interval, and u_0 is defined in the domain $\tilde{D} = F \times Q$ and satisfies the condition $u_0(x, q) \in I$ for any $x \in F$ and $q \in Q$.

Proposition 5. *Suppose that $f \in C^\infty(D)$, $u_0 \in C^\infty(\tilde{D})$, all partial derivatives of f and u_0 by x, u, and q are bounded in D and $\tilde{D}$, and for any $q \in Q$ there exists a unique solution $u : F \times [0, T] \times Q \to \mathbb{R}$ of Cauchy's problem* (80) *such that all its partial derivatives by x are bounded in $F \times [0, T] \times Q$. Then $u(\cdot\,, t, \cdot) \in C^\infty(F \times Q)$ for any $t \in [0, T]$.*

Proof. This is standard; we give it for the convenience of a reader. As is known, $u(\cdot\,, t, q) \in C^\infty(F)$ for any $q \in Q$, $t \in [0, T]$. We should prove the smooth dependence on q of the solution $u(x, t, q)$ and of all its partial derivatives by x for any fixed $t \in [0, T]$. We shall divide the proof into several steps.

Step 1: The continuous dependence of $u(x, t, q)$ in q. To show this, take $q_0 \in Q$ and denote by $v(x, t, q) = u(x, t, q) - u(x, t, q_0)$ and $v_0(x, q) = u_0(x, q) - u_0(x, q_0)$. Let us represent

$$f(x, u(x, t, q), q) - f(x, u(x, t, q_0), q_0) = F(x, t, q) v(x, t, q) + G(x, t, q) \cdot (q - q_0),$$

where

$$F(x,t,q) =: \int_0^1 \partial_u f(x,\, u(x,t,q_0) + \tau v(x,t,q),\, q_0 + \tau(q-q_0))\,d\tau$$
$$G(x,t,q) = \int_0^1 \mathrm{grad}_q f(x,\, u(x,t,q_0) + \tau v(x,t,q),\, q_0 + \tau(q-q_0))\,d\tau\,.$$

Then the function $v(x,t,q)$ is a solution of the Cauchy's problem:

$$\partial_t v = \Delta v + F(x,t,q)v + G(x,t,q)\cdot(q-q_0), \quad v_{\,|\,t=0} = G_0(x,q)\cdot(q-q_0),$$

where $G_0(x,q) = \int_0^1 \mathrm{grad}_q v_0(x,\, q_0 + \tau(q-q_0))\,d\tau$. Then by the maximum principle of Proposition 2,

$$|v(x,t,q)| \le w(t,q) \qquad \forall\,(x,t,q) \in F\times[0,T]\times Q, \tag{81}$$

where $w(t,q)$ is the solution of the following Cauchy's problem for the ODE:

$$\partial_t\, w = \bar F|w| + \bar G\,|q-q_0|, \qquad w(0,q) = \bar G_0\,|q-q_0| \tag{82}$$

with $\bar F = \sup\limits_{F\times[0,T]\times Q}|F|$, $\bar G = \sup\limits_{F\times[0,T]\times Q}|G|$ and $\bar G_0 = \sup\limits_{F\times Q}|G_0|$. Then from (81) and (82) we get

$$|v(x,t,q)| \le \big(\bar G_0 e^{\bar F t} + (e^{\bar F t}-1)(\bar G/\bar F)\big)|q-q_0|, \qquad (x,t,q)\in F\times[0,T]\times Q,$$

which implies the claim of Step 1.

Step 2: All the partial derivatives of $u(x,t,q)$ by x are continuous in q. Differentiating subsequently by x both sides of the equation and of the initial condition in (80), we have the following Cauchy's problems for $p_\alpha = \partial_x^\alpha u$ (α is the multi-index):

$$\begin{aligned}
\partial_t\, p_i &= \Delta p_i + \partial_u f(x,u(x,t,q),q)p_i + \partial_x^i f(x,u(x,t,q),q), \quad p_{i\,|\,t=0} = \partial_x^i u_0(x,q),\\
\partial_t\, p_{i,j} &= \Delta p_{i,j} + \partial_u f(x,u(x,t,q),q)p_{i,j} + \partial_u^2 f(x,u(x,t,q),q)p_i\,p_j\\
&\quad +\partial_u\partial_x^i f(x,u(x,t,q),q)p_j + \partial_x^{i,j} f(x,u(x,t,q),q),\\
&\quad p_{i,j\,|\,t=0} = \partial_x^{i,j}\, u_0(x,q),
\end{aligned} \tag{83}$$

and so on. Applying the claim of Step 1 to these Cauchy's problems, we prove the claim of Step 2.

Step 3: $u(x,t,q)$ is smooth with respect to q. Take $q_0 \in Q$ and consider the divided difference

$$\delta_{\mathbf y}(x,t,s) = \frac{1}{s}\,(u(x,t,q_0+s\mathbf y) - u(x,t,q_0)) \quad (\mathbf y\in\mathbb R^n,\ s\in\mathbb R).$$

Denote by $\delta^0_{\mathbf{y}}(x,s) = \frac{1}{s}(u_0(x,q_0+s\mathbf{y}) - u_0(x,q_0))$. As in Step 1, we obtain the Cauchy's problem for $\delta_{\mathbf{y}}$

$$\partial_t \delta_{\mathbf{y}} = \Delta\delta_{\mathbf{y}} + \tilde F(x,t,s)\delta_{\mathbf{y}} + \tilde G(x,t,s)\cdot\mathbf{y}, \quad \delta_{\mathbf{y}\,|\,t=0} = \delta^0_{\mathbf{y}}(x,s), \tag{84}$$

$$\tilde F(x,t,s) = \int_0^1 \partial_u f(x,u(x,t,q_0)+\tau(u(x,t,q_0+s\mathbf{y})-u(x,t,q_0))),\ q_0+s\mathbf{y})\,\mathrm{d}\tau,$$

$$\tilde G(x,t,s) = \int_0^1 \mathrm{grad}_q f(x,u(x,t,q_0)+\tau(u(x,t,q_0+s\mathbf{y})-u(x,t,q_0))),\ q_0+s\mathbf{y})\,\mathrm{d}\tau.$$

Applying to Cauchy's problem (84) the claim of Step 1, we conclude that $\delta_{\mathbf{y}}(x,t,s)$ is continuous by s at the point $s=0$, that is, there exists the directional derivative $d_{\mathbf{y}}(x,t,q_0) = \mathrm{grad}_q u(x,t,q_0)\cdot\mathbf{y} = \lim_{s\to 0}\delta_{\mathbf{y}}(x,t,s)$. Moreover, $d_{\mathbf{y}}(x,t,q)$ is the solution of the Cauchy's problem

$$\begin{aligned} \partial_t d_{\mathbf{y}} &= \Delta d_{\mathbf{y}} + \partial_u f(x,u(x,t,q),q)d_{\mathbf{y}} + \mathrm{grad}_q f(x,u(x,t,q)\cdot\mathbf{y}, \\ d_{\mathbf{y}\,|\,t=0} &= \mathrm{grad}_q u_0(x,q)\cdot\mathbf{y}. \end{aligned} \tag{85}$$

Applying the claim of Step 1 to this Cauchy's problem, we find that $d_{\mathbf{y}}(x,t,q) = \mathrm{grad}_q u(x,t,q)\cdot\mathbf{y}$ continuously depends on q for any $\mathbf{y}\in\mathbb{R}^n$. Thus, $u(x,t,q)$ is C^1-regular in q. Applying the above arguments to the Cauchy's problem (85), we conclude that $u(x,t,q)$ belongs to C^2 with respect to q. Finally, we prove by induction that $u(x,t,q)$ is smooth in q.

Step 4: Applying all the arguments of Step 3 to the Cauchy's problems (83) and so on, we prove that all derivatives of $u(x,t,q)$ in x smoothly depend on q.

□

References

1. T. Aubin: *Some nonlinear problems in Riemannian geometry*, Springer, 1998.
2. A. Candel and L. Conlon: *Foliations, I, II*, AMS, Providence, 2000.
3. B. Chow and D. Knopf: *The Ricci Flow: An Introduction*, AMS, 2004.
4. S. Hurder: Problem set, in "*Foliations-2012*", World Sci., Singapore 2013, pp. 235–255.
5. W. Kirsch, B. Simon: Approach to equilibrium for a forced Burgers equation. J. Evol. Equ. 1, No. 4 (2001), 411–419.
6. R. Langevin and P. Walczak: Conformal geometry of foliations, Geom. Dedicata 132 (2008), 135–178
7. R. Ponge and H. Reckziegel: Twisted products in pseudo-Riemannian geometry. Geom. Dedicata, 48 (1993), 15–25.
8. V. Rovenski: *Foliations on Riemannian Manifolds and Submanifolds*, Birkhäuser, 1998.
9. V. Rovenski and L. Zelenko: Prescribing the positive mixed scalar curvature of totally geodesic foliations, in "*Foliations-2012*", World Sci., Singapore 2013, pp. 163–203.

10. V. Rovenski and P. Walczak: *Topics in Extrinsic Geometry of Codimension-One Foliations*, Springer Briefs in Mathematics, Springer-Verlag, 2011.
11. V. Rovenski and R. Wolak: Deforming metrics of foliations. Centr. Eur. J. Math. 11, no. 6, (2013), 1039–1055.
12. D. Sullivan, A homological characterization of foliations consisting of minimal surfaces. Comm. Math. Helv., 54 (1979), 218–223
13. I. Vaisman: Conformal foliations, Kodai Math. J. 2 (1979), 26–37.
14. P. Walczak: An integral formula for a Riemannian manifold with two orthogonal complementary distributions. Colloq. Math. 58 (1990), 243–252.

The Partial Ricci Flow for Foliations

Vladimir Rovenski

Abstract We study the flow of metrics on a foliation (called the *Partial Ricci Flow*), $\partial_t g = -2\, r(g)$, where r is the partial Ricci curvature; in other words, for a unit vector X orthogonal to the leaf, $r(X, X)$ is the mean value of sectional curvatures over all mixed planes containing X. The flow preserves total umbilicity, total geodesy, and harmonicity of foliations. It is used to examine the question: Which foliations admit a metric with a given property of mixed sectional curvature (e.g., constant)? We prove local existence/uniqueness theorem and deduce the evolution equations (that are leaf-wise parabolic) for the curvature tensor. We discuss the case of (co)dimension-one foliations and show that for the warped product initial metric the solution for the normalized flow converges, as $t \to \infty$, to the metric with $r = \Phi\, \hat{g}$, where Φ is a leaf-wise constant.

Keywords Manifold • Foliation • Flow of metrics • Totally geodesic • Partial Ricci curvature • Conullity tensor • Parabolic differential equation • Warped product

Mathematics Subject Classifications (2010): Primary 53C12, Secondary 53C44

1 Introduction

We define the partial Ricci flow on foliations. It is proposed as the main tool to prescribe the partial Ricci and mixed curvature of a foliation (see Toponogov's question in what follows).

V. Rovenski (✉)
Mathematical Department, University of Haifa, Haifa, Israel
e-mail: rovenski@math.haifa.ac.il

V. Rovenski and P. Walczak (eds.), *Geometry and its Applications*, Springer Proceedings in Mathematics & Statistics 72, DOI 10.1007/978-3-319-04675-4_6,

1.1 Totally Geodesic Foliations

A Riemannian manifold may admit many kinds of geometrically interesting foliations (e.g., totally geodesic, totally umbilical, harmonic, and Riemannian). The problems of the existence and classification of metrics with a given geometry on foliations (first posed by H. Gluck in 1979 for geodesic foliations) were studied already in the 1970s when D. Sullivan provided a topological condition (called topological tautness) for a foliation, equivalent to the existence of a Riemannian metric making all the leaves minimal, see [5]. Several authors investigated whether on a given Riemannian manifold there exists a totally geodesic foliation, as well as the inverse problem of determining whether one can find a Riemannian metric on a foliated manifold with respect to which the foliation becomes totally geodesic, see [7, 8, 11] and a survey in [12]. Simple examples of totally geodesic foliations are parallel circles or winding lines on a flat torus, and a Hopf field of great circles on the sphere S^3. In the codimension-one case, totally geodesic foliations on closed nonnegatively curved space forms are completely understood: they are given by parallel hyperplanes in the case of a flat torus T^n and they do not exist for spheres S^n. If the codimension is greater than one, examples of geometrically distinct totally geodesic foliations are abundant.

Let (M^{n+p}, g) (where $n, p > 0$) be a connected Riemannian manifold with the Levi-Civita connection ∇, $\mathcal{F}$ a smooth p-dimensional foliation on M, and $\mathcal{D}$ its orthogonal n-dimensional distribution. We have the orthogonal splitting of the tangent bundle $T(M) = \mathcal{D}_{\mathcal{F}} + \mathcal{D}$, where $\mathcal{D}_{\mathcal{F}}$ consists of vectors tangent to the leaves. As usual, $R(X, Y, Z, V) = g(R(X, Y)Z, V)$ is the Riemannian curvature tensor, and $R(X, Y) = \nabla_Y\nabla_X - \nabla_X\nabla_Y + \nabla_{[X,Y]}$ is the curvature tensor. Thus, $R(X, Y) = \nabla^2_{Y,X} - \nabla^2_{X,Y}$, where $\nabla^2_{X,Y} := \nabla_X \circ \nabla_Y - \nabla_{\nabla_X Y}$ is the second covariant derivative. The sectional curvature of the plane $\sigma = X \wedge Y$ is $K_\sigma = R(X, Y, X, Y)/(g(X, X)g(Y, Y) - g(X, Y)^2)$.

The *mixed plane* is spanned by two vectors such that the first (second) vector is tangent (orthogonal) to a leaf. Let also *mixed curvatures* stand for the sectional curvatures of mixed planes. The mixed curvature of a foliated manifold regulates the deviation of leaves along the leaf geodesics. (The *geodesic deviation equation* involves the curvature tensor, which measures in mechanics the rate of relative acceleration of two particles moving forward on neighboring geodesics.)

Theorem 1 (see [6]). *Let (M^{n+p}, g) be foliated with complete totally geodesic leaves of dimension p. Denote by $\rho(n)-1$ the maximal number of point-wise linearly independent vector fields on a sphere S^{n-1}. If the sectional curvature of M has the same positive value for all mixed planes then*

$$p \leq \rho(n) - 1. \tag{1}$$

To the best of our knowledge, this is the unique theorem in Riemannian geometry, which involves the topological invariant $\rho(n)$, the *Adams number*; here,

$\rho\big((\text{odd})\,2^{4d+c}\big) = 8d + 2^c$, where $d \geq 0$ and $0 \leq c \leq 3$, see [1]. In the case of $p = 1$, the manifold M is foliated by complete geodesics. Theorem 1 prohibits the existence of a foliation of an even-dimensional manifold by geodesics with positive constant mixed curvatures, since $\rho(n) - 1 = 0$ for an odd n. Hopf's fiber bundle $\pi : S^3 \to S^2$ gives the simple example of such a foliation for the odd $n + p = 3$, where the sphere S^3 is equipped with the standard metric. Fibers of Hopf's bundle are closed geodesics (great circles). Theorem 1 has various applications to geometry of submanifolds, see survey in [11]. Among Toponogov's many important contributions to global Riemannian geometry is the following question, see [11, p. 30]:

Question 1. *Can Theorem 1 be generalized by replacing the hypothesis "all mixed curvatures are equal to a positive constant" with the weaker one: "all mixed curvatures are positive"?*

Although the question was posed in 1980s, it is still open for a closed foliated manifold M. The exactness of estimate (1) and necessity of more conditions when a foliation is given locally have been proven in [11]. The author solved the problem (i.e., Question 1) for the special case, when M^{n+p} is a ruled submanifold of a sphere (i.e., the leaves are the rulings).

One may try to attack Question 1 by deforming the metric in directions orthogonal to leaves. The candidate for such a deformation is the flow defined in the next section.

1.2 The Partial Ricci Tensor

Paper [4] (see also [15]) studies the new action on foliations; this is imitative of Einstein–Hilbert functional except that the scalar curvature is replaced by $\mathrm{Sc}_{\mathrm{mix}}$:

$$J_{\mathrm{mix}} : g \to \int_\Omega \mathrm{Sc}_{\mathrm{mix}}(g)\,\mathrm{d\,vol}_g\,.$$

Here Ω is a fixed relatively compact domain in M and

$$\mathrm{Sc}_{\mathrm{mix}} = \sum\nolimits_{i=1}^{p} \sum\nolimits_{a=1}^{n} R(\mathcal{E}_a, E_i, \mathcal{E}_a, E_i)$$

is the *mixed scalar curvature* (a function on M), see [11, 13, 20], where $\{E_i, \mathcal{E}_a\}_{i\leq p, a\leq n}$ is a local orthonormal frame on $T(M)$ adapted to $\mathcal{D}_\mathcal{F}$ and $\mathcal{D}$. In particular, $\mathrm{Sc}_{\mathrm{mix}} = \mathrm{Ric}(N, N)$ when one of the distributions is 1-dimensional and is spanned by a unit vector field N, see Sect. 4.1. For a foliated surface (M^2, g), i.e., $n = p = 1$, we obtain $\mathrm{Sc}_{\mathrm{mix}} = K$ – the Gaussian curvature.

An inspection of Euler–Lagrange equations of $J_{\rm mix}$ (called the mixed gravitational field equations) leads to a new kind of Ricci curvature, whose properties need to be further investigated.

The *partial Ricci curvature* is the symmetric $(0,2)$-tensor $r = r(g)$ defined as, see [13],

$$r(X,Y) = \operatorname{Tr}_{\mathcal{F}} R(X^\perp, \cdot, Y^\perp, \cdot) = \sum\nolimits_{i=1}^{p} R(X^\perp, E_i, Y^\perp, E_i), \qquad X,Y \in T(M), \tag{2}$$

where $^\perp$ is the orthogonal to $\mathcal{F}$ component of a vector. Definition (2) does not depend on the choice of $\{E_i\}_{i\le p}$; in other words, for a unit vector $X \in \mathcal{D}$, the quantity $r(X,X)$ is the mean value of sectional curvatures over all mixed planes containing X. The symmetric $(1,1)$-tensor

$$\operatorname{Ric}_{\mathcal{D}}(X) = \sum\nolimits_{i=1}^{p} \big(R(E_i, X^\perp)\, E_i\big)^\perp$$

(called the *partial Ricci tensor*) is dual to (2), i.e., $g(\operatorname{Ric}_{\mathcal{D}}(X), Z) = r(X,Z)$. Certainly, we have

$$\operatorname{Tr}_g r = \operatorname{Tr}(\operatorname{Ric}_{\mathcal{D}}) = \operatorname{Sc}_{\rm mix}.$$

For a 1-dimensional foliation (spanned by a unit vector N on a manifold) we have

$$r(X,Y) = R(X,N,Y,N), \qquad X,Y \in T(M) \tag{3}$$

and $\operatorname{Ric}_{\mathcal{D}} = R_N := R(N,\cdot)N$ the *Jacobi operator* for N.

The notion of the *$\mathcal{D}$-truncated* $(0,2)$*-tensor* will be helpful: $S(X,Y) = S(X^\perp, Y^\perp)$, $X,Y \in T(M)$. The tensor r provides the example of a $\mathcal{D}$-truncated symmetric $(0,2)$-tensor. Another useful example is the *$\mathcal{D}$-truncated metric tensor* $\hat g$, i.e., $\hat g(X,Y) = g(X^\perp, Y^\perp)$.

The author [13] studied the problem of prescribing r on a locally conformally flat foliated manifold (M,g), provided conditions for $(0,2)$-tensors S of a simple form (defined on M) to admit a metric $\tilde g$ conformal to g that solves the *partial Ricci equations* $r(g) = S$ (and Einstein-type equations, $r(g) = \frac{1}{p}\operatorname{Sc}_{\rm mix}\cdot \hat g$), and presented explicit solutions.

A geometric flow of metrics, g_t, on a manifold is a solution of a differential equation $\partial_t g = S(g)$, where the symmetric $(0,2)$-tensor $S(g)$ is usually related to some kind of curvature (e.g., the Ricci flow, see [2], and the mean curvature flow). The flows of metrics on foliations that depend on the second fundamental form of leaves are studied in [14–16]. In this chapter we study the flow of metrics (on foliations) and use it to examine the question: *Which foliations admit a metric with a given property of the partial Ricci curvature or mixed curvature* (*e.g., constant*)?

Definition 1. The *Partial Ricci Flow* (PRF) is a family of metrics $g_t,\ t \in [0,\varepsilon)$, satisfying the PDE

$$\partial_t g = -2\, r(g). \tag{4}$$

(The PRF on a 1-dimensional foliation was studied in [19]). The *normalized PRF* is defined by

$$\partial_t g = -2\, r(g) + 2\,\Phi\,\hat{g}, \tag{5}$$

where $\Phi : M \to \mathbb{R}$ is a leaf-wise constant. For a foliated surface, (5) reads

$$\partial_t g = -2\,(K - \Phi)\,\hat{g}.$$

Observe that $r(X,Y) = 0$ if either X or Y is tangent to $\mathcal{F}$. Thus, the PRF preserves the orthogonal distribution to $\mathcal{F}$, does not change the geometry of the leaves, and keeps them to be totally umbilical (totally geodesic) or minimal submanifolds, see Proposition 3 in Sect. 3.3.

Remark 1. The fixed points of (5) are metrics with $r = \Phi\,\hat{g}$ (examples are Hopf fibrations of odd-dimensional spheres). The author and Zelenko [17, 18] studied the $\mathcal{D}$-conformal flow

$$\partial_t g = -2\,(\mathrm{Sc}_{\mathrm{mix}}(g) - \Phi)\,\hat{g}\,, \tag{6}$$

i.e., 'Yamabe type' analogue to (5). For certain conditions, (6) admits a unique global solution g_t converging exponentially fast to a metric, whose $\mathrm{Sc}_{\mathrm{mix}}$ is a leaf-wise constant.

In the case of a general foliation, the topology of the leaf through a point can change dramatically with the point; this gives many difficulties in studying truncated flows of metrics and leaf-wise parabolic PDEs. Therefore, we assume, at least at the first stage of study (e.g., Theorem 2),

(a) *the leaves to be compact,* (b) *the manifold M to be fibered* instead of being foliated. (7)

Theorem 2. *Let $\mathcal{F}$ be a smooth foliation on a closed Riemannian manifold (M, g_0). Then the linearization of* (5) *at g_0 is a leaf-wise parabolic PDE; hence,* (5) *under assumptions* (7) *has a unique smooth solution g_t defined on a positive time interval* $[0, t_0)$.

We are going to study the PRF along with the same line as the classical Ricci flow is applied in the proof of the smooth $1/4$-pinching sphere theorem, see for example [2].

The author conjectured (in his project EU-FP7-P-2010-RG, No. 276919) the following:

Let $\mathcal{F}$ be a p-dimensional totally geodesic foliation on a closed Riemannian manifold (M^{n+p}, g). Assume all mixed curvatures to be sufficiently close to a positive constant. Then the PRF evolves the metric g to a limit metric whose mixed curvature is a positive function of a point.

The conjecture seems to be an analogue of the following result by C. Böhm and B. Wilking.

Theorem 3 (see Theorem 1.10 in [2]). *On a compact manifold the Ricci flow evolves a Riemannian metric with 2-positive curvature operator R (i.e., the sum of the first two eigenvalues of R is positive) to a limit metric with constant sectional curvature.*

Observe the following difference in statements of the conjecture and Theorem 3: the sectional curvature of the limit metric is constant in Theorem 3, while the mixed curvature can depend on a point in the conjecture. The difference is caused by the absence of Schur's lemma in the case of fiber bundles. Nevertheless, the statement of author's conjecture implies inequality (1).

Theorem 8 and Corollaries 6 and 7 (Sect. 4.3) confirm the conjecture for a special case of warped product metrics when the leaves are space forms.

2 Preliminaries

We survey the basic tensors of the extrinsic geometry of foliations, describe their behavior under $\mathcal{D}$-truncated variations of a metric, and find the $\mathcal{F}$-Laplacian of the curvature tensor.

2.1 *Basic Tensors of the Extrinsic Geometry of a Foliation*

The second fundamental tensor $h_{\mathcal{F}}$ of $\mathcal{F}$ is defined by $h_{\mathcal{F}}(N_1, N_2) = (\nabla_{N_1} N_2)^{\perp}$ $(N_i \in \mathcal{D}_{\mathcal{F}})$. The *Weingarten operator* $A^{\mathcal{F}}_X : \mathcal{D}_{\mathcal{F}} \to \mathcal{D}_{\mathcal{F}}$ is given by $g(A^{\mathcal{F}}_X(N_1), N_2) = g(h_{\mathcal{F}}(N_1, N_2), X)$. The mean curvature vector of $\mathcal{F}$ is given by $H_{\mathcal{F}} = \operatorname{Tr}_g h_{\mathcal{F}}$. A foliation $\mathcal{F}$ is called *totally umbilical, harmonic, totally geodesic*, if $h_{\mathcal{F}} = (H_{\mathcal{F}}/p)\, g_{|\mathcal{F}}$, $H_{\mathcal{F}} = 0$, and $h_{\mathcal{F}} = 0$, respectively.

Definition 2. The *conullity tensor* $C : \mathcal{D}_{\mathcal{F}} \times T(M) \to \mathcal{D}$ is defined by

$$C_N(X) = -(\nabla_X N)^{\perp}, \qquad N \in \mathcal{D}_{\mathcal{F}},\ X \in T(M). \tag{8}$$

In particular, $C_{N_1}(N_2) = -h_{\mathcal{F}}(N_1, N_2)$ when $N_1, N_2 \in \mathcal{D}_{\mathcal{F}}$. Hence, $C = 0$ if and only if $\mathcal{F}$ is totally geodesic and $\mathcal{D}$ is integrable with totally geodesic integral manifolds (in this case, by de Rham decomposition Theorem, M is locally the direct product).

The *second fundamental tensor* h and the *integrability tensor* T of $\mathcal{D}$ are given by

$$h(X,Y) = (1/2)\,(\nabla_X Y + \nabla_Y X)^{\mathcal{F}}, \quad T(X,Y) = (1/2)\,[X,\,Y]^{\mathcal{F}} \quad (X,Y \in \mathcal{D}). \tag{9}$$

Then $H = \operatorname{Tr}_g h$ is the mean curvature vector of $\mathcal{D}$. If $\mathcal{D}$ is integrable then $T = 0$, and if $\mathcal{F}$ is a Riemannian foliation then $h = 0$. The (self-adjoint) *Weingarten operator* $A_N : \mathcal{D} \to \mathcal{D}$ and the skew-symmetric operator $T^\sharp_N : \mathcal{D} \to \mathcal{D}$ related to $N \in \mathcal{D}_{\mathcal{F}}$ are dual to tensors h and T, respectively:

$$\begin{aligned} g(A_N(X),\,Y) &= g(h(X,Y),N),\\ g(T^\sharp(X),\,Y) &= g(T(X,Y),N)(X,Y \in \mathcal{D}). \end{aligned} \tag{10}$$

Let * be the conjugation of $(1,1)$-tensors on $\mathcal{D}$ with respect to g. We have the identities on $\mathcal{D}$

$$A_N = (C_N + C_N^*)/2, \qquad T^\sharp_N = (C_N - C_N^*)/2, \qquad C_N = A_N + T^\sharp_N. \tag{11}$$

For $N_1, N_2 \in \mathcal{D}_{\mathcal{F}}$, define the tensor $R_{N_1,N_2}(X) = \big(R(N_1, X^\perp)N_2\big)^\perp$ and the self-adjoint operator $R_N := R_{N,N}$. We have $\operatorname{Ric}_{\mathcal{D}}(X) = \sum_{i=1}^p R_{E_i}(X^\perp)$. Note that $\operatorname{div}_{\mathcal{F}} h := \sum_{i=1}^p g(\nabla_i\, h(\cdot,\,\cdot), E_i)$ is a symmetric bilinear form on $\mathcal{D}$. Define the symmetric tensors $\mathcal{A} = \sum_i A_i^2$ and $\mathcal{T} = \sum_i (T_i^\sharp)^2$.

The *deformation tensor* of a vector field Z is the symmetric part of ∇Z restricted to $\mathcal{D}$,

$$\operatorname{Def}_{\mathcal{D}}(Z)(X,Y) = (1/2)\,[g(\nabla_X Z, Y) + g(\nabla_Y Z, X)], \quad X,Y \in \mathcal{D}.$$

Lemma 1. *For a foliation $\mathcal{F}$ on (M,g) and any $X,Y \in \mathcal{D}$, $N_i \in \mathcal{D}_{\mathcal{F}}$, we have*

$$\begin{aligned} R(N_1,X,N_2,Y) &= g(((\nabla_{N_1} C)_{N_2} - C_{N_2}C_{N_1})(X),Y)\\ &\quad + g(((\nabla_X A^{\mathcal{F}})_Y - A^{\mathcal{F}}_Y A^{\mathcal{F}}_X)(N_1),N_2), \end{aligned} \tag{12}$$

$$\begin{aligned} r(X,Y) &= \operatorname{div}_{\mathcal{F}} h(X,Y) - g((\mathcal{A}+\mathcal{T})(X),Y)\\ &\quad + \operatorname{Def}_{\mathcal{D}}(H_{\mathcal{F}})(X,Y) - \operatorname{Tr}(A^{\mathcal{F}}_Y A^{\mathcal{F}}_X), \end{aligned} \tag{13}$$

$$\mathrm{Sc}_{\mathrm{mix}} = \operatorname{div}_{\mathcal{F}} H - \|h\|^2 + \|T\|^2 + \operatorname{div} H_{\mathcal{F}} + \|H_{\mathcal{F}}\|^2 - \|h_{\mathcal{F}}\|^2. \tag{14}$$

Proof. For (12) see [11, Lemma 2.25]. Note that

$$\begin{aligned}\sum\nolimits_i g((\nabla_X A^{\mathcal F})_Y(E_i),E_i) &= \sum\nolimits_i \nabla_X(g(A^{\mathcal F}_Y(E_i),E_i))\\ &= \nabla_X(g(\sum\nolimits_i h_{\mathcal F}(E_i,E_i),Y)) = g(\nabla_X H_{\mathcal F},Y).\end{aligned}$$

Denote $\operatorname{div}_{\mathcal F} C := \sum_i(\nabla_i C)_i$ and $\operatorname{div}_{\mathcal F} T^\sharp := \sum_i(\nabla_i T^\sharp)_i$. Tracing (12) on $\mathcal D_{\mathcal F}$ yields

$$r(X,Y) = g(\operatorname{div}_{\mathcal F} C(X),Y) - g(\sum\nolimits_i C_i^2(X),Y) + g(\nabla_X H_{\mathcal F},Y) - \operatorname{Tr}(A^{\mathcal F}_Y A^{\mathcal F}_X). \tag{15}$$

The symmetric part of above equation is (13). The antisymmetric companion of (13) is

$$g(\operatorname{div}_{\mathcal F} T^\sharp(X),Y) - \sum\nolimits_i g\big((A_i T_i^\sharp + T_i^\sharp A_i)(X),Y\big) = d_{\mathcal D} H_{\mathcal F}(X,Y), \tag{16}$$

where $d_{\mathcal D} H_{\mathcal F}(X,Y) = \frac12\,[g(\nabla_X H_{\mathcal F},Y) - g(\nabla_Y H_{\mathcal F},X)]$ is the antisymmetric part of $\nabla^\perp H_{\mathcal F}$, which is regarded as a 2-form. Tracing (13) on $\mathcal D$, we obtain (14). □

Remark 2. Note that (14) follows from the equality

$$\operatorname{div} H = \operatorname{div}_{\mathcal F} H - \|H\|^2 \tag{17}$$

and the known formula for complementary distributions $\mathcal D$ and $\mathcal D_{\mathcal F}$, see [20],

$$\mathrm{Sc}_{\rm mix} = \operatorname{div}(H+H_{\mathcal F}) + \|H\|^2 + \|H_{\mathcal F}\|^2 + \|T\|^2 + \|T_{\mathcal F}\|^2 - \|h\|^2 - \|h_{\mathcal F}\|^2\,. \tag{18}$$

One may find the norms of tensors using the orthonormal frame $\{E_i,\,\mathcal E_a\}_{i\le p,a\le n}$ as

$$\begin{aligned}&\|h_{\mathcal F}\|^2 = \sum\nolimits_{a,b}\|h_{\mathcal F}(\mathcal E_a,\mathcal E_b)\|^2, \quad \|h\|^2 = \sum\nolimits_{i,j}\|h(E_i,E_j)\|^2,\\ &|T\|^2 = \sum\nolimits_{i,j}\|T(E_i,E_j)\|^2.\end{aligned}$$

Corollary 1. *For a totally umbilical foliation $\mathcal F$ on (M,g) and any $X,Y\in\mathcal D$, $N_i\in\mathcal D_{\mathcal F}$ we have*

$$R_{N_1,N_2} = (\nabla_{N_1} C)_{N_2} - C_{N_2}C_{N_1} + g(N_1,N_2)\big(\nabla^\perp_\circ H^{\mathcal F} - g(H_{\mathcal F},\cdot)\,H_{\mathcal F}\big), \tag{19}$$

$$\begin{aligned}r(X,Y) = {}&\operatorname{div}_{\mathcal F} h(X,Y) - g(\mathcal A+\mathcal T)(X),Y) + \operatorname{Def}_{\mathcal D}(H_{\mathcal F})(X,Y)\\ &-p\,g(H_{\mathcal F},X)g(H_{\mathcal F},Y).\end{aligned} \tag{20}$$

For a harmonic foliation $\mathcal{F}$, we have

$$r = \operatorname{div}_{\mathcal{F}} h - (\mathcal{A} + \mathcal{T})^\flat - \operatorname{Tr}(A^{\mathcal{F}}_\circ A^{\mathcal{F}}_\circ),$$

$$\operatorname{Ric}_{\mathcal{D}} = (\operatorname{div}_{\mathcal{F}} h)^\sharp - \mathcal{A} - \mathcal{T} - \sum\nolimits_i h(A^{\mathcal{F}}_\circ(E_i), E_i), \tag{21}$$

$$(\operatorname{div}_{\mathcal{F}} T)^\sharp = \sum\nolimits_i (A_i T_i^\sharp + T_i^\sharp A_i). \tag{22}$$

For a totally geodesic foliation $\mathcal{F}$, we have

$$R_{N_1,N_2} = (\nabla_{N_1} C)_{N_2} - C_{N_2} C_{N_1}, \quad r = \operatorname{div}_{\mathcal{F}} h - (\mathcal{A} + \mathcal{T})^\flat,$$

$$\operatorname{Ric}_{\mathcal{D}} = (\operatorname{div}_{\mathcal{F}} h)^\sharp - \mathcal{A} - \mathcal{T}. \tag{23}$$

The symmetric and antisymmetric parts of $(23)_1$ *with* $N_1 = N_2 = N$ *are*

$$R_N = \nabla_N A_N - A_N^2 - (T_N^\sharp)^2, \qquad \nabla_N T_N^\sharp = A_N T_N^\sharp + T_N^\sharp A_N. \tag{24}$$

2.2 Time-Dependent Adapted Metrics

Denote by $\mathcal{M}$ the space of smooth Riemannian metrics on M such that the distribution $\mathcal{D}$ is orthogonal to $\mathcal{F}$. Elements of $\mathcal{M}$ are called $(\mathcal{D}, \mathcal{D}_{\mathcal{F}})$*-adapted metrics* (*adapted metrics*, in short).

Let $S(g)$ be a $\mathcal{D}$*-truncated* symmetric $(0,2)$-tensor on a foliated Riemannian manifold (M, g). Consider a family of adapted metrics g_t on a smooth manifold M (with $0 \le t < \varepsilon$) satisfying PDE

$$\partial_t g = S(g). \tag{25}$$

Since the difference of two connections is a tensor, $\partial_t \nabla^t$ is a $(1,2)$-tensor on (M, g_t) with the symmetry $(\partial_t \nabla^t)(X, Y) = (\partial_t \nabla^t)(Y, X)$. Recall the formula [2]:

$$2\, g_t((\partial_t \nabla^t)(X, Y), Z) = (\nabla^t_X S_t)(Y, Z) + (\nabla^t_Y S_t)(X, Z) - (\nabla^t_Z S_t)(X, Y) \tag{26}$$

for all $X, Y, Z \in \Gamma(T(M))$. If the vector fields $X = X(t)$, $Y = Y(t)$ are t-dependent then

$$\partial_t(\nabla^t_X Y) = (\partial_t \nabla^t)(X, Y) + \nabla_X(\partial_t Y) + \nabla_{\partial_t X} Y.$$

Let $S^\sharp : T(M) \to T(M)$ the $(1,1)$-tensor dual to S, i.e., $g(S^\sharp(X), Y) = S(X, Y)$.

Lemma 2. *Let the local $\mathcal{D}$-frame $\{\mathcal{E}_a\}$ evolve by* (25) *according to*

$$\partial_t \mathcal{E}_a = -(1/2)\, S^\sharp(\mathcal{E}_a)\,. \tag{27}$$

Then $\{\mathcal{E}_a(t)\}$ is a g_t-orthonormal frame of $\mathcal{D}$ for all t.

Proof. We have

$$\begin{aligned}\partial_t(g_t(\mathcal{E}_a,\mathcal{E}_\beta)) &= g_t(\partial_t\mathcal{E}_a(t),\mathcal{E}_\beta(t)) + g_t(\mathcal{E}_a(t),\partial_t\mathcal{E}_\beta(t)) + (\partial_t g_t)(\mathcal{E}_a(t),\mathcal{E}_\beta(t))\\ &= S_t(\mathcal{E}_a(t),\mathcal{E}_\beta(t)) - (1/2)\, g_t(S_t^\sharp(\mathcal{E}_a(t)),\mathcal{E}_\beta(t))\\ &\quad - (1/2)\, g_t(\mathcal{E}_a(t), S_t^\sharp(\mathcal{E}_\beta(t))) = 0. \end{aligned}$$

□

Lemma 3 (see [14]). *For* (25) *with* $\mathcal{D}$*-truncated tensor* S *and vectors* $X, Y \in \mathcal{D}$, $N \in \mathcal{D}_\mathcal{F}$ *we have*

$$2\, g(\partial_t h(X,Y),N) = -(\nabla_N S)(X,Y) + S(Y, C_N(X)) + S(X, C_N(Y)), \tag{28}$$

$$2\,\partial_t A_N = -\nabla_N S^\sharp + [A_N - T_N^\sharp,\ S^\sharp], \qquad \partial_t T_N^\sharp = -S^\sharp\, T_N^\sharp, \tag{29}$$

$$2\,\partial_t C_N = -\nabla_N S^\sharp + [C_N,\ S^\sharp] - 2\, T_N^\sharp\, S^\sharp, \tag{30}$$

$$2\,\partial_t H = -\nabla^\mathcal{F}(\mathrm{Tr}\ S^\sharp), \tag{31}$$

$$\partial_t h_\mathcal{F} = -S^\sharp \circ h_\mathcal{F}, \qquad \partial_t H_\mathcal{F} = -S^\sharp(H_\mathcal{F}). \tag{32}$$

Proof. Note that $\partial_t T = 0$. For all $X, Y \in \mathcal{D}$, using (26) and (9), we have

$$\begin{aligned}2\, g\big(\partial_t(\nabla^t_X Y),\ N\big) &= (\nabla^t_X S)(Y,N) + (\nabla^t_Y S)(X,N) - (\nabla^t_N S)(X,Y)\\ &= -(\nabla^t_N S)(X,Y) - S(Y,\nabla^t_X N) - S(X,\nabla^t_Y N).\end{aligned}$$

From this and symmetry of $\partial_t \nabla^t$, we have (28). Using (10), we then find

$$\begin{aligned}g(\partial_t A_N(X),Y) &= \partial_t g(h(X,Y),N) - (\partial_t g)(A_N(X),Y),\\ g(\partial_t T_N^\sharp(X),Y) &= -(\partial_t g)(T_N^\sharp(X),Y).\end{aligned}$$

The above, (11) and (28) yield (29). Using $\partial_t C_N = \partial_t A_N + \partial_t T_N^\sharp$ and (29), we obtain (30). Next, using Lemma 2, we deduce (31):

$$\begin{aligned}2\, g(\partial_t H, N) &= 2\sum\nolimits_a g(\partial_t(h(\mathcal{E}_a,\mathcal{E}_a)),N)\\ &= 2\sum\nolimits_a g(\partial_t h(\mathcal{E}_a,\mathcal{E}_a)) + 2\, h(\partial_t\mathcal{E}_a,\mathcal{E}_a)),\ N)\\ &= \sum\nolimits_a [-(\nabla_N S)(\mathcal{E}_a,\mathcal{E}_a) + 2\, S(C_N(\mathcal{E}_a),\mathcal{E}_a) - g(h(S^\sharp(\mathcal{E}_a),\mathcal{E}_a),\ N)]\\ &= -N(\mathrm{Tr}\ S^\sharp).\end{aligned}$$

Finally, from (26) we have

$$
\begin{aligned}
2\,g_t(\partial_t h_{\mathcal F}(N_1,N_2),X) &= g_t(\partial_t(\nabla^t_{N_1}N_2)+\partial_t(\nabla^t_{N_2}N_1),X)\\
&= (\nabla^t_{N_1}S)(X,N_2)+(\nabla^t_{N_2}S)(X,N_1)-(\nabla^t_X S)(N_1,N_2)\\
&= -S(\nabla^t_{N_1}N_2,X)+S(\nabla^t_{N_2}N_1,X)\\
&= -2\,S(h_{\mathcal F}(N_1,N_2),X).
\end{aligned}
$$

Hence, $2\,g_t(\partial_t h_{\mathcal F}(N_1,N_2),X) = -2\,g_t(S^\sharp\circ h_{\mathcal F}(N_1,N_2),X)$ for $X\in T(M)$, that is $(32)_1$. Since $\partial_t E_i = 0$, we have $(32)_2$

$$\partial_t H_{\mathcal F} = \sum\nolimits_i \partial_t h_{\mathcal F}(E_i,E_i) = -\sum\nolimits_i S^\sharp(h_{\mathcal F}(E_i,E_i)) = -S^\sharp(H_{\mathcal F}).\square$$

Corollary 2. *For (25), the tensors $\partial_t A_N$, $\partial_t T^\sharp_N$, where $N\in\mathcal{D}_{\mathcal F}$, and $\partial_t\,\mathrm{Ric}_{\mathcal D}$ may be not self-adjoint:*

$$
\begin{aligned}
&(\partial_t A_N)^* - \partial_t A_N = [S^\sharp, A_N],\ \ (\partial_t T^\sharp_N)^* + \partial_t T^\sharp_N = [T^\sharp_N, S^\sharp],\\
&(\partial_t\,\mathrm{Ric}_{\mathcal D})^* - \partial_t\,\mathrm{Ric}_{\mathcal D} = -[\mathrm{Ric}_{\mathcal D}, S^\sharp]. \qquad (33)
\end{aligned}
$$

Proof. From (29), formulae $(33)_{1,2}$ follow. Notice that

$$
\begin{aligned}
\partial_t r(X,Y) = \partial_t(g(\mathrm{Ric}_{\mathcal D}(X),Y)) &= S(\mathrm{Ric}_{\mathcal D}(X),Y)\\
&+ g((\partial_t\,\mathrm{Ric}_{\mathcal D})(X),Y)\quad (X,Y\in\mathcal{D}). \qquad (34)
\end{aligned}
$$

From this and symmetry of $\partial_t\, r$, equality $(33)_3$ follows. □

The metrics g_t on M in (25) are interpreted as a *natural bundle metric* on the *spatial tangent bundle* E, that is, the pull-back of $T(M)$ under the projection $M\times(0,\varepsilon)\to M,\ (q,t)\to q$, see [2]. The fiber of E over a point (q,t) is given by $E_{(q,t)} = T_q M$ and is endowed with the metric g_t.

A connection ∇ on a vector bundle E over M is a map $\nabla : \mathcal{X}(M)\times\Gamma(E)\to\Gamma(E)$, written as $(X,\sigma)\to\nabla_X\sigma$, such that, see [2],

1. ∇ is $C^\infty(M)$ linear in X: $\nabla_{f_1X_1+f_2X_2}\,\sigma = f_1\nabla_{X_1}\sigma + f_2\nabla_{X_2}\sigma$,
2. ∇ is $\mathbb{R}$-linear in σ: $\nabla_X(\lambda_1\sigma_1+\lambda_2\sigma_2) = \lambda_1\nabla_X\sigma_1+\lambda_2\nabla_X\sigma_2$, and
3. ∇ satisfies the product rule: $\nabla_X(f\sigma) = X(f)\,\sigma + f\,\nabla_X\sigma$.

A connection ∇ on a vector bundle E is said to be *compatiblewith a metric* g on E if for any $\xi,\eta\in\Gamma(E)$ and $X\in\mathcal{X}(M)$, we have $X(g(\xi,\eta)) = g(\nabla_X\xi,\eta) + g(\xi,\nabla_X\eta)$. Compatibility by itself is not enough to determine a unique connection. There is a natural connection $\tilde\nabla$ on E, which extends the Levi-Civita connection on $T(M)$. We need to specify only the covariant time derivative $\tilde\nabla_{\partial_t}$. Given any section X of the vector bundle E, we define $\tilde\nabla_{\partial_t}$ by

$$\tilde\nabla_{\partial_t}X = \partial_t X + (1/2)\,S^\sharp(X)\ \text{ for }\ X\in\mathcal{D},\qquad \tilde\nabla_{\partial_t}N = 0\ \text{ for }\ N\in\mathcal{D}_{\mathcal F}. \qquad (35)$$

Lemma 4. *The connection on E is compatible with the natural bundle metric:*

$$\tilde{\nabla}_{\partial_t} g = 0. \tag{36}$$

Proof. One may assume that $X, Y \in \mathcal{D}$ are constant in time. In this case, we have $\tilde{\nabla}_{\partial_t} X = \frac{1}{2} S^\sharp(X)$ and $\tilde{\nabla}_{\partial_t} Y = \frac{1}{2} S^\sharp(Y)$. Since $\partial_t g = S$, this and (25) imply (36):

$$(\tilde{\nabla}_{\partial_t} g)(X,Y) = \partial_t g(X,Y) - g(\tilde{\nabla}_{\partial_t} X, Y) - g(X, \tilde{\nabla}_{\partial_t} Y) = (\partial_t g)(X,Y) - S(X,Y) = 0. \quad \square$$

This connection is not symmetric: in general, $\tilde{\nabla}_{\partial_t} X \neq 0$, while $\tilde{\nabla}_X \partial_t = 0$ always for $X \in \mathcal{D}$. Clearly, the *torsion tensor* $\mathrm{Tor}(X,Y) := \tilde{\nabla}_X Y - \tilde{\nabla}_Y X - [X,Y]$ vanishes if both arguments are spatial; so, the only nonzero components are

$$\mathrm{Tor}(\partial_t, X) = \tilde{\nabla}_{\partial_t} X - \tilde{\nabla}_X \partial_t = (1/2)\, S^\sharp(X) \qquad (X \in \mathcal{D}).$$

However, each submanifold $M \times \{t\}$ is totally geodesic; so, computing derivatives of spatial tangent vector fields gives the same result as computing for sections of $T(M \times [0,\varepsilon))$. In particular, the corresponding Weingarten operators satisfy $\tilde{A}_N = A_N$.

Remark 3. Using connection (35), we also have

$$\begin{aligned} g((\tilde{\nabla}_{\partial_t} h)(X,Y), N) &= g(\partial_t h(X,Y) - h(\tilde{\nabla}_{\partial_t} X, Y) - h(X, \tilde{\nabla}_{\partial_t} Y),\ N) \\ &= -(1/2)\,(\nabla^t_N S)(X,Y), \\ (\tilde{\nabla}_{\partial_t} A_N)(X) &= (\partial_t A_N)(X) - A_N(\tilde{\nabla}_{\partial_t} X) \\ &= -(1/2)\,(\nabla^t_N S^\sharp)(X) - (1/2)\,[T^\sharp_N,\ S^\sharp]. \end{aligned}$$

If $\mathcal{D}$ is integrable then, see (29),

$$\tilde{\nabla}_{\partial_t} A_N = -(1/2)\, \tilde{\nabla}_N S^\sharp.$$

2.3 *The Leaf-Wise Laplacian of the Curvature Tensor*

In analogy with [2, Sect. 4.2.1], define the quadratic in the curvature tensor $B \in \Lambda^4_0(M)$ as

$$B(X,Y,V,Z) = \sum\nolimits_{i=1}^p \langle R(X,\,\cdot\,,Y,\,E_i),\ R(V,\,\cdot\,,Z,\,E_i)\rangle \quad \text{for all} \quad X,Y,V,Z \in T(M),$$

where $\{E_i\}_{i\le p}$ is a local orthonormal frame on $\mathcal{D}_\mathcal{F}$. Although generally we have $B(X,Y,Z,V)\ne B(Y,X,V,Z)$, the tensor B has some symmetries of the curvature tensor, as

$$B(X,Y,Z,V)=B(Z,V,X,Y). \tag{37}$$

The leaf-wise Laplacian is defined by $\Delta_\mathcal{F}=\operatorname{Tr}_\mathcal{F}(\nabla^2)=\sum_i\nabla^2_{i,i}$.

Proposition 1 (see [19] for $p=1$**).** *On a Riemannian manifold* (M,g) *endowed with a smooth foliation* $\mathcal{F}$*, the* $\mathcal{F}$*-Laplacian of the curvature tensor satisfies*

$$\begin{aligned}\Delta_\mathcal{F} R(X,Y,Z,V)&=\sum\nolimits_i[\nabla^2_{X,Z}\,R(Y,E_i,V,E_i)-\nabla^2_{Y,Z}\,R(X,E_i,V,E_i)\\&\quad+\nabla^2_{Y,V}\,R(X,E_i,Z,E_i)-\nabla^2_{X,V}\,R(Y,E_i,Z,E_i)]\\&\quad-\big(B(X,Y,Z,V)-B(X,Y,V,Z)-B(Y,X,Z,V)\\&\quad+B(Y,X,V,Z)-2\,B(Z,Y,V,X)\\&\quad+2\,B(Z,X,V,Y)\big)+\langle R(\cdot\,,Y,Z,V),\ \sum\nolimits_i R(X,E_i,\cdot\,,E_i)\rangle\\&\quad-\langle R(\cdot\,,X,Z,V),\ \sum\nolimits_i R(Y,E_i,\cdot\,,E_i)\rangle.\end{aligned} \tag{38}$$

Proof. Using $\nabla_i R(X,Y,Z,V)+\nabla_X R(Y,E_i,Z,V)+\nabla_Y R(E_i,X,Z,V)=0$ (the second Bianchi identity)—together with the linearity over $\mathbb{R}$ of ∇ on the space of tensor fields [2]—we find that

$$\begin{aligned}\Delta_\mathcal{F} R(X,Y,Z,V)&=\sum\nolimits_i\nabla^2_{i,i}R(X,Y,Z,V)\\&=\sum\nolimits_i[\nabla^2_{i,X}R(E_i,Y,Z,V)-\nabla^2_{i,Y}R(E_i,X,Z,V)].\end{aligned} \tag{39}$$

It suffices to express the first two terms on the rhs of (39) using lower order terms. To compute the first term on the rhs of (39), we transpose ∇_i and ∇_X,

$$\nabla^2_{i,X}R(E_i,Y,Z,V)=\nabla^2_{X,i}R(E_i,Y,Z,V)+(R(X,E_i)R)(E_i,Y,Z,V). \tag{40}$$

Using the second Bianchi identity $\nabla_i R(Z,V,E_i,Y)+\nabla_Z R(V,E_i,E_i,Y)+\nabla_V R(E_i,Z,E_i,Y)=0$, we transform the first term on the rhs of (40),

$$\nabla^2_{X,i}\,R(E_i,Y,Z,V)=\nabla^2_{X,Z}\,R(Y,E_i,V,E_i)-\nabla^2_{X,V}\,R(E_i,Y,E_i,Z). \tag{41}$$

Next, we transform the second term on the rhs of (40), using the identity

$$\begin{aligned}(R(X,Y)R)(Z,U,V,W)&=-R(R(X,Y)Z,U,V,W)-R(Z,R(X,Y)U,V,W)\\&\quad-R(Z,U,R(X,Y)V,W)-R(Z,U,V,R(X,Y)W)\end{aligned}$$

and noting that $R(X,Y)f = 0$ where $f = R(Z,U,V,W)$ (in our case, $R(X,E_i)(R(E_i,Y,Z,V)) = 0$)

$$
\begin{aligned}
(R(X,E_i)R)(E_i,Y,Z,V) &= -R(R(X,E_i)(E_i,Y,Z,V) - \ldots - R(E_i,Y,Z,R(X,E_i)V)\\
&= \langle R(X,E_i,\cdot,E_i),\ R(\cdot,Y,Z,V)\rangle\\
&\quad + \langle R(E_i,X,Y,\cdot)\ R(E_i,\cdot,Z,V)\rangle\\
&\quad + \langle R(E_i,X,Z,\cdot),\ R(E_i,Y,\cdot,V)\rangle\\
&\quad + \langle R(E_i,X,V,\cdot),\ R(E_i,Y,Z,\cdot)\rangle. \qquad (42)
\end{aligned}
$$

The first term on the rhs of (42) yields

$$\langle \sum\nolimits_i R(X,E_i,\cdot,E_i),\ R(\cdot,Y,Z,V)\rangle = -\langle R(Y,\cdot,Z,V),\ \sum\nolimits_i R(X,E_i,\cdot,E_i)\rangle.$$

We transform the second term on the rhs of (42), using the first Bianchi identity,

$$
\begin{aligned}
\sum\nolimits_i \langle R(E_i,X,Y,\cdot)\ R(E_i,\cdot,Z,V)\rangle &= -\sum\nolimits_i \langle R(\cdot,Y,X,E_i)\ R(Z,V,\cdot,E_i)\rangle\\
&= \sum\nolimits_i \big[\langle R(\cdot,Y,X,E_i)\ R(\cdot,Z,V,E_i)\rangle\\
&\quad + \langle R(\cdot,Y,X,E_i)\ R(V,\cdot,Z,E_i)\rangle\big]\\
&= B(Y,X,Z,V) - B(Y,X,V,Z).
\end{aligned}
$$

The third and the fourth terms on the rhs of (42) are transformed as

$$
\begin{aligned}
&\langle R(E_i,X,Z,\cdot),\ R(E_i,Y,\cdot,V)\rangle + \langle R(E_i,X,V,\cdot),\ R(E_i,Y,Z,\cdot)\rangle\\
&\quad = -\langle R(\cdot,Z,X,E_i),\ R(\cdot,V,Y,E_i)\rangle - \langle R(\cdot,V,X,E_i),\ R(\cdot,Z,Y,E_i)\rangle\\
&\quad = -B(Z,X,V,Y) + B(V,X,Z,Y).
\end{aligned}
$$

Hence, (42) takes the following form:

$$
\begin{aligned}
\sum\nolimits_i (R(X,E_i)R)(E_i,Y,Z,V) &= B(Y,X,Z,V) - B(Y,X,V,Z) - B(Z,X,V,Y)\\
&\quad + B(V,X,Z,Y)\\
&\quad - \langle R(Y,\cdot,Z,V),\ \sum\nolimits_i R(X,E_i,\cdot,E_i)\rangle. \qquad (43)
\end{aligned}
$$

Substituting expressions of (41) and (43) into (40), we have

$$
\begin{aligned}
\sum\nolimits_i \nabla^2_{i,X} R(E_i,Y,Z,V) &= \sum\nolimits_i \big[\nabla^2_{X,Z} R(Y,E_i,V,E_i) - \nabla^2_{X,V} R(Y,E_i,Z,E_i)\big]\\
&\quad - (B(Y,X,V,Z) - B(Y,X,Z,V) + B(Z,X,V,Y)\\
&\quad - B(V,X,Z,Y)) - \langle R(Y,\cdot,Z,V),\ \sum\nolimits_i R(X,E_i,\cdot,E_i)\rangle.
\end{aligned}
$$

Using symmetry $X \leftrightarrow Y$, we also have

$$\begin{aligned}\sum\nolimits_i \nabla^2_{i,Y} R(E_i, X, Z, V) = \sum\nolimits_i \big[\nabla^2_{Y,Z} R(X, E_i, V, E_i) - \nabla^2_{Y,V} R(X, E_i, Z, E_i)\big] \\ - \big(B(X, Y, V, Z) - B(X, Y, Z, V) + B(Z, Y, V, X) \\ - B(V, Y, Z, X)\big) - \langle R(X, \cdot, Z, V), \sum\nolimits_i R(Y, E_i, \cdot, E_i)\rangle.\end{aligned}$$

By the above, (39) reduces to

$$\begin{aligned}&\sum\nolimits_i \nabla^2_{i,i} R(X, Y, Z, V) \\ &= \sum\nolimits_i \big[\nabla^2_{X,Z} R(Y, E_i, V, E_i) - \nabla^2_{X,V} R(Y, E_i, Z, E_i) \\ &\quad -\nabla^2_{Y,Z} R(X, E_i, V, E_i) + \nabla^2_{Y,V} R(X, E_i, Z, E_i)\big] \\ &\quad +\langle R(\cdot, Y, Z, V), \sum\nolimits_i R(X, E_i, \cdot, E_i)\rangle \\ &\quad -\langle R(\cdot, X, Z, V), \sum\nolimits_i R(Y, E_i, \cdot, E_i)\rangle \\ &\quad - (B(Y, X, V, Z) - B(Y, X, Z, V) + B(Z, X, V, Y) - B(V, X, Z, Y) \\ &\quad -B(X, Y, V, Z) + B(X, Y, Z, V) - B(Z, Y, V, X) + B(V, Y, Z, X)).\end{aligned}$$

Using the symmetry (37) of B, from the above, we obtain (38). □

Remark 4. The distribution $\mathcal{D}$ (orthogonal to a foliation $\mathcal{F}$) will be called *averaged curvature-invariant* if $\sum_i R(E_i, \mathcal{D})E_i \subset \mathcal{D}$, where $\{E_i\}$ is a local orthonormal frame on $\mathcal{D}_\mathcal{F}$. This holds when $\mathcal{D}_\mathcal{F}$ is *curvature-invariant*, i.e., $R(X, Y)(\mathcal{D}_\mathcal{F}) \subset \mathcal{D}_\mathcal{F}$ for any $X, Y \in \mathcal{D}_\mathcal{F}$, see [11]; hence, the distribution orthogonal to a totally geodesic foliation $\mathcal{F}$ is (averaged) curvature-invariant (indeed, $R(E_i, X)E_i \in \mathcal{D}$ for $X \perp \mathcal{D}_\mathcal{F}$). Another example provide distributions orthogonal to foliations on space forms. For an averaged curvature-invariant distribution $\mathcal{D}$ and any vectors $X \in \mathcal{D}$, $Y \in T(M)$ we have

$$\sum\nolimits_i R(X, E_i, Y, E_i) = \sum\nolimits_i g(R(E_i, X)E_i, Y) = g(\mathrm{Ric}_\mathcal{D}(X), Y) = r(X, Y). \tag{44}$$

In this case, (38) reads

$$\begin{aligned}\Delta_\mathcal{F} R(X, Y, Z, V) = &\sum\nolimits_i [\nabla^2_{X,Z} R(Y, E_i, V, E_i) - \nabla^2_{Y,Z} R(X, E_i, V, E_i) \\ &+ \nabla^2_{Y,V} R(X, E_i, Z, E_i) - \nabla^2_{X,V} R(Y, E_i, Z, E_i)] \\ &- \big(B(X, Y, Z, V) - B(X, Y, V, Z) - B(Y, X, Z, V) \\ &+ B(Y, X, V, Z) - 2\, B(Z, Y, V, X) \\ &+ 2\, B(Z, X, V, Y)\big) + \langle R(\cdot, Y, Z, V), r(X, \cdot)\rangle \\ &- \langle R(\cdot, X, Z, V), r(Y, \cdot)\rangle.\end{aligned} \tag{45}$$

3 Main Results

In this section we prove local existence/uniqueness theorem and deduce the system of evolution equations (that are parabolic along the leaves) for the curvature and conullity tensors.

3.1 Short-Time Existence and Uniqueness for PRF

To linearize the differential operator $g \to -2\,r(g)$, see (4), on the space $\mathcal{M}$, we need the following.

Proposition 2 (see [2]). *Let g_t be a family of metrics on a manifold M such that $\partial_t\, g = S$. Then*

$$2\,\partial_t\, R(X,Y,Z,V) = \nabla^2_{X,V}\, S(Y,Z) + \nabla^2_{Y,Z}\, S(X,V) - \nabla^2_{X,Z}\, S(Y,V) - \nabla^2_{Y,V}\, S(X,Z) + S(R(X,Y)Z,\ V) - S(R(X,Y)V,\ Z). \tag{46}$$

Note that the first and second derivatives of a $(0,2)$-tensor S can be expressed as

$$\begin{aligned} \nabla_Z\, S(Y,V) &= Z(S(Y,V)) - S(\nabla_Z Y, V) - S(Y, \nabla_Z V), \\ \nabla^2_{X,Z}\, S(Y,V) &= \nabla_X(\nabla_Z\, S)(Y,V) - \nabla_{\nabla_X Z}\, S(Y,V) \\ &= \nabla_X(\nabla_Z\, S(Y,V)) - \nabla_Z\, S(\nabla_X Y, V) - \nabla_Z\, S(Y, \nabla_X V) \\ &\quad - \nabla_{\nabla_X Z}\, S(Y,V). \end{aligned} \tag{47}$$

Define the bilinear form $F_N : \mathcal{D} \times \mathcal{D} \to \mathbb{R}$ for $N \in \mathcal{D}_\mathcal{F}$ ($F_N = 0$ for totally geodesic foliations) by

$$F_N(Z,X) = g(((\nabla_Z\, A^\mathcal{F})_X - A^\mathcal{F}_X A^\mathcal{F}_Z)(N), N).$$

One may calculate,

$$\sum\nolimits_i F_{E_i}(Z,\, X) = g(\nabla_Z\, H_\mathcal{F}, X) - \mathrm{Tr}\,(A^\mathcal{F}_X A^\mathcal{F}_Z).$$

Note that

$$\sum\nolimits_i F_{E_i}(Z, S^\sharp(X)) = S(\nabla_Z\, H_\mathcal{F}, X) - \mathrm{Tr}\,(A^\mathcal{F}_{S^\sharp(X)} A^\mathcal{F}_Z).$$

Lemma 5. *Let (M,g) be a Riemannian manifold with a smooth foliation $\mathcal{F}$. Then the tensor r evolves by (25) (with a $\mathcal{D}$-truncated symmetric $(0,2)$-tensor $S(g)$) according to*

$$
\begin{aligned}
2\,\partial_t\, r(X,Z) &= -\Delta_{\mathcal F}\, S(X,Z) + \sum\nolimits_i \big[\nabla_i\, S(C_i(X),Z) + \nabla_i\, S(C_i(Z),X) \\
&\quad + S(C_i^2(X),Z) + S(C_i^2(Z),X) - 2\,S(C_i(X),C_i(Z))\big] \\
&\quad + S(\mathrm{Ric}_{\mathcal D}(Z),X) + S(\mathrm{Ric}_{\mathcal D}(X),Z) - \nabla_Z S(H_{\mathcal F},X) \\
&\quad - \nabla_X S(H_{\mathcal F},Z) - S(\nabla_X H_{\mathcal F},Z) \\
&\quad - S(\nabla_Z H_{\mathcal F},X) + \mathrm{Tr}\,(A^{\mathcal F}_{S^\sharp(X)} A^{\mathcal F}_Z).
\end{aligned} \tag{48}
$$

If, in addition, $\mathcal F$ is a totally geodesic foliation (i.e., $h_{\mathcal F}=0$) then

$$
\begin{aligned}
2\,\partial_t\, r(X,Z) &= -\Delta_{\mathcal F}\, S(X,Z) + \sum\nolimits_i \big[\nabla_i\, S(C_i(X),Z) + \nabla_i\, S(C_i(Z),X) \\
&\quad + S(C_i^2(X),Z) + S(C_i^2(Z),X) - 2\,S(C_i(X),C_i(Z))\big] \\
&\quad + S(\mathrm{Ric}_{\mathcal D}(Z),X) + S(\mathrm{Ric}_{\mathcal D}(X),Z),
\end{aligned} \tag{49}
$$

$$
\begin{aligned}
2\,\partial_t\, \mathrm{Ric}_{\mathcal D} &= -\Delta_{\mathcal F}\, S^\sharp + \sum\nolimits_i \big[(\nabla_i\, S^\sharp) C_i + C_i^*\nabla_i\, S^\sharp + S^\sharp\, C_i^2 + (C_i^*)^2 S^\sharp \\
&\quad - 2\, C_i^*\, S^\sharp\, C_i\big] + \mathrm{Ric}_{\mathcal D}\, S^\sharp - S^\sharp\, \mathrm{Ric}_{\mathcal D},
\end{aligned} \tag{50}
$$

$$
2\,\partial_t \mathrm{Sc}_{\mathrm{mix}}(g) = -\Delta_{\mathcal F}\,(\mathrm{Tr}\, S^\sharp) + 2\sum\nolimits_i \mathrm{Tr}\,((\nabla_i\, S^\sharp) C_i). \tag{51}
$$

Proof. Since the tensor r is $\mathcal D$-truncated, one may assume $X, Z \in \mathcal D$ and then calculate the time derivative $\partial_t\, r(X,Z) = \sum_i \partial_t\, R(X,E_i,Z,E_i)$. By Proposition 2 with $Y = V = E_i$, we then have

$$
\begin{aligned}
2\,\partial_t\, r(X,Z) &= S(\mathrm{Ric}_{\mathcal D}(X),Z) \\
&\quad + \sum\nolimits_i \big[\nabla^2_{X,i}\, S(E_i,Z) + \nabla^2_{i,Z}\, S(X,E_i) \\
&\quad - \nabla^2_{X,Z}\, S(E_i,E_i) - \nabla^2_{i,i}\, S(X,Z)\big].
\end{aligned} \tag{52}
$$

By definition (8), we have $(\nabla_X\, E_i)^\perp = -C_{E_i}(X)$, and we can take a local vector field X with the property $C_{E_i}(X) = -\nabla_i\, X$ at a fixed point $x \in M$. By the above and (47), for a $\mathcal D$-truncated symmetric $(0,2)$-tensor S we have $\nabla_Z\, S(X,E_i) = S(X,\, C_i(Z))$ and $\nabla_Z\, S(E_i,E_i) = 0$; hence,

$$
\begin{aligned}
\nabla^2_{X,i}\, S(E_i,Z) &= \nabla_X(\nabla_i\, S(E_i,Z)) - \nabla_i\, S(\nabla_X\, E_i,Z) - \nabla_i\, S(E_i,\nabla_X\, Z) \\
&\quad - \nabla_{\nabla_X\, E_i}\, S(E_i,Z) \\
&= \nabla_i\, S(C_i(X),Z) + S(C_i^2(X),Z) - \nabla_X\, S(\nabla_i\, E_i,Z) \\
&\quad - S(\nabla_X(\nabla_i\, E_i),Z), \\
\nabla^2_{i,Z}\, S(X,E_i) &= \nabla_i(\nabla_Z\, S(X,E_i)) - \nabla_Z\, S(\nabla_i X,E_i) - \nabla_Z\, S(X,\nabla_i\, E_i)
\end{aligned}
$$

$$
\begin{aligned}
&\quad - \nabla_{\nabla_i Z}\, S(X, E_i) \\
&= \nabla_i\, S(C_i(Z), X) - \nabla_Z\, S(\nabla_i\, E_i, X) \\
&\quad + S(\nabla_i (C_i(Z)), X) + S(C_i^2(Z), X) \\
&\overset{(12)}{=} \nabla_i\, S(C_i(Z), X) + S(C_i^2(Z), X) + S(R_{E_i}(Z), X) \\
&\quad - F_{E_i}(Z, S^\sharp(X)) - \nabla_Z S(\nabla_i\, E_i, X), \\
\nabla^2_{X,Z}\, S(E_i, E_i) &= \nabla_X(\nabla_Z\, S(E_i, E_i)) - 2\nabla_Z\, S(\nabla_X\, E_i, E_i) \\
&\quad - \nabla_{\nabla_X Z}\, S(E_i, E_i) \\
&= 2\, S(C_i(X), C_i(Z)).
\end{aligned}
$$

By $H_{\mathcal{F}} = \sum_i \nabla_i\, E_i$ and the above, (52) reduces to (48). □

Proof of Theorem 2. We will use variations of the form $g(t) = g_0 + t\, S$ with a $\mathcal{D}$-truncated symmetric $(0,2)$-tensor S. We will show that $\Delta_{\mathcal{F}}\, S_{ik}$ yields the principal symbol of order two, and other terms are of order less than two. By Lemma 5, the linearization of $-2\,r$ is the second-order differential operator (elliptic along the leaves)

$$
D(-2\,r)_{ik} = \Delta_{\mathcal{F}}\, S_{ik} + \tilde{S}_{ik},
$$

where $\tilde{S}_{ik}$ consists of the first- and zero-order terms. The result then follows from the theory of parabolic PDEs on vector bundles, see [2, Sect. 5.1], and assumption (7). □

3.2 Evolution of the Curvature Tensor Along the PRF

In this section we derive evolution equations for the Riemann curvature tensor, the partial Ricci tensor, and $\mathrm{Sc}_{\mathrm{mix}}$ along the PRF. These evolution equations for $p = 1$ were derived in [19].

Define the difference tensor $Q(X, Z; Y, V) = \big(\sum_i \nabla^2_{X,Z}\, R(Y, E_i, V, E_i)\big) - \nabla^2_{X,Z}\, r(Y, V)$.

Lemma 6. *For $Y, V \in \mathcal{D}$ we have*

$$
\begin{aligned}
Q(X, Z; Y, V) = \sum\nolimits_i \big[& \nabla_X R(Y, C_i(Z), V, E_i) + \nabla_X R(Y, E_i, V, C_i(Z)) \\
& + \nabla_Z R(Y, C_i(X), V, E_i) + \nabla_Z R(Y, E_i, V, C_i(X)) \\
& + R(Y, \nabla_X C_i(Z), V, E_i) + R(Y, E_i, V, \nabla_X C_i(Z)) \\
& - R(Y, C_i(Z), V, C_i(X)) - R(Y, C_i(X), V, C_i(Z)) \big].
\end{aligned}
$$

Proof. We calculate

$$\begin{aligned}\sum\nolimits_i \nabla^2_{X,Z} R(Y,E_i,V,E_i) &= \sum\nolimits_i \big[\nabla_X(\nabla_Z R)(Y,E_i,V,E_i) - \nabla_{\nabla_X Z} R(Y,E_i,V,E_i)\big]\\
&= \nabla_X\Big(\nabla_Z r(Y,V) - \sum\nolimits_i \big[R(Y,\nabla_Z E_i,V,E_i)\\
&\quad + R(Y,E_i,V,\nabla_Z E_i)\big]\Big) - \sum\nolimits_i \big[\nabla_Z R(\nabla_X Y,E_i,V,E_i)\\
&\quad + \nabla_Z R(Y,\nabla_X E_i,V,E_i) + \nabla_Z R(Y,E_i,\nabla_X V,E_i)\\
&\quad + \nabla_Z R(Y,E_i,V,\nabla_X E_i)\big] - \nabla_{\nabla_X Z} r(Y,V)\\
&\quad + \sum\nolimits_i \big[R(Y,\nabla_{\nabla_X Z} E_i,V,E_i)\\
&\quad + R(Y,E_i,V,\nabla_{\nabla_X Z} E_i)\big]\\
&= \nabla^2_{X,Z} r(Y,V) + \sum\nolimits_i \big[\nabla_X R(Y,C_i(Z),V,E_i)\\
&\quad + \nabla_X R(Y,E_i,V,C_i(Z))\\
&\quad + \nabla_Z R(Y,C_i(X),V,E_i) + \nabla_Z R(Y,E_i,V,C_i(X))\\
&\quad + R(Y,\nabla_X C_i(Z),V,E_i)\\
&\quad - R(Y,C_i(Z),V,C_i(X)) - R(Y,C_i(X),V,C_i(Z))\\
&\quad + R(Y,E_i,V,\nabla_X C_i(Z))\big]\end{aligned}$$

using

$$\begin{aligned}\sum\nolimits_i \nabla_Z R(Y,E_i,V,E_i) &= \nabla_Z r(Y,V) - \sum\nolimits_i \big[R(Y,\nabla_Z E_i,V,E_i)\\
&\quad + R(Y,E_i,V,\nabla_Z E_i)\big],\\
\nabla^2_{X,Z} r(Y,V) &= \nabla_X(\nabla_Z r(Y,V)) - \nabla_Z r(Y,\nabla_X V) - \nabla_Z r(\nabla_X Y,V)\\
&\quad - \nabla_{\nabla_X Z} r(Y,V).\end{aligned}$$

The above yields the claim. □

By Lemma 6, the tensor

$$\tilde{Q} = Q(X,Z;Y,V) - Q(Y,Z;X,V) + Q(Y,V;X,Z) - Q(X,V;Y,Z) \quad (53)$$

does not contain second-order derivatives when at least two vectors of $\{X,Y,Z,V\}$ belong to $\mathcal{D}$.

Remark 5. Using Gauss and Codazzi equations for submanifolds, one may study the remaining case and show that $\tilde{Q}$ does not contain the second-order derivatives when at most one vector belongs to $\mathcal{D}$. By Lemma 6, we find (when at least three vectors of $\{X,Y,Z,V\}$ belong to $\mathcal{D}$)

$$\begin{aligned}\tilde{Q} = \sum\nolimits_i \big[&\nabla_{C_i(Z)} R(X,Y,E_i,V) + \nabla_{C_i(Y)} R(X,E_i,Z,V) + \nabla_{C_i(X)} R(E_i,Y,Z,V)\\
&+ \nabla_{C_i(V)} R(X,Y,Z,E_i) + \nabla_i R(X,C_i(Y),Z,V) + \nabla_i R(C_i(X),Y,Z,V)\\
&+ \nabla_i R(X,Y,Z,C_i(V)) + \nabla_i R(X,Y,C_i(Z),V) + R(Y,\nabla_X C_i(Z),V,E_i)\\
&+ R(Y,E_i,V,\nabla_X C_i(Z)) - R(Y,C_i(Z),V,C_i(X)) - R(Y,C_i(X),V,C_i(Z))\\
&- R(Y,\nabla_X C_i(V),Z,E_i) - R(Y,E_i,Z,\nabla_X C_i(V)) + R(Y,C_i(V),Z,C_i(X))\\
&+ R(Y,C_i(X),Z,C_i(V)) + R(X,\nabla_Y C_i(V),Z,E_i) + R(X,E_i,Z,\nabla_Y C_i(V))\\
&- R(X,C_i(V),Z,C_i(Y)) - R(X,C_i(Y),Z,C_i(V)) - R(X,\nabla_Y C_i(Z),V,E_i)\\
&- R(X,E_i,V,\nabla_Y C_i(Z)) + R(X,C_i(Z),V,C_i(Y)) + R(X,C_i(Y),V,C_i(Z))\big].\end{aligned}$$

Theorem 4. *Let $\mathcal{F}$ be a smooth foliation on (M,g), and at least two vectors of $\{X,Y,Z,V\}$ belong to $\mathcal{D}$. Then the curvature tensor evolves by* (4) *according to a leaf-wise heat equation*

$$\begin{aligned}\partial_t R(X,Y,Z,V) = {}&\Delta_{\mathcal{F}} R(X,Y,Z,V) + B(X,Y,Z,V)\\
&- B(X,Y,V,Z) - B(Y,X,Z,V)\\
&+ B(Y,X,V,Z) - 2\,B(Z,Y,V,X) + 2\,B(Z,X,V,Y)\\
&- r(R(X,Y)V,\,Z) - r(R(X,Y)Z,\,V)\\
&- \langle R(\cdot,Y,Z,V),\, \sum\nolimits_i R(X,E_i,\cdot,E_i)\rangle\\
&+ \langle R(\cdot,X,Z,V),\, \sum\nolimits_i R(Y,E_i,\cdot,E_i)\rangle - \tilde{Q}\,. \qquad (54)\end{aligned}$$

For an averaged curvature-invariant distribution $\mathcal{D}$ this simplifies due to (45).

Proof. Applying (46) with $S = -2\,r$, we have

$$\begin{aligned}\partial_t R(X,Y,Z,V) = {}&\nabla^2_{X,Z}\, r(Y,V) - \nabla^2_{Y,Z}\, r(X,V) + \nabla^2_{Y,V}\, r(X,Z) - \nabla^2_{X,V}\, r(Y,Z)\\
&-r(R(X,Y)V,\,Z) - r(R(X,Y)Z,\,V). \qquad (55)\end{aligned}$$

Comparing (55) with (38), and using (53), completes the proof. □

Example 1. Let $\mathcal{F}$ be a one-dimensional foliation spanned by a unit vector field N. Then

$$B(X,Y,V,Z) = \langle R(X,\,\cdot\,,Y,\,N),\; R(V,\,\cdot\,,Z,\,N)\rangle \quad \text{for all} \quad X,Y,V,Z \in TM.$$

Formula (38) takes the form, see also [19],

$$\begin{aligned}\nabla^2_{N,N}\, R(X,Y,Z,V) = [&\nabla^2_{X,Z}\, R(Y,N,V,N) - \nabla^2_{Y,Z}\, R(X,N,V,N)\\
&+ \nabla^2_{Y,V}\, R(X,N,Z,N) - \nabla^2_{X,V}\, R(Y,N,Z,N)]\end{aligned}$$

$$
\begin{aligned}
&- \big(B(X,Y,Z,V) - B(X,Y,V,Z) - B(Y,X,Z,V)\\
&+ B(Y,X,V,Z) - 2\,B(Z,Y,V,X) + 2\,B(Z,X,V,Y)\big)\\
&+ \langle R(\cdot\,,Y,Z,V),\; r(X,\cdot)\rangle - \langle R(\cdot\,,X,Z,V),\; r(Y,\cdot)\rangle.
\end{aligned} \tag{56}
$$

Note that $Q(X,Z;Y,V) = \nabla^2_{X,Z} R(Y,N,V,N) - \nabla^2_{X,Z}\, r(Y,V)$. Hence, the curvature tensor evolves by (4) according to a heat-type equation along N-curves

$$
\begin{aligned}
\partial_t\, R(X,Y,Z,V) = \nabla^2_{N,N}\, R(X,Y,Z,V) &+ \big(B(X,Y,Z,V) - B(X,Y,V,Z)\\
&- B(Y,X,Z,V) + B(Y,X,V,Z) - 2\,B(Z,Y,V,X)\\
&+ 2\,B(Z,X,V,Y)\big) - \langle R(\cdot\,,Y,Z,V),\; r(X,\cdot)\rangle\\
&+ \langle R(\cdot\,,X,Z,V),\; r(Y,\cdot)\rangle - \tilde{Q}\,.
\end{aligned} \tag{57}
$$

Theorem 5. *Let $\mathcal{F}$ be a smooth foliation on (M,g). Then the tensor r evolves by* (4) *according to*

$$
\begin{aligned}
\partial_t\, r(X,Z) = \Delta_{\mathcal{F}}\, r(X,\, Z) &- 2\,r(X,\mathrm{Ric}_{\mathcal{D}}(Z)) - \sum\nolimits_i \big[\nabla_i\, r(C_i(X),Z)\\
&+ \nabla_i\, r(X,C_i(Z)) + r(X,C_i^{\,2}(Z)) + r(C_i^{\,2}(X),Z)\\
&+ 2\,r(C_i(X),C_i(Z))\big] - \mathrm{Tr}\,(A^{\mathcal{F}}_{\mathrm{Ric}_{\mathcal{D}}(X)} A^{\mathcal{F}}_Z)\\
&+ \nabla_X\, r(H_{\mathcal{F}},Z) + \nabla_Z\, r(H_{\mathcal{F}},X) + r(\nabla_X\, H_{\mathcal{F}},Z) + r(\nabla_Z\, H_{\mathcal{F}},X),
\end{aligned} \tag{58}
$$

where $X,Z \in \mathcal{D}$. We have $\mathrm{Tr}\,(A^{\mathcal{F}}_{\mathrm{Ric}_{\mathcal{D}}(X)} A^{\mathcal{F}}_Z) = (1/p)\,g(H_{\mathcal{F}},Z)\,\mathrm{Ric}_{\mathcal{D}}(H_{\mathcal{F}},X)$ *in* (58) *for a totally umbilical foliation $\mathcal{F}$, and for a totally geodesic foliation, we obtain*

$$
\begin{aligned}
\partial_t\, r(X,Z) &= \Delta_{\mathcal{F}}\, r(X,Z) - 2\,r(X,\mathrm{Ric}_{\mathcal{D}}(Z))\\
&\quad - \sum\nolimits_i \big[\nabla_i\, r(C_i(X),Z) + \nabla_i\, r(X,C_i(Z))\\
&\quad + r(X,C_i^{\,2}(Z)) + r(C_i^{\,2}(X),Z) - 2\,r(C_i(X),C_i(Z))\big], && (59)\\
\partial_t \mathrm{Ric}_{\mathcal{D}} &= \Delta_{\mathcal{F}}\, \mathrm{Ric}_{\mathcal{D}}\\
&\quad - \sum\nolimits_i \big[(\nabla_i\, R_{E_i})C_i\\
&\quad + (C_i)^*(\nabla_i\, R_{E_i}) + R_{E_i}\, C_i^{\,2} + (C_i^2)^*\, R_{E_i} - 2(C_i)^*\, R_{E_i}\, C_i\big], && (60)\\
\partial_t\, \mathrm{Sc}_{\mathrm{mix}}(g) &= \Delta_{\mathcal{F}}\, \mathrm{Sc}_{\mathrm{mix}}(g) - 2\sum\nolimits_i \big[\,\mathrm{Tr}\,(A_i \nabla_i\, R_{E_i}) + 2\,\mathrm{Tr}\,((T_i^\sharp)^2\, R_{E_i})\big]. && (61)
\end{aligned}
$$

Proof. Substituting $S = -2r$ into (48), we obtain (58), which for $h_{\mathcal{F}} = 0$ yields (59). By (59) and (34) we get (60). Tracing (60) and using $\partial_t(\operatorname{Tr}\operatorname{Ric}_{\mathcal{D}}) = \operatorname{Tr}(\partial_t \operatorname{Ric}_{\mathcal{D}})$ and $\operatorname{Tr}(T^\sharp \nabla_i R_{E_i}) = 0$ yield

$$\partial_t \mathrm{Sc}_{\mathrm{mix}}(g) = \Delta_{\mathcal{F}} \mathrm{Sc}_{\mathrm{mix}}(g) - 2\sum\nolimits_i \big[\operatorname{Tr}(C_i \nabla_i R_{E_i}) + 2\operatorname{Tr}(C_i T_i^\sharp R_{E_i})\big].$$

From this, the skew-symmetry of $T_i^\sharp$ and the property $\operatorname{Tr}(B_1 B_2) = \operatorname{Tr}(B_2 B_1)$, we obtain (61). □

We apply Uhlenbeck's trick (see [2]) to remove a group of terms in (54) with a 'change of variables.'

Corollary 3. *Let $\mathcal{F}$ be a totally geodesic foliation on a Riemannian manifold (M, g). Then the curvature tensor evolves by* (4) *according to*

$$\begin{aligned}\tilde\nabla_{\partial_t} R(X,Y,Z,V) &= \Delta_{\mathcal{F}} R(X,Y,Z,V) + B(X,Y,Z,V)\\ &\quad - B(X,Y,V,Z) - B(Y,X,Z,V)\\ &\quad + B(Y,X,V,Z) - 2\,B(Z,Y,V,X) + 2\,B(Z,X,V,Y) - \tilde Q,\end{aligned} \tag{62}$$

where $X,Y,Z,V \in \mathcal{D}$ and $\tilde Q$ is given in (53)*. The tensor r evolves by* (4) *according to*

$$\begin{aligned}\tilde\nabla_{\partial_t} r(X,Z) &= \Delta_{\mathcal{F}} r(X,Z) - \sum\nolimits_i \big[\nabla_i r(C_i(X),Z) + \nabla_i r(C_i(Z),X)\\ &\quad + r(C_i^2(X),Z) + r(C_i^2(Z),X) - 2\,r(C_i(X),C_i(Z))\big].\end{aligned} \tag{63}$$

Proof. Using definition $\tilde\nabla_{\partial_t} X = \partial_t X - \operatorname{Ric}_{\mathcal{D}}(X) = -\operatorname{Ric}_{\mathcal{D}}(X)$, we obtain

$$\begin{aligned}\tilde\nabla_{\partial_t} R(X,Y,Z,V) &= \partial_t R(X,Y,Z,V)\\ &\quad - R(-\tilde\nabla_{\partial_t} X,Y,Z,V) - \ldots - R(X,Y,Z,-\tilde\nabla_{\partial_t} V)\\ &= \partial_t R(X,Y,Z,V) + \langle R(\cdot,Y,Z,V),\ r(X,\cdot)\rangle\\ &\quad + \langle R(X,\cdot,Z,V),\ r(Y,\cdot)\rangle + \langle R(X,Y,\cdot,V),\ r(Z,\cdot)\rangle\\ &\quad + \langle R(X,Y,Z,\cdot),\ r(V,\cdot)\rangle.\end{aligned}$$

From this, (54), and (44), equation (62) follows. Similarly, (58) yields (63). □

3.3 Evolution by PRF of the Extrinsic Geometry

Proposition 3. *The normalized PRF* (5) *preserves the metric of $\mathcal{D}_{\mathcal{F}}$ and the orthogonality of vectors to $\mathcal{F}$. If $\mathcal{F}$ is either totally umbilical* (*totally geodesic*) *or harmonic foliation for $t = 0$ then it has the same property for all $t > 0$.*

Proof. Since $r(N,\cdot) = 0$ for any $N \in \mathcal{D}_{\mathcal{F}}$, (5) preserves the metric on $\mathcal{D}_{\mathcal{F}}$ and the scalar product $g(N, X)$ for any $X \in \mathcal{D}$. Hence, the normalized PRF preserves the distribution $\mathcal{D}$ orthogonal to $\mathcal{F}$.

By (32) with $S = -2\,r + 2\Phi\,\hat{g}$, we have

$$\partial_t h_{\mathcal{F}} = 2\,\mathrm{Ric}_{\mathcal{D}} \circ h_{\mathcal{F}} - 2\,\Phi\, h_{\mathcal{F}}, \qquad \partial_t H_{\mathcal{F}} = 2\,\mathrm{Ric}_{\mathcal{D}}(H_{\mathcal{F}}) - 2\,\Phi\, H_{\mathcal{F}}. \tag{64}$$

Hence, (5) preserves total geodesy and harmonicity of foliations. By (64) we have

$$\partial_t (h_{\mathcal{F}} - (1/p) H_{\mathcal{F}}\, g_{|\,\mathcal{F}}) = 2(\mathrm{Ric}_{\mathcal{D}} - \Phi\,\hat{\mathrm{id}}) \circ (h_{\mathcal{F}} - (1/p) H_{\mathcal{F}}\, g_{|\,\mathcal{F}}).$$

By the local theorem of existence and uniqueness of a solution to ODE, if $h_{\mathcal{F}} = (1/p) H_{\mathcal{F}}\, g_{|\,\mathcal{F}}$ (i.e., $\mathcal{F}$ is totally umbilical) for $t = 0$ then $h_{\mathcal{F}} = (1/p) H_{\mathcal{F}}\, g_{|\,\mathcal{F}}$ for all $t > 0$. □

By Corollary 2 with $S = -2\,r + 2\Phi\,\hat{g}$ and any $N \in \mathcal{D}_{\mathcal{F}}$, we have the symmetries of the PRF:

$$(\partial_t A_N)^* - \partial_t A_N = 2\,[A_N, \mathrm{Ric}_{\mathcal{D}}], \quad (\partial_t T^\sharp_N)^* + \partial_t T^\sharp_N = -2\,[T^\sharp_N, \mathrm{Ric}_{\mathcal{D}}],$$
$$(\partial_t\, \mathrm{Ric}_{\mathcal{D}})^* = \partial_t\, \mathrm{Ric}_{\mathcal{D}}\,.$$

Proposition 4. *Let $\mathcal{F}$ be a harmonic foliation. Then the tensor h and the mean curvature vector H of $\mathcal{D}$ evolve by* (5) *according to*

$$\begin{aligned}\partial_t h(X,Y) &= \nabla^{\mathcal{F}} \mathrm{div}_{\mathcal{F}}\, h(X,Y) - \mathrm{div}_{\mathcal{F}}\big(h(X, C_\circ(Y)) + h(C_\circ(X), Y)\big) + \Phi\, h(X,Y)\\ &\quad + g([C_\circ^*(\mathcal{A}+\mathcal{T}) + (\mathcal{A}+\mathcal{T})C_\circ - \nabla^{\mathcal{F}}(\mathcal{A}+\mathcal{T})](X), Y)\\ &\quad + \mathrm{Tr}\,(A^{\mathcal{F}}_{C_\circ(Y)} A^{\mathcal{F}}_X + A^{\mathcal{F}}_{C_\circ(X)} A^{\mathcal{F}}_Y) - \nabla^{\mathcal{F}}\, \mathrm{Tr}\,(A^{\mathcal{F}}_Y A^{\mathcal{F}}_X),\end{aligned} \tag{65}$$

$$\partial_t H = \nabla^{\mathcal{F}}(\mathrm{div}_{\mathcal{F}}\, H) + \nabla^{\mathcal{F}}(\|T\|^2 - \|h\|^2 - \|h_{\mathcal{F}}\|^2). \tag{66}$$

The tensors A_N, $T^\sharp_N$, and $C_N = A_N + T^\sharp_N$ evolve by (5) *according to*

$$\partial_t A_N = \nabla_N\, \mathrm{Ric}_{\mathcal{D}} + [T^\sharp_N - A_N,\ \mathrm{Ric}_{\mathcal{D}}], \qquad \partial_t T^\sharp_N = 2(\mathrm{Ric}_{\mathcal{D}} - \Phi\,\hat{\mathrm{id}})\, T^\sharp_N, \tag{67}$$

$$\partial_t C_N = \nabla_N\, \mathrm{Ric}_{\mathcal{D}} + [\mathrm{Ric}_{\mathcal{D}}, C_N] + 2\, T^\sharp_N\, \mathrm{Ric}_{\mathcal{D}} - 2\,\Phi\, T^\sharp_N. \tag{68}$$

Proof. From (28) (with $S = -2r + 2\Phi\,\hat{g}$) we have (65). From (31) with $S = -2r + 2\Phi\,\hat{g}$ we have

$$\partial_t H = \nabla^{\mathcal{F}} \mathrm{Sc}_{\mathrm{mix}}(g_t). \tag{69}$$

Substituting $\mathrm{Sc}_{\mathrm{mix}}$ from (18) into (69) and using (17) (or tracing (65)) yield (66). Indeed, from (29) and (30) (with $S = -2r + 2\Phi\,\hat{g}$), we obtain (67) and (68). □

4 Examples

In this section we show that PRF preserves several classes of foliations and prove existence/uniqueness of global leaf-wise smooth solution metrics convergence of a solution under certain conditions.

4.1 Totally Geodesic Foliations of Codimension One (n = 1)

Let $\mathcal{F}$ be a codimension-one totally geodesic foliation on a Riemannian manifold (M^{p+1}, g) with a unit normal vector field N. Then M is locally (globally, if M is simply connected and the leaves are complete) isometric to a product manifold $F^p \times \mathbb{R}$, with a *twisted* product metric $\mathrm{dx}^2 + \varphi^2 \mathrm{dy}^2$, where dx^2 is a fixed metric on F^p and $\varphi \in C^\infty(F^p \times \mathbb{R})$, see [9, Theorem 1]. The mean curvature vector of N-curves is $H = -\nabla^{\mathcal{F}} \log \varphi$. Note that $T = 0$, $H_{\mathcal{F}} = 0$, and $\mathrm{Sc}_{\mathrm{mix}} = \mathrm{Ric}(N, N)$, and (18) reads

$$\mathrm{Ric}(N, N) = \mathrm{div}_{\mathcal{F}}\, H - \|H\|^2 = \mathrm{div}\, H. \tag{70}$$

Since $r = \mathrm{Ric}(N, N)\,\hat{g}$, PRF (5) with a leaf-wise constant $\Phi : M \to \mathbb{R}$ reduces to (6),

$$\partial_t g = -2\,(\mathrm{Ric}(N, N) - \Phi)\,\hat{g}. \tag{71}$$

The spectrum of $\Delta_{\mathcal{F}}$ on a compact leaf F is an infinite sequence of isolated real eigenvalues $0 = \lambda_0 < \lambda_1 \le \lambda_2 \le \dots$ counting their multiplicities, and $\lim_{j\to\infty} \lambda_j = \infty$. One may fix in $L_2(F)$ an orthonormal basis of corresponding eigenfunctions $\{e_j\}$, i.e., $-\Delta_{\mathcal{F}}(e_j) = \lambda_j e_j$, and $e_0 = \mathrm{const} > 0$.

One may compare the next result with [18, Theorem 4].

Theorem 6. *Let $\mathcal{F}$ be a totally geodesic foliation of codimension one with simply connected leaves and a unit normal N on a Riemannian manifold (M, g_0), and let assumptions* (7) *are satisfied. Then* (71) *has a unique global smooth solution g_t $(t \ge 0)$. If $\Phi = 0$ then as $t \to \infty$, the metrics g_t converge with the exponential rate λ_1 to the limit smooth metric $\bar{g}$ and*

$$\overline{\mathrm{Ric}}(N, N) = 0, \qquad \bar{H} = 0.$$

Proof. From (66) with $T = 0$, $h_{\mathcal{F}} = 0$, and $\|h\| = \|H\|$, we obtain the Burgers-type equation for H

$$\partial_t H + \nabla^{\mathcal{F}}(\|H\|^2) = \nabla^{\mathcal{F}}(\operatorname{div}_{\mathcal{F}} H). \tag{72}$$

For any leaf (fiber) F there is a simply connected neighborhood $U_F \simeq F \times \mathbb{R}$ such that $H_0 = -\nabla^{\mathcal{F}} \log u_0$ for a smooth function $u_0 > 0$ on U_F. One may take $H = -\nabla^{\mathcal{F}} \log u$, where the function $u(t, x) > 0$ obeys the heat equation (see also [17, Proposition 2])

$$\partial_t u = \Delta_{\mathcal{F}}\, u, \qquad u(0, \cdot) = u_0. \tag{73}$$

The Cauchy's problem (73) has a unique global (smooth for $t > 0$) solution and $\lim\limits_{t\to\infty} u = \bar u > 0$ is a leaf-wise constant. Denote by $(\cdot\,, \cdot)_0$ and $\|\cdot\|_0$ the inner product and the norm in $L_2(F)$. Using Fourier series representation

$$u = (u_0, e_0)_0\, e_0 + e^{-\lambda_1 t} \sum\nolimits_{j>1} e^{(\lambda_1-\lambda_j)t} (u_0, e_j)_0\, e_j\,,$$

we find $\nabla^{\mathcal{F}} u = e^{-\lambda_1 t} \sum_{j>1} e^{(\lambda_1-\lambda_j)t} (u_0, e_j)_0\, \nabla^{\mathcal{F}} e_j$. Since the series above converge absolutely and uniformly with exponential rate, and $(u_0, e_0)_0 > 0$, we have $\lim\limits_{t\to\infty} u = (u_0, e_0)_0\, e_0 > 0$ is leaf-wise constant and $\lim\limits_{t\to\infty} \nabla^{\mathcal{F}} u = 0$, see [17, Proposition 4]. Hence, (72) admits a unique smooth solution H and

$$\lim_{t\to\infty} H = \lim_{t\to\infty} \frac{\nabla^{\mathcal{F}} u}{u} = \lim_{t\to\infty} \frac{e^{-\lambda_1 t} \sum_{j>1} e^{(\lambda_1-\lambda_j)t} (u_0, e_j)_0\, \nabla^{\mathcal{F}} e_j}{(u_0, e_0)_0\, e_0 + e^{-\lambda_1 t} \sum_{j>1} e^{(\lambda_1-\lambda_j)t} (u_0, e_j)_0\, e_j} = 0.$$

Note that $\operatorname{div}_{\mathcal{F}} H = \operatorname{div}_{\mathcal{F}}(\frac{\nabla^{\mathcal{F}} u}{u}) = \frac{1}{u}\Delta_{\mathcal{F}}\, u - |\nabla^{\mathcal{F}} u|^2/u^2 \to 0$ as $t \to \infty$. By (70), this corresponds to smooth functions $\operatorname{Ric}_t(N, N)$, and

$$\lim_{t\to\infty} \operatorname{Ric}_t(N, N) = \lim_{t\to\infty} (\operatorname{div}_{\mathcal{F}} H - \|H\|^2) = 0,$$

where convergence is exponential. Then we recover the metrics g_t $(t \geq 0)$ from (71). If $\Phi = 0$ then g_t converge exponentially to a smooth limit metric $\bar g$. □

4.2 *Geodesic Foliations* $(p = 1)$

Let $\mathcal{F}$ be a one-dimensional foliation spanned by a unit vector field N. Denote by $C := C_N$ the conullity tensor, $T^\sharp := T^\sharp_N$ the integrability tensor, and $A := A_N$ the Weingarten operator of $\mathcal{F}$. Then (19) and (20) hold, where $H_{\mathcal{F}} = \nabla_N N$. The above and the equality $\operatorname{div} N = -\operatorname{Tr} A$ provide

$$N(\operatorname{Tr} A) = \operatorname{Tr}(A^2) + \operatorname{Tr}((T^\sharp)^2) + \operatorname{Ric}(N, N) - \operatorname{div} H_{\mathcal{F}}. \tag{74}$$

Here, $\operatorname{Tr} A = \operatorname{Tr}_g h = H$ is the mean curvature of $\mathcal{D}$. Note that $\|T^\sharp\|^2 = -\operatorname{Tr}((T^\sharp)^2)$. By Theorem 5 with $p = 1$ and Lemma 3 with $p = 1$ and $S = -2\,r$, we have the following (see also [19]).

Proposition 5. *Let N be a unit geodesic vector field. Then the curvature evolves by* (4) *according to*

$$\partial_t\, r(X,Y) = \nabla^2_{N,N}\, r(X,Y) - \nabla_N\, r(C(X),Y) - \nabla_N\, r(X,C(Y)) \tag{75}$$
$$- r(C^2(X),Y) - r(X,C^2(Y)) + 2\,r(C(X),C(Y)) - 2\,r(X,R_N(Y)),$$
$$\partial_t\, R_N = \nabla^2_{N,N}\, R_N - (\nabla_N R_N)C - C^*\nabla_N R_N - R_N C^2 - (C^*)^2 R_N$$
$$+\,2\,C^* R_N C, \tag{76}$$
$$\partial_t\, \mathrm{Ric}_N = N(N(\mathrm{Ric}_N)) - 2\operatorname{Tr}(A\,\nabla_N R_N) - 4\operatorname{Tr}((T^\sharp)^2 R_N). \tag{77}$$

For (5)*, we also have*

$$\partial_t C = \nabla_N(\nabla_N C) - (C + C^*)\,\nabla_N C - (C - C^*)C^2 - 2\,\Phi\, T^\sharp, \tag{78}$$
$$\partial_t T^\sharp = 2\,(\nabla_N A)T^\sharp - 2\,A^2 T^\sharp - 2\,(T^\sharp)^3 - 2\,\Phi\, T^\sharp, \tag{79}$$
$$\partial_t A = \nabla_N(\nabla_N A) - 2\,A\,\nabla_N A + [A^2,\ T^\sharp] - 2\,(T^\sharp)^2 A - 2\,T^\sharp A T^\sharp, \tag{80}$$
$$\partial_t H = \nabla_N(\nabla_N H) - \nabla_N \operatorname{Tr}(A^2) - 4\operatorname{Tr}((T^\sharp)^2 A). \tag{81}$$

Using (80) and definition (35), we obtain the following.

Corollary 4 ([19]). *Let N be a unit geodesic vector field with integrable orthogonal distribution. Then the Weingarten operator A evolves by* (4) *according to*

$$\tilde\nabla_{\partial_t} A = \nabla_N(\nabla_N A) - \nabla_N(A^2). \tag{82}$$

By the existence/uniqueness Theorem 2, we have the following.

Proposition 6 ([19]). *If $\nabla_N N = 0$ and $A = 0$ at $t = 0$ then flow* (5) *preserves these properties.*

Next theorem deals with a geodesic Riemannian foliation such that $T \neq 0$. Examples are provided by Hopf fibrations of odd-dimensional spheres.

Theorem 7. *Let $\mathcal{F}$ be a geodesic foliation spanned by a unit vector field N on (M, g_0). Suppose that $\mathcal{F}$ is a Riemannian foliation ($A = 0$) and the orthogonal distribution $\mathcal{D}$ is nowhere integrable ($T \neq 0$). If $r \le \Phi\,\hat g$ and $r_{|\mathcal{D}} > 0$ at $t = 0$ then* (5) *admits a unique solution g_t ($t \in \mathbb{R}$) such that $\lim_{t\to-\infty} R_N(t) = \Phi\,\widehat{\mathrm{id}}$ and $\lim_{t\to\infty} R_N(t) = 0$.*

Proof. By Proposition 6, we have $A = 0$ for $t \ge 0$; hence, $C = T^\sharp$. By (24), $\nabla_N T^\sharp = 0$ and $R_N = -(T^\sharp)^2 \ge 0$. This yields $\nabla_N R_N = 0$, $\nabla_N r = 0$, and $N(\mathrm{Ric}_N) = 0$ for $t \geq 0$; hence, (75)–(80) reduce to ODEs in the variable t. By (50)–(51) with $p = 1$ and $S^\sharp = -2\,R_N + 2\,\Phi\,\hat{\mathrm{id}}$, we obtain

$$\partial_t\,R_N = -R_N(T^\sharp)^2 - (T^\sharp)^2 R_N - 2\,T^\sharp R_N T^\sharp - 4\,\Phi R_N = 4\,R_N(R_N - \Phi\,\hat{\mathrm{id}}),$$

$$\partial_t\,\mathrm{Ric}_N = -4\,\mathrm{Tr}\,((T^\sharp)^2 R_N) - 4\,\Phi\,\mathrm{Ric}_N = 4\,\mathrm{Tr}\,(R_N^2) - 4\,\Phi\,\mathrm{Ric}_N \geq \frac{4}{n}\,(\mathrm{Ric}_N)^2 - 4\,\Phi\,\mathrm{Ric}_N,$$

see also (76)–(77). One may show that (5) preserves positive Ric_N. By Proposition 4, we have

$$\partial_t T^\sharp = -2\,T^\sharp\big((T^\sharp)^2 + \Phi\,\hat{\mathrm{id}}\big). \tag{83}$$

In our case $r_{|\mathcal{D}} > 0$, the dimension n should be even. (Indeed, if n is odd then the skew-symmetric operator $T^\sharp$ has zero eigenvalues; hence, R_N also has zero eigenvalues.)

Let $\mu_i(t) > 0$ be the eigenvalue and $e_i(t)$ the eigenvector of $R_N(t)$ under flow (5). Then

$$\partial_t e_i = (\mu_i - \Phi)e_i, \qquad \partial_t \mu_i = 4\mu_i(\mu_i - \Phi).$$

Hence, the PRF preserves the directions of eigenvectors of R_N. Furthermore, if $\Phi \geq \mu_i(0) > 0$ then $\lim_{t\to-\infty} \mu_i(t) = \Phi$ and $\lim_{t\to\infty} \mu_i(t) = 0$. □

4.3 Warped Products

Definition 3 (see [9]). Let (B^p, dx^2) and $(\bar{M}^n, \bar{g})$ be Riemannian manifolds, and $\varphi \in C^\infty(B)$ a positive function. The *warped product* $B \times_\varphi \bar{M}$ is the manifold $M = B \times \bar{M}$ with the metric $g = \mathrm{dx}^2 + \varphi^2(x)\,\bar{g}$. The *fibers* $\{x\} \times \bar{M}$ of a warped product are totally umbilical, while the *leaves* $B \times \{y\}$ compose a totally geodesic foliation $\mathcal{F}$ on M. (Examples are rotational symmetric metrics, i.e., $\bar{M}$ is a unit sphere, which appear on rotation surfaces in space forms.)

Recall that *Hessian* of a function on a Riemannian manifold is defined by $\mathrm{Hess}_\varphi\,X = \nabla_X \nabla\varphi$.

Lemma 7. *Let $M = B \times_\varphi \bar{M}$ be the warped product. Then $r = -(\Delta_{\mathcal{F}}\,\varphi/\varphi)\,\hat{g}$, and the following conditions are equivalent:*

K_{mix} *depends on a point on M only* $\Leftrightarrow$ $\mathrm{Hess}_\varphi^{\mathcal{F}} = -\lambda\varphi\,\mathrm{id}_{\mathcal{F}}$ *for some function* $\lambda : \bar{M} \to \mathbb{R}$.

Proof. For a warped product $B \times_\varphi \bar{M}$ we have $h_{\mathcal{F}} = 0$, $T = 0$, and

$$h = -\nabla^{\mathcal{F}}(\log\varphi)\,\hat{g}, \quad H = -n\nabla^{\mathcal{F}}(\log\varphi) \qquad \text{when} \quad \varphi \neq 0.$$

The sectional curvature is $K(X,N) = -\frac{1}{\varphi}\,N(N(\varphi))$ for any unit vectors $N \in \mathcal{D}_{\mathcal{F}}$ and $X \in \mathcal{D}$, when $\varphi \neq 0$, see [9]. Note that $g(\mathrm{Hess}^{\mathcal{F}}_{\varphi}(N), N) = N(N(\varphi))$. Using definitions $\Delta_{\mathcal{F}}\,\varphi = \sum_i E_i(E_i(\varphi))$ and (2) we then get

$$r = -(\Delta_{\mathcal{F}}\,\varphi/\varphi)\,\hat{g}, \quad \mathrm{Ric}_{\mathcal{D}} = -(\Delta_{\mathcal{F}}\,\varphi/\varphi)\,\hat{\mathrm{id}} \qquad (\text{when } \varphi \neq 0).$$

From the above the last claim follows. □

One may apply the existence/uniqueness Theorem 2 to conclude that (5) preserves total umbilicity of foliations with integrable orthogonal distribution. Thus we have the following.

Corollary 5. *Flow* (5) *preserves warped product metrics.*

Let us look at what happens when B has a boundary (e.g., B is a ball in $\mathbb{R}^p$) and $\varphi > 0$ in the interior of B. By the maximum principle, see [3, Sect. 3.73], the problem $\Delta_{\mathcal{F}}\,u = 0$, $u_{|\,\partial B} = 0$ has only zero solution; hence, $\lambda = 0$ is not the eigenvalue. Let $0 < \lambda_1 \leq \ldots \leq \lambda_i \ldots$ be eigenvalues and $\{e_i\}_{1 \leq i < \infty}$ the unit L_2-norm eigenfunctions of the eigenvalue problem $-\Delta_{\mathcal{F}}\,e_i = \lambda_i e_i$ in B and $e_i = 0$ on ∂B. Note that λ_1 has multiplicity 1 and one may assume $e_1 > 0$ in the interior of B.

Assume that $\mu(t,x) := \varphi(t,x)_{|\,\partial B}$ is twice continuously differentiable in t, and there exist limits

$$\lim_{t\to\infty} \mu(t,x) = \tilde{\mu}(x), \quad \lim_{t\to\infty} \partial_t \mu(t,x) = 0, \quad \lim_{t\to\infty} \partial_t^2 \mu(t,\cdot) = 0 \tag{84}$$

for a smooth nonnegative function $\tilde{\mu} : \partial B \to \mathbb{R}$ uniformly with respect to $x \in \partial B$.

We shall study when for the warped product initial metric on $M = B^p \times \bar{M}^n$ the solution of (5) converges to one with leaf-wise constant partial Ricci curvature (see [19] for $p = 1$).

Theorem 8. *Let the warped product metrics* $g_t = \mathrm{dx}^2 + \varphi_t^2(x)\,\bar{g}$ *on* $B^p \times \bar{M}^n$ *solve* (5), *and any of conditions* (i)–(iii) *are satisfied:*

(i) $\Phi < 0$ *and* $(84)_{1,2}$, $\quad (ii)$ $0 \leq \Phi < \lambda_1$, $p < 4$, *and* (84),

(iii) $\Phi = \lambda_1$, $p < 4$, (84) *and*

$$\tilde{\mu} \equiv 0, \quad \int_0^\infty \nu(\tau)\,\mathrm{d}\tau < \infty, \quad \nu(t) := \max\{\|\mu(t,\cdot) - \tilde{\mu}\|_{C^0(\partial B)},\ \|\partial_t \mu(t,\cdot)\|_{C^0(\partial B)}\}. \tag{85}$$

Then g_t *exist for all* $t \geq 0$, *as* $t \to \infty$, g_t *converge in the* C^0*-norm uniformly on* $B \times \bar{M}^n$ *to the limit metric* $g_\infty = \mathrm{dx}^2 + \varphi_\infty^2(x)\,\bar{g}$ *with* $r(g_\infty) = \Phi\,\bar{g}$. *Moreover,* (5) *has a global single point attractor for cases* (i) *and* (ii), *but for* (iii) *the limit metric* g_∞ *depends on initial and boundary conditions.*

Proof. If a family of warped product metrics $g_t = \mathrm{dx}^2 + \varphi^2(t,x)\bar{g}$ solve (5) on M, then, by Lemma 7, $\partial_t(\varphi^2)\,\hat{g} = 2(\Delta_{\mathcal{F}}\,\varphi/\varphi)\,\hat{g} + 2\,\Phi\,\varphi^2\,\hat{g}$. This yields the leaf-wise parabolic Cauchy's problem (with Dirichlet boundary conditions) for the warping function φ,

$$\partial_t \varphi = \Delta_{\mathcal{F}}\,\varphi + \Phi\,\varphi, \quad \varphi(0,\cdot) = \varphi_0, \quad \varphi(t,\cdot)_{|\,\partial B} = \mu(t,\cdot). \tag{86}$$

Linear problem (86) has a unique classical solution $\varphi : [0,\infty) \times B \to \mathbb{R}$. By Lemma 8, φ converges, as $t \to \infty$, to a stationary state, i.e., to a solution $\tilde\varphi : B \to \mathbb{R}$ of the problem

$$-\Delta_{\mathcal{F}}\tilde\varphi = \Phi\,\tilde\varphi, \qquad \tilde\varphi_{|\,\partial B} = \tilde\mu. \quad \square \tag{87}$$

Example 2. For $p = 1$, i.e., $B = [0,l]$ and $\lambda_j = (\pi\, j/l)^2$, the solution of (87) has the form

$$\tilde\varphi(x) = \begin{cases} \frac{\tilde\mu_1 \sin(\sqrt{\Phi}\,x) + \tilde\mu_0 \sin(\sqrt{\Phi}\,(l-x))}{\sin(\sqrt{\Phi}\,l)} & \text{if } 0 < \Phi < \lambda_1, \\ \tilde\mu(0) + (\tilde\mu_1 - \tilde\mu_0)(x/l) & \text{if } \Phi = 0, \\ \frac{\tilde\mu_1 \sinh(\sqrt{-\Phi}\,x) + \tilde\mu_0 \sinh(\sqrt{-\Phi}\,(l-x))}{\sinh(\sqrt{-\Phi}\,l)} & \text{if } \Phi < 0. \end{cases}$$

For $\Phi = \lambda_1$, problem (87) is solvable when $\tilde\mu(l) = \tilde\mu(0) = 0$; in this resonance case, the solutions are $\tilde\varphi(x) = a\sin(\pi x/l) + \tilde\mu(0)\cos(\pi x/l)$, where $a > 0$ is an arbitrary constant.

To formulate Lemma 8, we need some notations. Let $U : [0,\infty)\times B \to \mathbb{R}$ solves the Dirichlet problem on B, where t plays role of a parameter,

$$\Delta_{\mathcal{F}} U = 0, \qquad U_{|\,\partial B} = \mu(t,\,\cdot) - \tilde\mu.$$

Then the function $v(t,x) := \varphi(t,x) - \tilde\varphi(x) - U(t,x)$ on $[0,\infty)\times B$ solves the Cauchy's problem

$$\partial_t v = \Delta_{\mathcal{F}}\, v + \Phi\, v + f, \quad v(0,\,\cdot) = v_0, \quad v(t,\,\cdot)_{|\,\partial B} = 0,$$

where $v_0 = \varphi_0 - \tilde\varphi - U(0,\,\cdot)$ and $f = \Phi U - \partial_t U$. Similar problems can be studied for Neumann boundary conditions. Consider Fourier series $f(t,x) = \sum_{j=1}^{\infty} f_j(t)\, e_j(x)$ and $v_0(x) = \sum_{j=1}^{\infty} v_j^0\, e_j(x)$, where $f_j = \int_B f(\cdot,s)\, e_j(s)\,\mathrm{ds}$ and $v_j^0 = \int_B v_0(s)\, e_j(s)\,\mathrm{ds}$. Recall that $\int_B e_i(s)e_j(s)\,\mathrm{ds} = \delta_{ij}$.

Lemma 8 (see [18]). *Let the function φ on B^p solve* (86) *with $\Phi \le \lambda_1$. If $\Phi > \lambda_1$ then $\varphi(t,x)$ diverges as $t \to \infty$. Otherwise, $\varphi(t,x)$ converges in the C^0-norm uniformly on B to the limit*
(i) $\tilde\varphi$, see (87)*, when $\Phi < 0$ and* $(84)_{1,2}$ *hold,* *(ii) $\tilde\varphi$, when $0 \le \Phi < \lambda_1$, $p < 4$, and* (84) *hold,*
(iii) $\varphi_\infty := (v_1^0 + \int_0^\infty f_1(\tau)\,\mathrm{d}\tau)\, e_1$ when $\Phi = \lambda_1$, $p < 4$, (84) *and conditions* (85) *on ∂B hold.*

For $p = 1$, Lemma 8, cases *(*ii*)* and *(*iii*)* hold under assumption $(84)_{1,2}$ only, see [19].

Corollary 6 (see [19]). *Let the warped product metrics $g_t = \mathrm{dx}^2 + \varphi_t^2(x)\,\bar g$ on $M = [0,l] \times \bar M$ solve* (5) *and* (84) *with $B = [0,l]$ hold. If $\Phi < (\pi/l)^2$ (see* *conditions (i) and (ii) of Theorem 8) then g_t converge uniformly for $x \in [0, l]$ to the limit metric g_∞, whose mixed sectional curvature is Φ.*

The following result will be used in the proof of Corollary 7.

Theorem A ([10]). *Let (B^p, g) be a compact Riemannian manifold with totally geodesic boundary. Assume that there exists a function $\varphi \neq \text{const}$ satisfying $\text{Hess}_\varphi = -k^2\varphi\,\text{id}$ on B, $\varphi = 0$ on ∂B, for a positive number k. Then (B^p, g) is isometric to the upper hemisphere of radius $1/k$ in $\mathbb{R}^{p+1}$.*

Corollary 7. *Let conditions of Theorem 8(iii) hold for (5) with $\Phi = \lambda_1$, $p < 4$, and let (B^p, dx^2) be a hemisphere of radius $\sqrt{p/\lambda_1}$ in $\mathbb{R}^{p+1}$. Then the mixed curvature of the metric g_∞ is constant:*

$$K(N, X) = \Phi/p \qquad \text{(for any unit vectors } N \in \mathcal{D}_\mathcal{F},\ X \in \mathcal{D}). \tag{88}$$

Proof. By Theorem 8 (ii), φ_∞ is proportional to e_1. The Hessian of first eigenfunction e_1 (of Laplacian on a hemisphere of curvature $k > 0$) satisfies the condition $\text{Hess}^\mathcal{F}_{e_1} = -k\,e_1\,\text{id}_\mathcal{F}$, see Theorem A. Hence, $\text{Hess}^\mathcal{F}_\varphi$ is also proportional to the identity mapping. Thus, (88) holds. □

Acknowledgements Supported by the Marie-Curie actions grant EU-FP7-P-2010-RG, No. 276919.

References

1. J. Adams: Vector fields on spheres, *Annals of Math.* 75 (1962), 603–632.
2. B. Andrews B., and C. Hopper: *The Ricci flow in Riemannian geometry*, LNM 2011, Springer, 2011.
3. T. Aubin: *Some nonlinear problems in Riemannian geometry*, Springer, 1998.
4. E. Barletta, S. Dragomir, V. Rovenski, and M. Soret: Mixed gravitational field equations on globally hyperbolic spacetimes, Class. Quantum Grav. 30, (2013), 26 pp.
5. A. Candel, and L. Conlon: *Foliations, I–II*, AMS, Providence, 2000.
6. D. Ferus: Totally geodesic foliations. Math. Ann. 188 (1970), 313–316.
7. E. Ghys: Classification of totally geodesic foliations of codimension one. Comment. Math. Helv., 58, no. 4 (1983) 543–572.
8. B. McKay: The Blaschke conjecture and great circle fibrations of spheres. Amer. J. Math. 126 (2004), no. 5, 1155–1191.
9. R. Ponge, and H. Reckziegel: Twisted products in pseudo-Riemannian geometry, Geom. Dedicata 48 (1993), 15–25.
10. R. Reilly: Applications of the Hessian operator in a Riemannian manifold, Ind. Univ. Math. J. 26, 1977, 459–472.
11. V. Rovenski: *Foliations on Riemannian manifolds and submanifolds*, Birkhäuser, 1998.
12. V. Rovenski: *Foliations, submanifolds and mixed curvature*. J. Math. Sci., New York, 2000, 99 (6), 1699–1787.
13. V. Rovenski: On solutions to equations with partial Ricci curvature, to be published in J. of Geometry and Physics (see also ArXiv:1010.2986 v2.)

14. V. Rovenski: Extrinsic geometric flows on codimension-one foliations, J. of Geom. Analysis, Vol. 23, No. 3 (2013), 1530–1558.
15. V. Rovenski, and P. Walczak: *Topics in extrinsic geometry of codimension-one foliations*, Springer Briefs in Mathematics, Springer-Verlag, 2011.
16. V. Rovenski and R. Wolak: Deforming metrics of foliations, Centr. Eur. J. Math. 11, no. 6, (2013), 1039–1055.
17. V. Rovenski and L. Zelenko: Prescribing the positive mixed scalar curvature of totally geodesic foliations, in "*Foliations-2012*", World Sci., Singapore 2013, pp. 163–203.
18. V. Rovenski and L. Zelenko: Prescribing the mixed scalar curvature of a foliation, pp. 83–123 (This Volume) Proceedings "Geometry and its Applications", Springer 2014.
19. V. Rovenski, and V. Sharafutdinov: The partial Ricci flow on one-dimensional foliations. ArXiv:1308.0985, preprint, 18 pp. 2013.
20. P. Walczak: An integral formula for a Riemannian manifold with two orthogonal complementary distributions. Colloq. Math. 58 (1990), 243–252.

Osculation in General: An Approach

Paweł G. Walczak

Abstract This is an essay about osculation, that is, the tangency of highest order of different types, of different objects (hyperplanes, spheres, cyclides, and others), and arbitrary hypersurfaces.

Keywords Hypersurface • Tangency type • Osculation

Mathematics Subject Classification: Primary 53A07, Secondary 53A04, 53A05

Introduction

Everyone knows what is an osculating plane, circle, or sphere for a curve in the three-dimensional space, or the osculating sphere for a surface. These notions belong to foundations of the classical differential geometry of curves and surfaces, are similar but differ a bit. The dimensions of the osculating object (plane, circle, sphere) and of the object being osculated (curve, surface) are related differently ($\leq$, $\geq$, $=$). The order of tangency varies depending on the case and the way of tangency differs. Either these objects are tangent "globally" (that is, along the whole space tangent to one of them) of highest rank or the direction of tangency diminishes with rank growing. In general, a sphere osculating a surface is tangent (of order one) in all directions but it is tangent of order two just in one particular direction.

In the literature, one can find several articles about other objects (conics, quadrics, helices, Dupin cyclides, etc.) osculating either curves or surfaces ([1, 4, 6, 7, 13] etc.). Upon discussion with colleagues at several universities, we realized that one can imagine completely different and incomparable approaches to the order and

P.G. Walczak (✉)
Wydział Matematyki i Informatyki Uniwersytet Łódzki, Łódź, Poland
e-mail: pawelwal@math.uni.lodz.pl

V. Rovenski and P. Walczak (eds.), *Geometry and its Applications*, Springer Proceedings in Mathematics & Statistics 72, DOI 10.1007/978-3-319-04675-4_7,

type of tangency and to osculation of different objects of arbitrary dimension and codimension. Our goal here is to introduce and discuss one of possible approaches "in general," that is, for an arbitrary submanifold and arbitrary (to some extend) family of submanifolds, candidates for osculating objects, of a given manifold. Our chapter is not a real research paper, it has a character of an essay discussing notions which belong to the so-called *folklore* but can be useful and are rather difficult to find in the literature in large generality.

Since the notion of tangency is local, we shall work with hypersurfaces (of arbitrary dimension and codimension) in $\mathbb{R}^N$ but one can observe that the problem has also a global aspect: the candidates for osculating objects (as spheres, circles, or planes) are global so, one can consider them as global submanifolds of a manifold equipped with a geometric (for example, Riemannian) structure satisfying a given extrinsic geometric property (for example, being totally geodesic, umbilical or so). It seems also that osculating objects of different sorts can be useful for computer graphics or computer-aided geometric design (CADG).

Here, we work at generic points; however, we are aware of the fact that there is a large interest in "super-osculation" at special points which comes from the classical Four Vertex Theorem [12] and several its generalizations ([5, 14], and the bibliographies therein), results on *sextactic* and 3-*extactic* points ([10] and the bibliography therein), and other related problems.

We assume that all our objects are as smooth as needed, say differentiable of class C^{2013} or more, if necessary.

1 Type and Order of Tangency

Let Σ and S be two hypersurfaces of $\mathbb{R}^N$ given respectively by the equation $F = 0$, $F : \mathbb{R}^N \to \mathbb{R}^n$ being a submersion, and by a parametrization ϕ, $S = \phi(\mathbb{R}^k)$, $\phi : \mathbb{R}^k \to \mathbb{R}^N$ being an immersion. (Certainly, one may consider F and ϕ defined locally, on open sets, but our approach does not diminish generality.) Let $f = F \circ \phi$ and $p = \phi(0) \in S \cap \Sigma$.

Consider a sequence $\mathcal{V} = (V_1, V_2, V_3, \dots)$ of linear subspaces V_j of $\mathbb{R}^k$ such that $V_{j+1} \subset V_j$ for all $j \in \mathbb{N}$. Certainly, $V_j = V_{j+1}$ for all j large enough. We shall say that Σ and S are *tangent of type* $\mathcal{V}$ at p whenever $f(0) = 0$,

$$df(0)|V_1 = 0 \quad \text{and} \quad d^j f(0)| \odot^{j-1} V_{j-1} \otimes V_j = 0, \quad \text{for all} \quad j > 1, \tag{1}$$

where $\odot^j V$ denotes the jth symmetric power of a vector space V. Roughly speaking, condition (1) says that all the directional derivatives at 0 of order j in the direction of an arbitrary sequence of j vectors, $j - 1$ of them belonging to V_{j-1} and one to V_j, are equal to 0.

Remark. Condition (1) can be modified to either

$$d^j f(0)| \odot^j V_j = 0, \quad j \in \mathbb{N} \tag{2}$$

or

$$d^j f(0)| \odot^{j-k_j} V_j \otimes \odot^{k_j} V_{j-1} = 0, \quad j \in \mathbb{N}, \tag{3}$$

where $k = (k_1, k_2, \ldots)$ is a sequence of integers satisfying $0 \le k_j \le j$ for all j, providing other approaches to tangency and osculation. These conditions are not discussed here; anyway, one can imagine situations where such types of tangency could be of some interest.

If $r = \max\{j \in \mathbb{N}; V_j \neq \{0\}\}$, then we shall say that the *rank* of the tangency of Σ and S equals r.

Finally, if $d = (d_1, d_2, d_3, \ldots)$ is a nonincreasing sequence of nonnegative integers, then we shall say that Σ and S are *tangent of type* d and *rank* r at p if $r = \max\{j \in \mathbb{N}; d_j > 0\}$ and there exists a system $\mathcal{V}$ as above which satisfies (1) and the equalities $\dim V_j = d_j$ for all j.

Standard results of differential calculus imply directly that the above definitions are correct, that is, independent of the maps F and ϕ describing hypersurfaces Σ and S.

2 Osculation

Let $S = \phi(\mathbb{R}^k)$ be as before while $\mathfrak{S} = \{\Sigma_\lambda, \lambda \in \Lambda\}$, Λ being an open subset of $\mathbb{R}^m$, be a family of hypersurfaces given by $F_\lambda = 0$, $F_\lambda : \mathbb{R}^N \to \mathbb{R}^n$ being submersions. Assume that the family $\mathfrak{S}$ is *smooth*, that is, the map $\mathbb{R}^m \times \mathbb{R}^N \ni (\lambda, x) \mapsto F_\lambda(x) \in \mathbb{R}^n$ is smooth. Fix $p = \phi(0) \in S$ and denote by Λ_p (resp., by $\mathfrak{S}_p$) the family of all $\lambda \in \Lambda$ for which $\Sigma_\lambda \in \mathfrak{S}$ passes through p (resp., the family of all Σ_λ with $\lambda \in \Lambda_p$).

Fix, as before, a decreasing sequence $d = (d_1, d_2, d_3, \ldots)$ of nonnegative integers and denote by r_λ, $\lambda \in \Lambda_p$, the rank of tangency at p of type d of Σ_λ and S. An element Σ_{λ_0} with $\lambda_0 \in \Lambda_p$ and $r_{\lambda_0} = \max\{r_\lambda; \lambda \in \Lambda_p\}$ will be called *d osculating* S at p.

First, let us collect some of the simplest examples of objects osculating curves. In this case, the only possible types of tangencies are of the form $d(r) = (d_1, d_2, \ldots)$, where $d_j = 1$ for $j \le r$ and $d_j = 0$ for all $j > r$, and r is a fixed natural number.

Example 1.

(1) Generically, given a curve $\Gamma : s \mapsto \gamma(s)$ in $\mathbb{R}^N$, the affine r-dimensional, $r < N$, hyperplane P through $\gamma(s)$ spanned by the vectors $\gamma'(s), \ldots, \gamma^{(r)}(s)$ is tangent to Γ of order r and there is no r-hyperplane tangent to Γ of higher order, so P is $d(r)$-osculating in our sense.

(2) Since the dimension of the space of all the $(N-1)$-dimensional spheres in $\mathbb{R}^N$ passing through a given point p_0 equals N (a sphere like that is uniquely determined by its center), the best order of tangency of such a sphere with a generic curve equals N as well, and N is the order of tangency of the $(N-1)$-sphere osculating a curve. More generally, consider the space of all $(r-1)$-spheres in $\mathbb{R}^N$ passing through p_0; its dimension equals $r(N-r+1)$ (indeed, every such sphere is uniquely determined by the affine $(m+1)$-hyperplane H containing the sphere and a point of H (the center), and the dimension of the Grassmannian G_r^N equals $r(N-r)$) and the system (1) consists of $r \cdot (N-r)$ equations when we think about the tangency of order r. Therefore, the $(r-1)$-sphere osculating a generic curve should have order of tangency r. In particular, circles osculating generic curves in $\mathbb{R}^N$ are tangent of order 2 for any N.

(3) Consider now the space $\mathfrak{H}$ of all helices in $\mathbb{R}^3$ passing through a given point. A helix like that is determined by the position of the cylinder containing it, its position on the cylinder and two constants, and its curvature and torsion. Therefore, the dimension of our space $\mathfrak{H}$ equals 5. For $d = d(r)$, system of equations (1) contains $2r$ members; therefore, one can expect that generically the best order of tangency of a helix and a curve is only 2, the same as for osculating circles even if the space of helices is definitely richer than that of circles. However, one may expect the existence of a one-parameter family of osculating helices and, in fact, this is the case: this fact was known already to Olivier [13] in the first half of nineteenth century.

(4) If $\mathfrak{S}_0^s$ is the space of all planar algebraic curves of degree s passing through the origin, then $\dim \mathfrak{S}_0^s = (s^2 + 3s - 2)/2$ (the dimension of the space of all polynomials P of two variables, of degree s, with $P(0,0) = 0$, modulo multiplication by constants), system (1) reduces—after suitable normalization of the coefficients of the polynomials—to a linear system of r, r being the order of tangency under consideration, and equations with $(s^2 + 3s - 2)/2$ variables which for a generic curve through the origin has solutions whenever $r \leq (s^2 + 3s - 2)/2$ and the solution is unique when one has equality in the preceding inequality. This means that $r = (s^2 + 3s - 2)/2$ is, generically, the order of tangency of the algebraic curve of degree s osculating a planar curve. In particular, for $s = 1$ we get $r = 1$, the order of tangency of the tangent line, while for $s = 2$ we get $r = 4$, the order of tangency of an osculating conic.

Now, let us turn our attention to surfaces. In this case, the only possible types of tangency are $d(r_1, r_2) = (d_1, d_2, \ldots)$, where $d_j = 2$ for $1 \leq j \leq r_1$, $d_j = 1$ for $r_1 < d_j \leq r_1 + r_2$ and $d_j = 0$ for all $j > r_1 + r_2$.

Example 2.

(1) If $S = \phi(U)$, $U \subset \mathbb{R}^2$, is a surface is $\mathbb{R}^N$ with parametrization ϕ as in Sect. 1, then, given $u \in \mathbb{R}^k$, any hyperplane H spanned by all the derivatives $(\partial^{i+j}\phi/(\partial^i x_1 \partial^j x_2))(u)$ with $i + j \leq r_1$ and, for example, $(\partial^k \phi/\partial^k x_1)(u)$, $r_1 < k \leq r_1 + r_2$ (and possibly other vectors of $\mathbb{R}^N$) is tangent to S of

type $d(r_1, r_2)$ at $p = \phi(u)$ and $\dim H = r_2 + (r_1^2 + 3r_1)/2$; if $N - n = r_2 + (r_1^2 + 3r_1)/2$, then—given a one-dimensional subspace of the plane tangent to S at p—the $(N - n)$-hyperplane H_0 spanned exactly by all these derivatives is the unique one tangent to S of type $d(r_1, r_2)$, no better tangency is possible in this dimension; so, generically one has a one-dimensional family of $(N - n)$-dimensional hyperplanes osculating S in this type. For example, if $N > 6$, then one can find such a family of six-dimensional hyperplanes osculating S of types either $d(1, 4)$ or $d(2, 1)$.

(2) Consider a surface S in $\mathbb{R}^N$ and the space of all $(N - 1)$-dimensional spheres passing through a given point, say the origin o, of S.

(2.1) In the best known case of $N = 3$, condition (1) for degree 1 implies obviously that the center of any sphere tangent to S at o lies on the normal line through o; then one can observe that generically tangency of type $d(2, 0)$ is impossible to obtain but that of type $d(1, 1)$ may happen when the space (here, line) $V_2 \subset T_oS$ is chosen in the direction of the eigenvalue of the Weingarten operator $A = -\nabla N$, N being a unit normal. And, generically, this is optimal, that is, the order of tangency of osculating spheres of type $d(1, 1)$ is 2. Observe also that generic surfaces intersect their osculating spheres along curves, so they are $(2, 1, 1, 1, \ldots)$-tangent in the sense of (2) to these spheres and the rank of this type of osculation is infinite.

(2.2) If $N = 4$, then, for any direction ν orthogonal to the surface at the reference point one has the corresponding Weingarten operator A^ν, its eigenvalues, eigenvectors, and principal directions. Generically, all the corresponding osculating spheres are tangent to S of type $d(1, 1)$ and, therefore, a generic surface has at a generic point a one-parameter family of osculating spheres of this type.

(2.3) Next, if $N = 5$, then condition (1) for $d = (2, 2, 0 \ldots)$ reduces to a system of five (in fact, linear) equations in five variables and generically has a unique solution while a higher order of tangency cannot occur; therefore, a generic surface in $\mathbb{R}^5$ has the unique osculating sphere of type $d(2, 0)$. In the same way, a generic surface S in $\mathbb{R}^N$, where $N = k(k + 3)/2$ and $k \geq 2$, admits, at a generic point p, the sphere which $d(k, 0)$-osculates S at p.

(3) Keep S as before and take as $\mathfrak{S}$ the space of all the circles passing through a given point, say o. Each circle Σ can be expressed as the intersection of $N - 1$ spheres Σ_i, $i = 1, \ldots, N - 1$ of dimension $N - 1$. Expressing S locally in the form

$$x_j = a_j x_1^2 + b_j x_1 x_2 + c_j x_2^2 + d_j x_1^3 + e_j x_1^2 x_2 + f_j x_1 x_2^2 + g_j x_2^3 + H.O.T., \qquad j = 3, \ldots, N, \tag{4}$$

(where "H. O. T." reads as "higher order terms") and substituting the right hand side of the above into the equations

$$\sum_{i=1}^{N}(x_i^2 - 2\mu_i^j x_i) = 0, \quad j = 1, \dots, N-1, \tag{5}$$

we learn that generically there are exactly three directions (called, according to our knowledge, *Laguerre directions*) $x_2 = t_i x_1$, $i \leq 3$, for which S and Σ are tangent of order $d(3) = (1, 1, 1, 0, 0, \dots)$ for suitable choice of coordinates μ^i_j of centers of the spheres Σ_j. Better tangency of a circle and a sphere is, generically, impossible, so order of tangency of a circle osculating (in our sense) a surface in $\mathbb{R}^N$ is 3 and does not depend on N.

(4) Again, keep S as before with $N = 3$ and take as $\mathfrak{S}$ the space of all the *Dupin cyclides*, that is, conformal images of tori, cylinders, and cones of revolution, passing through the origin. By Fialkov results (see [9] and [2]), given a nonumbilical point p of S, S can be mapped by a unique Möbius transformation (mapping p to the origin o) to the position

$$z = \frac{1}{2}(x^2 - y^2) + \frac{1}{6}(\theta_1 x^3 + \theta_2 y^3) + \frac{1}{24}(ax^4 + 4bx^3y + 6\Psi x^2y^2 + 4cxy^3 + dy^4) + H.O.T., \tag{6}$$

where θ_1 and θ_2 are conformal principal curvatures of S, coefficients a, b, c, d depend on θ_i's and their derivatives, and Ψ is another local conformal invariant determining, together with θ_i's and two conformally invariant 1-forms ω_1, ω_2, the conformal type of S. For a Dupin cyclid C, (6) reduces to

$$z = \frac{1}{2}(x^2 - y^2) + \frac{1}{8}(x^4 - y^4) + \frac{1}{6}\Psi_C x^2 y^2 + H.O.T., \tag{7}$$

Ψ_C being the corresponding conformal invariant (at o) of C. Comparing (6) and (7) one can observe (see [1]) that generically (at a nonumbilical point where one of the conformal principal curvatures, say θ_2, is different from 0) a surface and the Dupin cyclid can be strongly tangent of type $d(2, 2)$ with $V_1 = T_o S$, the space $V_2 \subset T_o S$ being determined by the condition $y = tx$ with $t = -\sqrt[3]{\theta_1/\theta_2}$ and ψ_C being suitably chosen (to satisfy the relation $\frac{1}{24}(a + 4bt + 6\Psi t^2 + 4ct^3 + dt^4) = \frac{1}{8}(1 - t^4) + \frac{1}{6}\Psi_C t^2$). Again, a better tangency cannot be expected; so, the Dupin cyclid described above can be considered as the one which osculates S at p in our sense.

3 Existence

Given a sequence $d = (d_1, d_2, \dots)$ as before, we shall denote by $\mathrm{Gr}(l, d)$ the space of all the sequences $\mathcal{V} = (V_1, V_2, \dots)$ of linear subspaces of $\mathbb{R}^l$ such that $V_{j+1} \subset V_j$

and $\dim V_j = d_j$ for all j. Certainly, $\mathrm{Gr}(l,d)$ has a natural structure of a manifold. Its dimension $D(l,d)$ equals

$$D(l,d) = d_1(l - d_1) + d_2(d_1 - d_2) + d_3(d_2 - d_3) + \dots. \tag{8}$$

On the other hand, due to the symmetry of mixed derivatives and the well-known equality

$$\dim \odot^j V = \binom{d+j-1}{j} \quad \text{when } \dim V = d,$$

in this situation, the number $\tilde{D}(d)$ of independent equations in system (1) equals

$$\begin{aligned}\tilde{D}(d) &= d_1 + (2d_1 - d_2 + 1)d_2/2 \\ &\quad + \sum_{j=3}^{s} \left(\binom{d_{j-1}+j-1}{j} - \binom{d_{j-1}-d_j+j-1}{j} \right),\end{aligned} \tag{9}$$

where $s = \max\{j; d_j > 0\}$.

Consider S and $\mathfrak{S}$ as in Sect. 2 and let $l = \min(N - n, k)$.

Certainly, if the family $\mathfrak{S}$ is small, one cannot expect existence of members of $\mathfrak{S}$ tangent of order r to an arbitrary hypersurface S. Therefore, we will consider families which are "large enough" in the following sense: $\mathfrak{S}$ is called (d,r)-*complete* whenever the map

$$\begin{aligned}\Lambda \times \mathbb{R}^N \times \mathrm{Gr}(l,d) &\ni (\lambda, p, \mathcal{V}) \\ &\mapsto (F_\lambda(p), dF_\lambda(p)|V_1, \dots, d^r F_\lambda| \otimes^{r-1} V_{r-1} \otimes V_r)\end{aligned} \tag{10}$$

is surjective: its values are sequences of symmetric multilinear maps which can be identified with elements of suitable powers of $\mathbb{R}^n$ and, taken all together, constitute elements of the space $\mathbb{R}^{n(1+\tilde{D}(d))}$. Certainly, such situation may occur only when

$$m + N + D(l,d) \geq n(1 + \tilde{D}(d)). \tag{11}$$

Set

$$\Delta(d,r) = m + D(l,d) - n \cdot (1 + \tilde{D}(d(r))), \tag{12}$$

where $d(r) = (d_1, d_2, \dots, d_r, 0, 0, \dots)$. With this terminology, using standard Explicit Function Theorem, we can summarize our discussion in the following.

Proposition 1. *If the family $\mathfrak{S}$ is (d,r)-complete and $\Delta(d,r) \geq 0$ while $\Delta(d,r+1) < 0$, then for a generic S and $p \in S$ there exists a $\Delta(d,r)$-dimensional space of elements of $\mathfrak{S}$ which d-osculate S at p.*

One can verify that this general observation agrees with what we said in Examples 1 and 2. For instance,

- In Example 1(1) one has $n = N - r$, $m = (r+1)(N-r)$ and $\tilde{D}(d(r)) = r$, so $\Delta(d(r), r) = 0$,
- In the case of $(r-1)$-dimensional spheres considered in Example 1(2) one has $m = (N - r + 1)(r + 1)$ and $\Delta(d(r), r) = 0$ as before,
- In Example 2(1) one has $m = n(N - n + 1)$, and $D(2, d) = 1$ and $\tilde{D}(d) = r_2 + (r_1^2 + 3r_1)/2$ when $d = d(r_1, r_2) = (2, \dots, 2, 1, \dots, 1, 0, 0, \dots)$, so $\Delta(d, r) = 1$ when $N - n = r_2 + (r_1^2 + 2r_1)/2$ and $r = r_1 + r_2$,

and so on.

4 Final Remarks

Certainly, one can imagine and consider a variety of situations different from those described in Examples 1 and 2. Let us comment briefly about some of them, those which are currently of some interest for us.

4.1 Dupin Cyclides and Curves

Analogously to the situation studied in [1] and described here in Example 2 (4), one can search for Dupin cyclides osculating curves in $\mathbb{R}^3$. Since every Dupin cyclid is characterized up to a Möbius transformation by the value of Ψ, the dimension of the Möbius group in $\mathbb{R}^3$ equals 10 and for a Dupin cyclid C there exists a two-parameter family of Dupin transformations preserving C, the dimension of the space of all the Dupin cylides equals 9 and the expected order of best tangency of a Dupin cyclid and a curve equals 8. At the moment, we are not able to describe the Dupin cyclid osculating a surface as we have done for surfaces. This should be possible if one could calculate more terms in the canonical equation of a curve:

$$\begin{aligned} y &= x^3/6 + (2Q - T^2)x^5/120 + H.O.T., \\ z &= Tx^4/24 + T'x^5/(120\sqrt{\nu}) + H.O.T., \end{aligned} \tag{13}$$

where $\nu = \sqrt{(\kappa')^2 + \kappa^2\tau^2}$ while κ and τ are, respectively, the standard curvature and torsion of our curve. Analogously to what we said in Example 2 (4) about surfaces, a generic curve Γ together with a generic point $p \in \Gamma$ can be transformed to the position (13) by a unique Möbius transformation sending p to the origin o, see [8] and [2]. In (13), the quantities Q and T are called, respectively, the *conformal curvature* and the *conformal torsion* of Γ. The quantities Q and T can be expressed in terms of the curvature κ and torsion τ of Γ, and their derivatives, for example

$$T = (2(\kappa')^2\tau + \kappa^2\tau^3 + \kappa\kappa'\tau' - \kappa\kappa''\tau)/\nu^{5/2}. \tag{14}$$

Comparing (14) either with classical textbooks on differential geometry of curves or, for instance, with [15], on can see that a generic curve is spherical if and only if its torsion T vanishes identically and this holds when the osculating sphere along the curve is constant. A similar result should hold for curves on Dupin cyclides: one should be able to find a condition expressed in terms of Q, T and, perhaps, their derivatives equivalent to "*Dupin cyclidity*" of a curve. Curves located on Dupin cylides should be interesting for CAGD.

4.2 Algebraic Varieties

One should be able to generalize Example 1 (4) to the case of algebraic varieties of arbitrary degree. Given a finite system $s = (s_1, \dots, s_n)$ of natural numbers, one can consider the space $\mathfrak{S}(s)$ of all algebraic varieties given by

$$P_1 = 0, \dots, P_n = 0, \tag{15}$$

P_i being a polynomial of degree s_i of N real variables. Calculating the dimension of the space $\mathfrak{S}(s)$ seems to be an exercise in combinatorics and should allow to find the degree and dimension of the space of varieties from $\mathfrak{S}(s)$ which d-osculate a given hypersurface $S \subset \mathbb{R}^N$. Certainly, there is a problem of regularity of the space of solutions of (15) at the point of tangency to S. A complex version of this situation can be also considered.

4.3 Isoparametric Hypersurfaces

Any Dupin cyclid is conformally equivalent to a surface with constant principal curvatures in a three-dimensional space form $M^3(c)$. Such a surface provides an example of an isoparametric hypersurface: a codimension-1 hypersurface $\Sigma \subset M^N(c)$ is *isoparametric* whenever its principal curvatures are constant. Such hypersurfaces were studied already by Elli Cartan in 1930s but still are of great interest, see, among the others, [3, 11] and the bibliographies therein. Therefore, given a space form $M^N(c)$, it seems to be interesting to consider the space $\mathfrak{S}(l, s)$, $s = (s_1, \dots, s_l)$ of all its isoparametric hypersurfaces which have l distinct principal curvatures of multiplicities $s_1, \dots, s_l$ and, given an arbitrary submanifold $S \subset M^N(c)$ search for osculating elements of $\mathfrak{S}(l, s)$. For $l = 1$ the problem reduces to the one for umbilical hypersurfaces, in fact spheres, and this was discussed to some extent in our examples of Sect. 2.

One can generalize this problem once again replacing isoparametric hypersurfaces by isoparametric submanifolds of arbitrary (fixed) codimension: a submanifold N of $M^N(c)$ is said to be *isoparametric* whenever its normal bundle νN is flat and its principal curvatures corresponding to parallel sections of νN are constant.

Acknowledgements Supported by the Polish NSC grant 6065/B/H03/2011/40.

References

1. A. Bartoszek, P. Walczak, Sz. Walczak, *Dupin cyclides osculating surfaces*, preprint, 2012.
2. G. Cairns, R. W. Sharpe and L. Webb. *Conformal invariants for curves in three dimensional space forms*, Rocky Mountain J. Math. **24** (1994), 933–959.
3. T. E. Cecil, Q. Chi, and G. R. Jensen, *Isoparametric hypersurfaces with four principal curvatures*, Ann. of Math. **166** (2007), 1–76.
4. G. Darboux, *Sur le contact des courbes et des surfaces*, Bull. sci. math. et astr., **4** (1880), 348–384.
5. D. DeTurck, H. Gluck, D. Pomerleano, D. S. Vick, *The Four Vertex Theorem and Its Converse*, Notices of the Amer. Math. Soc. **54** (2007), 192–207.
6. J. Drach, *Sur les lignes d'osculation quadrique des surfaces*, C. R. Acad. Sci. Paris **224** (1947), 309–312.
7. J. Drach, *Détermination des lignes d'osculation quadrique (lignes de Darboux) sur les surfaces cubiques. Lignes asymptotiques de la surface de Bioche*, C. R. Acad. Sci. Paris **226** (1948), 1561–1564.
8. A. Fialkov, *The conformal geometry of curves*, Trans. Amer. Math. Soc. **51** (1942), 435–501.
9. A. Fialkov, *Conformal differential geometry of a subspace*, Trans. Amer. Math. Soc. **56** (1944), 309–433.
10. D. Fuchs, S. Tabachnikov, *Mathematical Omnibus*, Amer. Math.Soc., Providence 2007.
11. R. Miyaoka, *Isoparametric hypersurfaces with (g,m)=(6,2)*, Ann. of Math. **177** (2013), 53–110.
12. S. Mukhopadhyaya, *New methods in the geometry of a plane arc*, Bull. Calcutta Math. Soc. **1** (1909), 21–27.
13. T. Olivier, J. École Polytechnique, **15** (1835), 61–91, 252–263.
14. R. Uribe-Vargas, *Four-vertex theorems in higher-dimensional spaces for a larger class of curves than theconvex ones*, C. R. Acad. Sci. Paris St'er. I Math. **330** (2000), 1085–1090.
15. Y.-C. Wong, *On an explicit charcterization of spherical curves*, Proc. Amer. Math. Soc. **34** (1972), 239–242.

Stability of Left-Invariant Totally Geodesic Unit Vector Fields on Three-Dimensional Lie Groups

Alexander Yampolsky

Abstract We consider the problem of stability or instability of unit vector fields on three-dimensional Lie groups with left-invariant metric which have totally geodesic image in the unit tangent bundle with the Sasaki metric with respect to classical variations of volume. We prove that among non-flat groups only $SO(3)$ of constant curvature $+1$ admits stable totally geodesic submanifolds of this kind. Restricting the variations to left-invariant (i.e., equidistant) ones, we give a complete list of groups which admit stable/unstable unit vector fields with totally geodesic image.

Mathematics Subject Classifications (2010): 53C40, 53B21

Introduction

Let (M, g) be a Riemannian manifold and ξ a unit tangent vector field on M. Then ξ can be considered as a local or global (if exists) immersion $\xi : M \to T_1(M)$ into the unit tangent bundle. The Sasaki metric $\tilde{g}$ on $T(M)$ gives rise to the metric on $T_1(M)$ and hence on $\xi(M)$. In this way $(\xi(M), \tilde{g})$ gets definite intrinsic and extrinsic geometry. Particularly, a unit vector field is said to be minimal or totally geodesic if $\xi(M)$ is a minimal or totally geodesic submanifold in $(T_1(M), \tilde{g})$. From the variation theory viewpoint, a minimal unit vector field is a stationary point of the first local normal variation of the volume functional of $\xi(M)$. In other words, ξ is a *minimal unit vector field* if the mean curvature vector of $\xi(M) \subset (T_1(M), \tilde{g})$

A. Yampolsky (✉)
Department of Geometry, Faculty of Mechanics and Mathematics, Kharkiv Karazin National University, Svobody Sq. 4, 61022, Kharkiv, Ukraine
e-mail: a.yampolsky@karazin.ua

V. Rovenski and P. Walczak (eds.), *Geometry and its Applications*, Springer Proceedings in Mathematics & Statistics 72, DOI 10.1007/978-3-319-04675-4_8,

vanishes; ξ is a *totally geodesic unit vector field* if all the second fundamental forms of $\xi(M) \subset (T_1(M), \tilde{g})$ vanish. We refer to this kind of minimality as *the classical.*

A different type of volume variations and hence the minimality for a given unit vector field was proposed in [12] and developed in [9,10]. Denote by $\mathfrak{X}^1(M)$ a space of all smooth unit vector fields on M. The variation of ξ within $\mathfrak{X}^1(M)$ gives rise to variation of $\xi(M)$ and hence the volume functional $\mathrm{Vol}_\xi : \mathfrak{X}^1(M) \to \mathbb{R}$. We call this type of variations *the field variations.* A unit vector field ξ is called *minimal*, if ξ is a stationary point of the latter functional. It was proved that this definition of minimality is equivalent to the classical one, i.e. minimal unit vector field gives rise to minimal immersion $\xi : M \to T_1(M)$. The minimality condition in a meaning of [10] was expressed in terms of a special 1-form. A number of examples of minimal unit vector fields by using this 1-form [3–5, 8, 13, 14] (the list is not complete) was constructed. In the case of three-dimensional Lie group G with the left-invariant metric, K. Tsukada and L. Vanhecke managed to find a list of all *minimal* left-invariant unit vector fields [17]. It was proved that each minimal left-invariant unit vector field on three-dimensional unimodular Lie group is an eigenvector of the Ricci operator.

A. Borisenko [1] was the first who asked on unit vector fields with *totally geodesic image* in the unit tangent bundle of Riemannian manifold. The author solved the problem in two-dimensional case [19] and has extracted the subclass of totally geodesic fields on three-dimensional Lie groups by equalizing to zero the whole second fundamental form [21]. As a result, it was proved that each totally geodesic left-invariant unit vector field on three-dimensional unimodular Lie group is the unit eigenvector of the Ricci operator of G with eigenvalue 2, if exists.

The *second variation formula* for the $\xi(M)$-volume functional with respect to the field variation was obtained in [11] and is very complicated to handle with. That is why only little number of results concerning stability/instability are known. Particularly, a minimal unit vector field on two-dimensional Riemannian manifold is always stable with respect to the field variations [11]. In application to the Hopf vector field on the unit 3-sphere, it was also proved that it is minimal and stable [11]. Remark that the Hopf vector field is a totally geodesic one as well as the unit characteristic vector field of the Sasakian structure [18].

On the other hand, there is a well-known formula for the second variation of volume [16] which allows to check stability/instability of minimal submanifold in the Riemannian space with respect to local (or global, if admissible) normal variations of the submanifold. We refer to this kind of stability as *classical.* This kind of stability/instability is different from the one considered in [11] because the normal variation of the $\xi(M)$ gives rise to the wider class of the field variations. Namely, the variation field can be non-orthogonal to ξ.

In some cases, the normal variation of the minimal submanifold $\xi(M) \subset T_1(M)$ is probably equivalent to the field variation of minimal unit vector field. The case of totally geodesic left-invariant unit vector field on the three-dimensional Lie group with the left-invariant metric gives a corresponding example. In [14], the authors tried to check stability/instability of left-invariant unit vector fields from Tsukada-Vanhecke list [17]. They have constructed the left-invariant variations of minimal unit left-invariant vector field on compact quotient of unimodular three-dimensional Lie groups which produce instability with respect to the field variations.

In this chapter, we check the list of all totally geodesic left-invariant unit vector fields on three-dimensional Lie group G with the left-invariant metric and provide stability or instability conditions for them with respect to classical normal variations of domains in $\xi(G) \subset T_1(G)$. In the case of unimodular groups, we conduct a complete proof for their compact quotients for the sake of simplicity.

The main result (Theorem 2.2) says that *only* S^3 *of constant curvature* $+1$ *admits classically stable totally geodesic left-invariant unit vector field.* We also give a list of left invariant *totally geodesic* unit vector fields on unimodular three-dimensional Lie groups with the left-invariant metric which are stable/unstable with respect to classical *left-invariant* variations (Theorem 2.4).

1 Preliminaries

The definition of the Sasaki metric is based on the bundle projection differential $\pi_* : TT(M) \to T(M)$ and the connection map $\mathcal{K} : TT(M) \to T(M)$ [7]. For any $\tilde{X}, \tilde{Y} \in T_{(q,\xi)}T(M)$, we have

$$\tilde{g}(\tilde{X}, \tilde{Y}) = g(\pi_* \tilde{X}, \pi_* \tilde{Y}) + g(\mathcal{K}\tilde{X}, \mathcal{K}\tilde{Y}).$$

By definition, the *vertical* distribution $\mathcal{V}_{(q,\xi)} = \ker \pi_*$ and the *horizontal* one $\mathcal{H}_{(q,\xi)} = \ker \mathcal{K}$. Then $T_{(q,\xi)}T(M) = \mathcal{V}_{(q,\xi)} \oplus \mathcal{H}_{(q,\xi)}$ and the horizontal and vertical distributions are mutually orthogonal with respect to $\tilde{g}$.

The *horizontal and vertical lifts* of a vector field X on the base are defined as the unique vector fields X^h and X^v on $T(M)$ such that

$$\pi_* X^h = X,\ \pi_* X^v = 0,$$
$$\mathcal{K}X^h = 0, \quad \mathcal{K}X^v = X.$$

The h and v lifts of a tangent frame on M form a lifted frame on $T(M)$. As concerns the unit tangent bundle, the lifted frame on $T_1(M)$ at $(q,\xi) \in T_1(M)$ is formed by h lift and the *tangential lift* [2] of the frame on M. The latter is defined by

$$X^{tan} = X^v - g(X,\xi)\xi^v.$$

Evidently, $X^{tan} = X^v$ for all X from the orthogonal complement of the "vector part" of a point (q,ξ). *We use this fact without special comments.*

Denote by $\mathfrak{X}(M)$ the Lie algebra of smooth vector fields on M and by $\mathfrak{X}_{\xi^\perp}(M)$ the orthogonal complement of a unit vector field ξ in $\mathfrak{X}(M)$. If ξ is a *unit vector field* on M, then it can be considered as a mapping $\xi : M \to T_1(M)$. Then its differential ξ_* sends a vector field $X \in \mathfrak{X}(M)$ into $T\xi(M)$, by [20],

$$\xi_* X = X^h + (\nabla_X \xi)^{tan} = X^h + (\nabla_X \xi)^v,$$

where ∇ means the Riemannian connection of (M, g).

In what follows, we use the notion of the *Nomizu operator* $A_\xi : \mathfrak{X}(M) \to \mathfrak{X}_{\xi\perp}(M)$ given by

$$A_\xi X = -\nabla_X \xi.$$

Denote by A^t_ξ a conjugate Nomizu operator defined by $g(A_\xi X, Y) = g(X, A^t_\xi Y)$. Then one can define the tangent $\xi_* : \mathfrak{X}(M) \to T\xi(M)$ and the normal $\nu : \mathfrak{X}(M) \to T^\perp\xi(M)$ mappings by

$$\begin{aligned} \xi_*(X) &= X^h - (A_\xi X)^{tan} = X^h - (A_\xi X)^v, \\ \nu(Y) &= (A^t_\xi Y)^h + Y^{tan}. \end{aligned} \tag{1}$$

Then there are local orthonormal frames $(e_1, \dots, e_n) \in \mathfrak{X}(M)$ and $(f_1, \dots, f_{n-1}) \in \mathfrak{X}_{\xi\perp}$ such that

$$A_\xi e_i = \sigma_i f_i, \quad A^t_\xi f_i = \sigma_i e_i,$$

where $\sigma_i \geq 0$ are *the singular values* of the linear operator A_ξ. In fact, e_i are the eigenvectors of the symmetric linear operator $A^t_\xi A_\xi$ and its eigenvalues are the squares of the singular values.

By dimension reasons, there is at least one local unit vector field e_0 such that $A_\xi e_0 = 0$. Then

$$\begin{aligned} \tilde e_\alpha &= \tfrac{\xi_*(e_\alpha)}{|\xi_*(e_\alpha)|} = \tfrac{1}{\sqrt{1+\sigma_\alpha^2}}(e^h_\alpha - \sigma_\alpha f^v_\alpha),\ \tilde e_n = e^h_0, \\ \tilde n_\alpha &= \tfrac{\nu(f_\alpha)}{|\nu(f_\alpha)|} = \tfrac{1}{\sqrt{1+\sigma_\alpha^2}}(\sigma_\alpha e^h_\alpha + f^v_\alpha) \quad \alpha = 1, \dots, n-1 \end{aligned} \tag{2}$$

form the tangent and normal framing over $\xi(M) \subset T_1(M)$. We call this framing the *singular* one. If ξ is a geodesic unit vector field, i.e., $A_\xi \xi = 0$, then one can always put $\tilde e_n = \xi^h$.

Let $\tilde n = \frac{\nu(Z)}{|\nu(Z)|}$ be a unit normal vector field on $\xi(M)$ and $F \subset M$ be a domain with a compact closure. Denote by $\tilde N = w\tilde n$ a local normal variation vector field, where $w : F \to \mathbb{R}$ is a smooth function such that $w|_{\partial \bar F} = 0$. Suppose $\xi(M)$ is minimal. Then the formula for second variation of the volume in application to our case takes the form

$$\delta^2(\mathrm{Vol}_\xi) = \int_{\xi(F)} \Big(||\tilde\nabla^\perp \tilde N||^2 - (\widetilde{\mathrm{Ric}}(\tilde N) + ||\tilde S_{\tilde N}||^2) \Big) dV_\xi,$$

where $\tilde\nabla^\perp$ means the covariant derivative in the normal bundle of $\xi(M)$, $\widetilde{\mathrm{Ric}}(\tilde N)$ is the partial Ricci curvature and $\tilde S$ is the shape operator of $\xi(M)$.

In the case of *compact orientable* M and *totally geodesic* $\xi(M)$, the formula takes a simpler form, namely

$$\delta^2 \mathrm{Vol}_\xi = \int_{\xi(M)} \sum_{i=1}^{n} \left(||\tilde{\nabla}^{\perp}_{\tilde{e}_i} \tilde{N}||^2 - w^2 \tilde{K}(\tilde{e}_i, \tilde{n}) \right) dV_\xi.$$

Finally remark that

$$dV_\xi = \sqrt{\det(I + A_\xi^t A_\xi)}\, dV := L^{1/2}\, dV,$$

where dV is the volume element of the base manifold. That is why one can rewrite the formula of the second variation as follows

$$\delta^2 \mathrm{Vol}_\xi = \int_M \sum_{i=1}^{n} \left(||\tilde{\nabla}^{\perp}_{\tilde{e}_i} \tilde{N}||^2 - w^2 \tilde{K}(\tilde{e}_i, \tilde{n}) \right) L^{1/2}\, dV. \tag{3}$$

In next sections, we simplify this formula in the case of three-dimensional Lie groups with the left-invariant metric.

2 Three-dimensional Unimodular Lie Groups with the Left-Invariant Metric

Let ξ be a unit left-invariant vector field on the three-dimensional Lie group G with the left-invariant Riemannian metric. The group G is unimodular if and only if there is a discrete subgroup Γ acting on G by left translations free and properly discontinuous such that the left quotient $\Gamma \backslash G$ is compact [15]. The $\Gamma \backslash G$ is a compact Riemannian manifold with the same curvature properties as G. The descended unit vector field has the same properties concerning minimality, harmonicity, etc. as the one on G [14].

For each three-dimensional unimodular Lie group G, there is an orthonormal frame e_1, e_2, e_3 such that [15]

$$[e_2, e_3] = \lambda_1 e_1, \quad [e_3, e_1] = \lambda_2 e_2, \quad [e_1, e_2] = \lambda_3 e_3. \tag{4}$$

We will refer to this frame as to the *canonical* one. This frame consists of *eigenvectors* of the Ricci curvature operator. Each frame vector field is a *Killing* one and hence *geodesic*. The *Levi-Civita connection* coefficients on G can be easily found, namely, with $\mu_i = \frac{1}{2}(\lambda_1 + \lambda_2 + \lambda_3) - \lambda_i$, the frame covariant derivatives take the form $\nabla_{e_i} e_k = \mu_i\, e_i \times e_k$. It is also well known that the *principal Ricci curvatures* are $\rho_i = 2\mu_j \mu_k$ and *the basic sectional curvatures* are $k_{ij} := g(R(e_i, e_j)e_j, e_i) = \frac{1}{2}(\rho_i + \rho_j - \rho_k)$, where i, j, k are all different.

The constants $\lambda_1, \lambda_2, \lambda_3$ define the topological structure of G in the following sense:

Signs of $\lambda_1, \lambda_2, \lambda_3$	Associated Lie group
$+, +, +$	$SO(3)$
$+, +, -$	$SL(2, \mathbb{R})$
$+, +, 0$	$E(2)$
$+, 0, -$	$E(1, 1)$
$+, 0, 0$	Nil^3 (Heisenberg group)
$0, 0, 0$	$\mathbb{R} \oplus \mathbb{R} \oplus \mathbb{R}$

The class of left-invariant totally geodesic unit vector fields on three-dimensional unimodular Lie group G can be described as the eigenvectors of the Ricci operator associated with the eigenvalue 2, if exists [21]. Namely, where $\mathcal{S}$ stands for vector

ρ_1	ρ_2	ρ_3	μ_1	μ_2	μ_3	ξ
0	0	0	0	0	0	$\mathcal{S}$
0	0	0	$\neq 0$	0	0	$\pm e_1, \cos t\, e_2 + \sin t\, e_3$
0	0	0	0	$\neq 0$	0	$\pm e_2, \cos t\, e_1 + \sin t\, e_3$
0	0	0	0	0	$\neq 0$	$\pm e_3, \cos t\, e_1 + \sin t\, e_2$
	2					$\pm e_2$
		2				$\pm e_3$
2	2					$\cos t\, e_1 + \sin t\, e_2$
2		2				$\cos t\, e_1 + \sin t\, e_3$
	2	2				$\cos t\, e_2 + \sin t\, e_3$
2	2	2				$\mathcal{S}$

fields of the form $\xi = \cos t \cos s\, e_1 + \cos t \sin s\, e_2 + \sin t\, e_3$ with fixed parameters t and s. Analysis of the above table yields the following result [21].

Theorem 2.1. *Let G be a three-dimensional unimodular Lie group with a left-invariant metric. Let $\{e_i, i = 1, 2, 3\}$ be the canonical frame of its Lie algebra. Set for definiteness $\lambda_1 \geq \lambda_2 \geq \lambda_3$. Then the totally geodesic left-invariant unit vector fields on (a compact quotient of) G are the following:*

G or $\Gamma\backslash G$	*Conditions on* $\lambda_1, \lambda_2, \lambda_3$	ξ
SO(3)	$\lambda_1 = \lambda_2 = \lambda_3 = 2$	$\cos t \cos s\, e_1 + \cos t \sin s\, e_2 + \sin t\, e_3$
	$\lambda_1 = \lambda_2 = \lambda > \lambda_3 = 2$	$\pm e_3$
	$\lambda_1 = \lambda_2 = \lambda > 2 > \lambda_3 = \lambda - \sqrt{\lambda^2 - 4}$	$\cos t\, e_1 + \sin t\, e_2$
	$\lambda_1 = 2 > \lambda_2 = \lambda_3 = \lambda > 0$	$\pm e_1$
	$\lambda_1 = \lambda + \sqrt{\lambda^2 - 4} > \lambda = \lambda_2 = \lambda_3 > 2$	$\cos t\, e_2 + \sin t\, e_3$
	$\lambda_1 > \lambda_2 > \lambda_3 > 0$, $\lambda_m^2 - (\lambda_i - \lambda_k)^2 = 4$	$\pm e_m$ *(i,k,m=1,2,3)*
$SL(2, R)$	$\lambda_3^2 - (\lambda_1 - \lambda_2)^2 = 4$	$\pm e_3$
	$\lambda_1^2 - (\lambda_2 - \lambda_3)^2 = 4$	$\pm e_1$
$E(2)$	$\lambda_1 = \lambda_2 > 0,\ \lambda_3 = 0$	$\pm e_3$, $\cos t\, e_1 + \sin t\, e_2$
	$\lambda_1^2 - \lambda_2^2 = 4, \lambda_1 > \lambda_2 > 0,$ $\lambda_3 = 0$	$\pm e_1$
$E(1,1)$	$\lambda_3^2 - \lambda_1^2 = 4,$ $\lambda_1 > 0, \lambda_2 = 0,\ \lambda_3 < 0$	$\pm e_3$
	$\lambda_1^2 - \lambda_3^2 = 4,$ $\lambda_1 > 0, \lambda_2 = 0,\ \lambda_3 < 0$	$\pm e_1$
Nil^3	$\lambda_1 = 2,\ \lambda_2 = 0, \lambda_3 = 0$	$\pm e_1$
$R \oplus R \oplus R$	$\lambda_1 = \lambda_2 = \lambda_3 = 0$	$\cos t \cos s\, e_1 + \cos t \sin s\, e_2 + \sin t\, e_3$

where t and s are arbitrary fixed parameters.

For any left invariant vector field $\xi = x_1 e_1 + x_2 e_2 + x_3 e_3$, we have $\nabla_{e_i}\xi = \mu_i\, e_i \times \xi$ and as a consequence, with respect to the canonical frame, we have

$$A_\xi = \begin{pmatrix} 0 & -\mu_2 x_3 & \mu_3 x_2 \\ \mu_1 x_3 & 0 & -\mu_3 x_1 \\ -\mu_1 x_2 & \mu_2 x_1 & 0 \end{pmatrix}. \tag{5}$$

To calculate the integrand in (3), we need some Lemmas.

Lemma 2.1. *Let $\xi := e_m$ be a totally geodesic left-invariant unit vector field on (compact quotient of) unimodular three-dimensional Lie group G with a left-invariant metric. Then the normal bundle connection coefficients of $\xi(G)$ with respect to framing* (1) *are*

$$\tilde{\gamma}^i_{j|s} = -\frac{1}{2} k_{ij} \delta_{sm}, \qquad (i < j) \neq m,$$

where k_{ij} means the sectional curvature of G along $e_i \wedge e_j$.

Proof. We will conduct the proof for $\xi = e_3$. Observe that since ξ is supposed totally geodesic, the principal Ricci curvature $\rho_3 = 2$ and hence $\mu_1\mu_2 = \frac{1}{2}\rho_3 = 1$. From (5) we get

$$A_\xi = \begin{pmatrix} 0 & -\mu_2 & 0 \\ \mu_1 & 0 & 0 \\ 0 & 0 & 0 \end{pmatrix}, \quad A_\xi^t = \begin{pmatrix} 0 & \mu_1 & 0 \\ -\mu_2 & 0 & 0 \\ 0 & 0 & 0 \end{pmatrix}, \quad A_\xi^t A_\xi = \begin{pmatrix} \mu_1^2 & 0 & 0 \\ 0 & \mu_2^2 & 0 \\ 0 & 0 & 0 \end{pmatrix}.$$

Hence, the e_1, e_2 can be taken as the vectors of singular frame. Since

$$A_\xi e_1 = \mu_1 e_2, \quad A_\xi e_2 = -\mu_2 e_1,$$

we may put $\sigma_1 = \mu_1$, $\sigma_2 = \mu_2$ and take $f_1 = e_2$, $f_2 = -e_1$. Then the framing (2) takes the form

$$\tilde{e}_1 = \left(\frac{1}{\sqrt{1+\mu_1^2}}e_1\right)^h - \left(\frac{\mu_1}{\sqrt{1+\mu_1^2}}e_2\right)^v, \quad \tilde{e}_2 = \left(\frac{1}{\sqrt{1+\mu_2^2}}e_2\right)^h + \left(\frac{\mu_2}{\sqrt{1+\mu_2^2}}e_1\right)^v,$$

$$\tilde{e}_3 = (e_3)^h \tag{6}$$

$$\tilde{n}_1 = \left(\frac{\mu_1}{\sqrt{1+\mu_1^2}}e_1\right)^h + \left(\frac{1}{\sqrt{1+\mu_1^2}}e_2\right)^v, \quad \tilde{n}_2 = \left(\frac{\mu_2}{\sqrt{1+\mu_2^2}}e_2\right)^h - \left(\frac{1}{\sqrt{1+\mu_2^2}}e_1\right)^v. \tag{7}$$

Recall, that $\tilde{\gamma}^i_{j|s} := \tilde{g}(\tilde{\nabla}_{\tilde{e}_s}\tilde{n}_j, \tilde{n}_i)$ and in our case we only need to calculate $\tilde{\gamma}^2_{1|s}$. To do this we use Kowalski-type formulas from [2], namely

$$\tilde{\nabla}_{X^h}Y^h = (\nabla_X Y)^h - \tfrac{1}{2}(R(X,Y)\xi)^{tan}, \quad \tilde{\nabla}_{X^h}Y^{tan} = (\nabla_X Y)^{tan} + \tfrac{1}{2}(R(\xi,Y)X)^h,$$

$$\tilde{\nabla}_{X^{tan}}Y^h = \tfrac{1}{2}(R(\xi,X)Y)^h, \qquad \tilde{\nabla}_{X^{tan}}Y^{tan} = -g(Y,\xi)X^{tan}.$$

Then

$$\tilde{\nabla}_{X_1^h+X_2^{tan}}(Y_1^h+Y_2^{tan}) = \Big(\nabla_{X_1}Y_1 + \frac{1}{2}R(\xi,Y_2)X_1 + \frac{1}{2}R(\xi,X_2)Y_1\Big)^h + \Big(\nabla_{X_1}Y_2 - \frac{1}{2}R(X_1,Y_1)\xi - g(Y_2,\xi)X_2\Big)^{tan}$$

Straightforward calculations show that the curvature tensor components are of the form

	e_1	e_2	e_3
$R(e_1,e_2)\bullet$	$-k_{12}e_2$	$k_{12}e_1$	0
$R(e_1,e_3)\bullet$	$-k_{13}e_3$	0	$k_{13}e_1$
$R(e_2,e_3)\bullet$	0	$-k_{23}e_3$	$k_{23}e_2$

where $k_{ij} = \frac{1}{2}(\rho_i + \rho_j - \rho_m)$ $(i \neq j \neq m \neq i)$ are basic sectional curvatures.

Using this formulas, we obtain easily $\tilde{\nabla}_{\tilde{e}_1}\tilde{n}_1 = ((*)e_3)^{tan} = 0$, $\tilde{\nabla}_{\tilde{e}_2}\tilde{n}_1 = (*)e_3^h$ and hence $\tilde{\gamma}^2_{1|1} = \tilde{\gamma}^2_{1|2} = 0$. Finally,

$$\tilde{\nabla}_{\tilde{e}_3}\tilde{n}_1 = \frac{1}{2}\frac{k_{12}}{\sqrt{1+\mu_1^2}}e_2^h - \frac{1}{2}\frac{2\mu_3 - \mu_1 k_{13}}{\sqrt{1+\mu_1^2}}\,e_1^v.$$

As $\mu_1\mu_2 = 1$, we can simplify

$$\begin{aligned}\frac{2\mu_3 - \mu_1 k_{13}}{\sqrt{1+\mu_1^2}} &= \frac{2\mu_3 - \mu_1(\mu_1\mu_2 + \mu_2\mu_3 - \mu_1\mu_3)}{\sqrt{1+\mu_1^2}}\\ &= \frac{2\mu_3 - (\mu_1 + \mu_3 - \mu_1^2\mu_3)}{\sqrt{1+\mu_1^2}}\\ &= \frac{\mu_3 + \mu_1^2\mu_3 - \mu_1}{\sqrt{1+\mu_1^2}} = \frac{\mu_2\mu_3 + \mu_1\mu_3 - \mu_2\mu_3}{\mu_2\sqrt{1+\mu_1^2}} = \frac{k_{12}}{\sqrt{1+\mu_2^2}},\end{aligned}$$

$$\frac{k_{12}}{\sqrt{1+\mu_1^2}} = \frac{\mu_2 k_{12}}{\sqrt{1+\mu_2^2}}.$$

So, we have

$$\tilde{\nabla}_{\tilde{e}_3}\tilde{n}_1 = \frac{1}{2}k_{12}\tilde{n}_2,$$

hence $\tilde{\gamma}^2_{1|3} = \frac{1}{2}k_{12}$. In the cases of $\xi = e_1$ and $\xi = e_2$ the calculations are similar. □

The partial Ricci curvature $\widetilde{\mathrm{Ric}}(\tilde{N}) = w^2\sum_{i=1}^n \tilde{K}(\tilde{e}_i, \tilde{n})$, where $\tilde{e}_i$ are the vectors of orthonormal frame tangent to $\xi(M)$, can be calculated by using the formula for the sectional curvature of $T_1(M)$. Namely, if $\tilde{X} = X_1^h + X_2^{tan}$ and $\tilde{Y} = Y_1^h + Y_2^{tan}$ are orthonormal, then [6]

$$\begin{aligned}\tilde{K}(\tilde{X},\tilde{Y}) = &\langle R(X_1,Y_1)Y_1, X_1\rangle - \frac{3}{4}\|R(X_1,Y_1)\xi\|^2\\ &+ \frac{1}{4}\|R(\xi,Y_2')X_1 + R(\xi,X_2')Y_1\|^2 + 3\langle R(X_1,Y_1)Y_2', X_2'\rangle\\ &- \langle R(\xi,X_2')X_1, R(\xi,Y_2')Y_1\rangle + \|X_2'\|^2\|Y_2'\|^2 - \langle X_2', Y_2'\rangle^2\\ &+ \langle(\nabla_{X_1}R)(\xi,Y_2')Y_1, X_1\rangle + \langle(\nabla_{Y_1}R)(\xi,X_2')X_1, Y_1\rangle,\end{aligned} \tag{8}$$

where $X_2' = X_2 - g(X_2, \xi)\xi$, $Y_2' = Y_2 - g(Y_2, \xi)\xi$, R and ∇ are the curvature tensor and Riemannian connection of the base manifold (M, g) respectively.

So, to find the partial Ricci curvature of $\xi(G)$, we need the covariant derivatives of the curvature tensor. One can find them by standard calculations.

Lemma 2.2. *Let (e_1, e_2, e_3) be the canonical left-invariant frame on (compact quotient of) a three-dimensional unimodular Lie group with a left-invariant metric. Then the covariant derivatives of the curvature tensor are of the form*

	$(\nabla_\bullet R)(e_1, e_2)e_1$	$(\nabla_\bullet R)(e_1, e_2)e_2$	$(\nabla_\bullet R)(e_1, e_2)e_3$
e_1	$\mu_1(\rho_3 - \rho_2)e_3$	0	$-\mu_1(\rho_3 - \rho_2)e_1$
e_2	0	$\mu_2(\rho_3 - \rho_1)e_3$	$-\mu_2(\rho_3 - \rho_1)e_2$
e_3	0	0	0

	$(\nabla_\bullet R)(e_1, e_3)e_1$	$(\nabla_\bullet R)(e_1, e_3)e_2$	$(\nabla_\bullet R)(e_1, e_3)e_3$
e_1	$\mu_1(\rho_3 - \rho_2)e_2$	$-\mu_1(\rho_3 - \rho_2)e_1$	0
e_2	0	0	0
e_3	0	$\mu_3(\rho_2 - \rho_1)e_3$	$-\mu_3(\rho_2 - \rho_1)e_2$

	$(\nabla_\bullet R)(e_2, e_3)e_1$	$(\nabla_\bullet R)(e_2, e_3)e_2$	$(\nabla_\bullet R)(e_2, e_3)e_3$
e_1	0	0	0
e_2	$\mu_2(\rho_3 - \rho_1)e_2$	$-\mu_2(\rho_3 - \rho_1)e_1$	0
e_3	$\mu_3(\rho_2 - \rho_1)e_3$	0	$-\mu_3(\rho_2 - \rho_1)e_1$

where ρ_i are the principal Ricci curvatures and μ_i are the connection coefficients.

Now we can calculate the partial Ricci curvature with respect to arbitrary normal vector field for totally geodesic $\xi(G)$.

Lemma 2.3. *Let $\xi = e_m$ be a totally geodesic unit vector field on a (compact quotient of) three-dimensional unimodular Lie group G with a left-invariant metric. The partial Ricci curvature of $\xi(G)$ with respect to arbitrary normal vector field $\tilde{N} = h_i\tilde{n}_i + h_j\tilde{n}_j$ $(i \neq j \neq m \neq i)$ is of the form*

$$\widetilde{\mathrm{Ric}}(\tilde{N}) = \frac{1}{4}k_{ij}(h_i^2 + h_j^2) + \left(1 - \frac{\rho_j^2}{4}\right)h_i^2 + \left(1 - \frac{\rho_i^2}{4}\right)h_j^2 ,$$

where k_{ij} is a basic $e_i \wedge e_j$ sectional curvature and ρ_i are the principal Ricci curvatures.

Proof. We will conduct the proof for $\xi = e_3$, since the other cases are similar. Take the $\xi(G)$ tangent and normal framing according to (6) and (7). Then the arbitrary normal vector field $\tilde{N}$ can be expressed by

$$\tilde{N} = \left(\frac{h_1\mu_1}{\sqrt{1+\mu_1^2}}e_1 + \frac{h_2\mu_2}{\sqrt{1+\mu_2^2}}e_2\right)^h + \left(-\frac{h_2}{\sqrt{1+\mu_2^2}}e_1 + \frac{h_1}{\sqrt{1+\mu_1^2}}e_2\right)^v.$$

Observe that if $\tilde{X}$ is of unit length and orthogonal to $\tilde{Y}$, then $|\tilde{Y}|^2\tilde{K}(\tilde{X},\tilde{Y})$ could be calculated by (8) assuming that Y_1 and Y_2 are the components of the non-normalized vector. Keeping this, put

$$Y_1 = \frac{h_1\mu_1}{\sqrt{1+\mu_1^2}}e_1 + \frac{h_2\mu_2}{\sqrt{1+\mu_2^2}}e_2, \quad Y_2 = -\frac{h_2}{\sqrt{1+\mu_2^2}}e_1 + \frac{h_1}{\sqrt{1+\mu_1^2}}e_2.$$

To calculate $\tilde{K}(\tilde{e}_1,\tilde{N})$, put

$$X_1 = \frac{1}{\sqrt{1+\mu_1^2}}e_1, \qquad X_2 = \frac{-\mu_1}{\sqrt{1+\mu_1^2}}e_2.$$

The MAPLE calculations yield:

$$\begin{aligned}
\langle R(X_1,Y_1)Y_1,X_1\rangle &= \frac{\mu_2^2k_{12}}{(1+\mu_1^2)(1+\mu_2^2)}h_2^2\Big|_{\mu_1\mu_2=1}\\
&= \frac{\mu_1^2\mu_3+\mu_3-\mu_1}{\mu_1(1+\mu_1^2)^2}h_2^2,
\end{aligned}$$

$$||R(X_1,Y_1)\xi||^2 = 0,$$

$$\begin{aligned}
||R(\xi,Y_2)X_1 + R(\xi,X_2)Y_1||^2 &= \frac{(k_{13}+\mu_1\mu_2k_{23})^2}{(1+\mu_1^2)(1+\mu_2^2)}h_2^2\Big|_{\mu_1\mu_2=1}\\
&= \frac{4\mu_1^2}{(1+\mu_1^2)^2}h_2^2,
\end{aligned}$$

$$\begin{aligned}
\langle R(X_1,Y_1)Y_2,X_2\rangle &= -\frac{\mu_1\mu_2k_{12}}{(1+\mu_1^2)(1+\mu_2^2)}h_2^2\Big|_{\mu_1\mu_2=1}\\
&= -\frac{\mu_1(\mu_1^2\mu_3+\mu_3-\mu_1)}{(1+\mu_1^2)^2}h_2^2,
\end{aligned}$$

$$\langle R(\xi,X_2)X_1, R(\xi,Y_2)Y_1\rangle = 0,$$

$$\|X_2\|^2\|Y_2\|^2 - \langle X_2, Y_2\rangle^2 = \frac{\mu_1^2}{(1+\mu_1^2)(1+\mu_2^2)} h_2^2\Big|_{\mu_1\mu_2=1}$$
$$= \frac{\mu_1^4}{(1+\mu_1^2)^2} h_2^2,$$

$$\langle(\nabla_{X_1} R)(\xi, Y_2)Y_1, X_1\rangle = \frac{\rho_3(\rho_2-\rho_3)}{2(1+\mu_1^2)(1+\mu_2^2)} h_2^2\Big|_{\rho_3=2,\mu_1\mu_2=1}$$
$$= \frac{2\mu_1^2(\mu_1\mu_3-1)}{(1+\mu_1^2)^2} h_2^2,$$

$$\langle(\nabla_{Y_1} R)(\xi, X_2)X_1, Y_1\rangle = -\frac{\mu_2^2\rho_3(\rho_1-\rho_3)}{2(1+\mu_1^2)(1+\mu_2^2)} h_2^2\Big|_{\rho_3=2,\mu_1\mu_2=1}$$
$$= \frac{2(\mu_1-\mu_3)}{\mu_1(1+\mu_1^2)^2} h_2^2.$$

After substitution into (8), we get

$$|\tilde N|^2 \tilde K(\tilde e_1, \tilde N) = \left(1 - \frac{1}{2}\rho_1\right) h_2^2.$$

After similar calculations, one can find

$$|\tilde N|^2 \tilde K(\tilde e_2, \tilde N) = \left(1 - \frac{1}{2}\rho_2\right) h_1^2,$$
$$|\tilde N|^2 \tilde K(\tilde e_3, \tilde N) = \frac{1}{4} k_{12}^2 (h_1^2 + h_2^2) + \left(\frac{\rho_2}{2} - \frac{\rho_2^2}{4}\right) h_1^2 + \left(\frac{\rho_1}{2} - \frac{\rho_1^2}{4}\right) h_2^2.$$

It follows then

$$\widetilde{\mathrm{Ric}}(\tilde N) = \frac{1}{4} k_{12}^2 (h_1^2 + h_2^2) + \left(1 - \frac{\rho_2^2}{4}\right) h_1^2 + \left(1 - \frac{\rho_1^2}{4}\right) h_2^2,$$

which completes the proof. □

The following Lemma is the principal one.

Lemma 2.4. *Let $\xi = e_m$ be a totaly geodesic left-invariant unit vector field on a three-dimensional non-flat compact quotient of a unimodular Lie group with a left-invariant metric. Then the integrand in the second volume variation formula* (3) *can be reduced to*

$$W(h,h) := \left[\frac{e_i(h_i)^2}{1+\mu_i^2} - \frac{2k_{ij}}{\lambda_m} e_i(h_i)e_j(h_j) + \frac{e_j(h_j)^2}{1+\mu_j^2} + \frac{e_i(h_j)^2}{1+\mu_i^2} + \frac{2k_{ij}}{\lambda_m} e_i(h_j)e_j(h_i) \right.$$
$$\left. + \frac{e_j(h_i)^2}{1+\mu_j^2} + e_m(h_i)^2 + e_m(h_j)^2 + \left(\frac{\rho_j^2}{4} - 1\right)h_i^2 + \left(\frac{\rho_i^2}{4} - 1\right)h_j^2 \right] |\lambda_m|, \quad (9)$$

where $i \neq j \neq m \neq i$, ρ_i and ρ_j are the principal Ricci curvatures, k_{ij} are the basic sectional curvatures of G and h_i are the variation functions.

Proof. We will conduct the calculations for the case $m = 3$. First of all observe that $\xi = e_m$ is the unit Ricci eigenvector of eigenvalue $\rho_3 = 2$, which means that $\mu_1\mu_2 = 1$ and hence

$$L = \det(I + A_\xi^t A_\xi) = 1 + \mu_1^2 + \mu_2^2 + \mu_1^2\mu_2^2 = 2 + \mu_1^2 + \mu_2^2 = (\mu_1 + \mu_2)^2 = \lambda_3^2.$$

Therefore, dV_ξ is a constant multiple of dV, namely $dV_\xi = |\lambda_3| dV$. Take the $\xi(G)$ tangent and normal framing according to (6) and (7). Put $\tilde{N} = h_1\tilde{n}_1 + h_2\tilde{n}_2$. To calculate $|\tilde{\nabla}^{\perp}_{\tilde{e}_i}\tilde{N}|^2$, observe that

$$\begin{aligned}
\tilde{\nabla}^{\perp}_{\tilde{e}_i}\tilde{N} &= \langle\langle \tilde{\nabla}_{\tilde{e}_i}\tilde{N}, \tilde{n}_1\rangle\rangle\, \tilde{n}_1 + \langle\langle \tilde{\nabla}_{\tilde{e}_i}\tilde{N}, \tilde{n}_2\rangle\rangle\, \tilde{n}_2 \\
&= \tilde{e}_i(h_1)\tilde{n}_1 + \tilde{e}_i(h_2)\tilde{n}_2 + h_2\langle\langle \tilde{\nabla}_{\tilde{e}_i}\tilde{n}_2, \tilde{n}_1\rangle\rangle\, \tilde{n}_1 \\
&\quad + h_1\langle\langle \tilde{\nabla}_{\tilde{e}_i}\tilde{n}_1, \tilde{n}_2\rangle\rangle\, \tilde{n}_2 \\
&= \tilde{e}_i(h_1)\tilde{n}_1 + \tilde{e}_i(h_2)\tilde{n}_2 + h_2\tilde{\gamma}^1_{2|i}\tilde{n}_1 + h_1\tilde{\gamma}^2_{1|i}\tilde{n}_2.
\end{aligned}$$

By Lemma 2.1, we have

$$\tilde{\nabla}^{\perp}_{\tilde{e}_1}\tilde{N} = \tilde{e}_1(h_1)\tilde{n}_1 + \tilde{e}_1(h_2)\tilde{n}_2, \quad \tilde{\nabla}^{\perp}_{\tilde{e}_2}\tilde{N} = \tilde{e}_2(h_1)\tilde{n}_1 + \tilde{e}_2(h_2)\tilde{n}_2,$$
$$\tilde{\nabla}^{\perp}_{\tilde{e}_3}\tilde{N} = \left(\tilde{e}_3(h_1) - \frac{1}{2}k_{12}h_2\right)\tilde{n}_1 + \left(\tilde{e}_3(h_2) + \frac{1}{2}k_{12}h_1\right)\tilde{n}_2.$$

Therefore

$$\sum_{i=1}^{3} ||\tilde{\nabla}^{\perp}_{\tilde{e}_i}\tilde{N}||^2 = \sum_{i=1}^{3}\left(\tilde{e}_i(h_1)^2 + \tilde{e}_i(h_2)^2\right) + k_{12}\left(\tilde{e}_3(h_2)h_1 - \tilde{e}_3(h_1)h_2\right) + \frac{1}{4}k_{12}^2(h_1^2 + h_2^2).$$

Since h_i are the functions on the base manifold, we have $\tilde{e}_\alpha(h_\sigma) = \frac{1}{\sqrt{1+\mu_\alpha^2}} e_\alpha(h_\sigma)$ and $\tilde{e}_3(h_\sigma) = e_3(h_\sigma)$, where $(\alpha, \sigma = 1, 2)$. Hence,

$$\sum_{i=1}^{3} ||\tilde{\nabla}^{\perp}_{\tilde{e}_i} \tilde{N}||^2 = \sum_{\alpha=1}^{2} \frac{1}{1+\mu_\alpha^2}\big(e_\alpha(h_1)^2 + e_\alpha(h_2)^2\big)$$
$$+ e_3(h_1)^2 + e_3(h_2)^2 + k_{12}\big(e_3(h_2)h_1 - e_3(h_1)h_2\big)$$
$$+ \frac{1}{4}k_{12}^2(h_1^2 + h_2^2).$$

Since G is compact, by the divergence theorem

$$\int_G div(X)dV = 0$$

for any vector field X. For $X = h_1h_2e_3$, we have

$$div(h_1h_2e_3) = g(\text{grad}(h_1h_2), e_3) = e_3(h_1)h_2 + e_3(h_2)h_1$$

and hence

$$\int_G \Big(e_3(h_2)h_1 - e_3(h_1)h_2\Big)dV = 2\int_G e_3(h_2)h_1dV.$$

Analyzing the table in Theorem 2.1 one can observe that in all cases (except $E(2)$ and T^3 with flat metric) the totaly geodesic e_i corresponds to $\lambda_i \neq 0$. Therefore, we can continue as

$$2\int_{G/\Gamma} e_3(h_2)h_1dV = \frac{2}{\lambda_3}\int_G [e_1, e_2](h_2)h_1\, dV.$$

Expand

$$h_1[e_1, e_2](h_2) = h_1e_1(e_2(h_2)) - h_1e_2(e_1(h_2))$$
$$= e_1(h_1e_2(h_2)) - e_1(h_1)e_2(h_2) - e_2(h_1e_1(h_2)) + e_2(h_1)e_1(h_2).$$

Since G is compact and boundaryless, after applying the Stokes formula we get

$$\int_G (e_3(h_2)h_1 - e_3(h_1)h_2)dV = \frac{2}{\lambda_3}\int_G (e_2(h_1)e_1(h_2) - e_1(h_1)e_2(h_2))dV.$$

Hence,

$$\int_{\xi(G)} \sum_{i=1}^{3} ||\tilde{\nabla}^{\perp}_{\tilde{e}_i} \tilde{N}||^2 dV_\xi = \int_G \Big(\frac{e_1(h_1)^2}{1+\mu_1^2} - \frac{2k_{12}}{\lambda_3}e_1(h_1)e_2(h_2) + \frac{e_2(h_2)^2}{1+\mu_2^2} + \frac{e_1(h_2)^2}{1+\mu_1^2}$$
$$+ \frac{2k_{12}}{\lambda_3}e_1(h_2)e_2(h_1) + \frac{e_2(h_1)^2}{1+\mu_2^2} + e_3(h_1)^2 + e_3(h_2)^2$$
$$+ \frac{1}{4}k_{12}^2(h_1^2 + h_2^2)\Big)|\lambda_3|dV.$$

Taking into account the result of Lemma 2.3, we obtain

$$\delta^2\mathrm{Vol}_\xi = \int_G \left(\frac{e_1(h_1)^2}{1+\mu_1^2} - \frac{2k_{12}}{\lambda_3} e_1(h_1)e_2(h_2) + \frac{e_2(h_2)^2}{1+\mu_2^2} + \frac{e_1(h_2)^2}{1+\mu_1^2} + \frac{2k_{12}}{\lambda_3} e_1(h_2)e_2(h_1)\right.$$
$$\left. + \frac{e_2(h_1)^2}{1+\mu_2^2} + e_3(h_1)^2 + e_3(h_2)^2 + \left(\frac{\rho_2^2}{4} - 1\right) h_1^2 + \left(\frac{\rho_1^2}{4} - 1\right) h_2^2\right) |\lambda_3|\, dV.$$

The other cases can be treated in a similar way. □

Remark 1. It is worthwhile to mention that if $\mu_1 = \mu_2 = \mu_3 = 1$, then $\rho_1 = \rho_2 = \rho_3 = 2$ and the integrand (9) up to multiple 2 is the same as in [11] obtained for the Hopf vector field on $S^3(1)$. In this case, we deal with $SO(3)$ of constant curvature $+1$ which is isometric to $S^3(1)$. The left-invariant unit vector field corresponds the Hopf vector field on $S^3(1)$. So we can conclude that in this case the second variation of volume with respect to the field variation is equal to a half of classical second variation of volume. Therefore, the Hopf vector field is stable with respect to both types of variations. The stability the Hopf vector field with respect to the field variations was proved in [11] and in [18] for the classical treatment.

From Lemma 2.4 we immediately conclude the following.

Theorem 2.2. *Let ξ be a left-invariant unit vector field on a compact quotient of a non-flat three-dimensional unimodular Lie group G with a left-invariant metric. Then $\xi(\Gamma\backslash G)$ is a stable totally geodesic submanifold in $T_1(\Gamma\backslash G)$ if and only if $G = SO(3)$ of constant curvature $+1$ and ξ is arbitrary left-invariant.*

Proof. Let $\xi = e_m$ be totally geodesic. Then $\rho_m = 2$ and to be left-invariant stable, the other Ricci curvatures must satisfy

$$|\rho_i| \geq 2 \quad \text{or, equivalently,} \quad |\mu_m \mu_j| \geq 1$$

and

$$|\rho_j| \geq 2 \quad \text{or, equivalently,} \quad |\mu_i \mu_m| \geq 1.$$

To be generally stable, both quadratic expressions involving derivatives must be positively semi-definite. The latter condition is equivalent to

$$\frac{k_{ij}^2}{\lambda_m^2} \leq \frac{1}{(1+\mu_i^2)(1+\mu_j^2)}.$$

Since $\rho_m = 2\mu_i\mu_j = 2$, we have $(1+\mu_i^2)(1+\mu_j^2) = (2+\mu_i^2+\mu_j^2) = (\mu_i+\mu_j)^2 = \lambda_m^2$. As a result, $|k_{ij}| \leq 1$. Observe that $k_{ij} = \mu_i\mu_m + \mu_m\mu_j - 1$ and hence the equality $|k_{ij}| \leq 1$ is equivalent to

$$0 \leq \mu_m(\mu_i + \mu_j) \leq 2 \quad \text{or} \quad 0 \leq \frac{\mu_m}{\mu_i}(1+\mu_i^2) \leq 2 \quad \text{or} \quad 0 \leq \frac{\mu_m}{\mu_i} \leq \frac{2}{1+\mu_i^2}.$$

Evidently, all connection coefficients have to be of the same sign. Therefore, the classical stability take place if

$$\mu_m\mu_i \geq 1, \quad \frac{\mu_m}{\mu_i} \geq 1, \quad 0 \leq \frac{\mu_m}{\mu_i} \leq \frac{2}{1+\mu_i^2}.$$

The possible solutions of the system satisfy $\mu_1 = \mu_2 = \mu_3 = \pm 1$. Taking into account the signs of λ_i, we obtain a unique solution $\mu_1 = \mu_2 = \mu_3 = 1$ which means that the base manifold is $SO(3)$ of constant curvature $+1$ and hence ξ is arbitrary left-invariant.

If the system is inconsistent, then the totally geodesic submanifold $\xi(G)$ is unstable. Indeed, if say $\rho_i < 2$, then in the case of compact quotient one can take $h_i = 0$, $h_m = 0$, and $h_j = const \neq 0$ and we get $W(h,h) < 0$ over whole compact quotient. If $\rho_i \geq 2$ and $\rho_j \geq 2$ but $|k_{ij}| > 1$, then both quadratic expressions that involve derivatives of h_i and h_j in $W(h,h)$ are not positively semi-definite. By taking $h_3 = 0$ and h_1, h_2 sufficiently small with derivatives making the quadratic expressions negative, we obtain negative $W(h,h)$ at least over some domain $F \subset \Gamma\backslash G$. □

The proof of Lemma 2.4 essentially uses non-flatness of the group. If the group is flat, then the second classical variation of volume for the unit vector field with totally geodesic image is much simpler.

Theorem 2.3. *Let ξ be a unit vector field on a compact quotient of a flat three-dimensional unimodular Lie group G with a left-invariant metric. Then*

- *If $G = E(2)$ and ξ is a parallel unit vector field on $E(2)$, then $\xi(\Gamma\backslash G)$ is a stable totally geodesic submanifold;*
- *If $G = E(2)$ and ξ is in integrable distribution orthogonal to the parallel vector field on $E(2)$, then $\xi(\Gamma\backslash G)$ is an unstable totally geodesic submanifold;*
- *If $G = R \oplus R \oplus R$ and is arbitrary left-invariant, then $\xi(T^3)$ is a stable totally geodesic submanifold.*

Proof. We have flat $E(2)$ if $\lambda_1 = \lambda_2 = a > \lambda_3 = 0$. In this case $\mu_1 = 0, \mu_2 = 0, \mu_3 = a$ and $\rho_1 = \rho_2 = \rho_3 = 0$. The field $\xi = e_3$ is the field of unit normals of the integrable orthogonal distribution $\xi^\perp$. In this case, $\widetilde{\mathrm{Ric}}(\tilde N) = 0$ and we have $\xi(G)$ **stable** totally geodesic submanifold in $T_1(G)$.

As concerns the field $\xi = \cos t e_1 + \sin t e_2$, rotating the frame in $e_1 \wedge e_2$ plane, we may always put $\xi = e_1$ without loss of generality. Then

$$A_\xi = \begin{pmatrix} 0 & 0 & 0 \\ 0 & 0 & -a \\ 0 & 0 & 0 \end{pmatrix}, \quad A_\xi^t = \begin{pmatrix} 0 & 0 & 0 \\ 0 & 0 & 0 \\ 0 & -a & 0 \end{pmatrix}.$$

The tangent frame consists of

$$\tilde e_1 = e_1^h, \quad \tilde e_2 = e_2^h, \quad \tilde e_3 = \frac{1}{\sqrt{1+a^2}}(e_3^h + a e_2^v).$$

The normal frame on $\xi(G)$ consists of

$$\tilde{n}_2 = \frac{1}{\sqrt{1+a^2}}(-ae_3^h + e_2^v), \quad \tilde{n}_3 = e_3^v.$$

For the field of normal variation $\tilde{N} = h_2\tilde{n}_2 + h_3\tilde{n}_3$, we have

$$\widetilde{\mathrm{Ric}}(\tilde{N}) = \frac{a^2}{1+a^2}h_3^2.$$

The normal connection of $\xi(G)$ is flat and hence, by choosing the variation with constant h_1 and h_2, we get

$$W(h,h) = -\frac{a^2}{1+a^2}h_3^2,$$

which means that $\xi(G)$ is **unstable** totally geodesic submanifold in $T_1(G)$.

In the case of $R \oplus R \oplus R$, the compact quotient is flat torus T^3. Each left-invariant field is parallel and therefore, the $\xi(T^3)$ is **stable** totally geodesic submanifold. □

Considering the field of normal variation of $\xi(G)$ for $\xi = e_3$, namely,

$$\tilde{N} = \left(\frac{h_1\mu_1}{\sqrt{1+\mu_1^2}}e_1 + \frac{h_2\mu_2}{\sqrt{1+\mu_2^2}}e_2\right)^h + \left(-\frac{h_2}{\sqrt{1+\mu_2^2}}e_1 + \frac{h_1}{\sqrt{1+\mu_1^2}}e_2\right)^v,$$

one can observe that this field generates two variations of the field ξ in a meaning of [11], namely

$$Z_1 = \pi_*(\tilde{N}) = \frac{h_1\mu_1}{\sqrt{1+\mu_1^2}}e_1 + \frac{h_2\mu_2}{\sqrt{1+\mu_2^2}}e_2, \quad Z_2 = \mathcal{K}(\tilde{N}) = -\frac{h_2}{\sqrt{1+\mu_2^2}}e_1 + \frac{h_1}{\sqrt{1+\mu_1^2}}e_2.$$

If h_1 and h_2 are non-constant, then these variations exclude $\xi(G)$ from the class of submanifolds in $T_1(G)$, generated by the left-invariant unit vector fields. This fact justifies the following definition.

Definition 2.1. Let ξ be left-invariant unit vector field on a Lie group G with a left-invariant metric. The normal variation vector field $\tilde{N}$ on $\xi(G) \subset T_1(G)$ is called *left-invariant*, if $Z_1 = \pi_*(\tilde{N})$ and $Z_2 = \mathcal{K}(\tilde{N})$ are left-invariant vector fields on G.

If we restrict the variations to the left-invariant ones, we obtain a wider class of classically stable totally geodesic unit vector fields.

Theorem 2.4. *Let G be a three-dimensional unimodular Lie group with a left-invariant metric. Let (e_1, e_2, e_3) be the canonical frame of its Lie algebra. Set for definiteness $\lambda_1 \geq \lambda_2 \geq \lambda_3$. Then stable/unstable with respect to left-invariant variations totally geodesic submanifolds generated by unit left-invariant vector field ξ on compact quotient of G are the following.*

G *or* $\Gamma\backslash G$	*Ricci principal curvatures*	ξ	*left-invariant stability or instability*
SO(3)	$\rho_1 = \rho_2 = \rho_3 = 2$	$\mathcal{S}$	*stable*
	$\rho_1 = \rho_2 > \rho_3 = 2$	$\pm e_3$	*stable*
	$\rho_1 = \rho_2 = 2 > \rho_3$	$\cos t\, e_1 + \sin t\, e_2$	*unstable*
	$\rho_1 = 2 > \rho_2 = \rho_3$	$\pm e_1$	*unstable*
	$\rho_1 > \rho_2 = \rho_3 = 2$	$\cos t\, e_2 + \sin t\, e_3$	*stable*
	$\rho_1 = 2 > \rho_2 > \rho_3$	$\pm e_1$	*unstable*
	$\rho_1 > \rho_2 = 2 > \rho_3$	$\pm e_2$	*unstable*
	$\rho_1 > \rho_2 > \rho_3 = 2$	$\pm e_3$	*stable*
$\Gamma\backslash SL(2,R)$	$\rho_3 = 2 > -2 > \rho_2 > \rho_1$	$\pm e_3$	*unstable*
	$\rho_1 = 2 > -2 > \rho_2 > \rho_3$	$\pm e_1$	*stable*
$\Gamma\backslash E(2)$	$\rho_1 = \rho_2 = \rho_3 = 0,$	$\pm e_3,$	*stable*
	$\mu_1 = \mu_2 = 0, \mu_3 > 0$	$\cos t\, e_1 + \sin t\, e_2$	*unstable*
	$\rho_1 = 2 > \rho_3 > \rho_2 = -2$	$\pm e_1$	*unstable*
$\Gamma\backslash E(1,1)$	$\rho_3 = 2 > \rho_1 = -2 > \rho_2$	$\pm e_3$	*stable*
	$\rho_1 = 2 > \rho_2 = -2 > \rho_3$	$\pm e_1$	*stable*
$\Gamma\backslash Nil^3$	$\rho_1 = 2 > \rho_2 = \rho_3 = -2$	$\pm e_1$	*stable*
T^3	$\rho_1 = \rho_2 = \rho_3 = 0,$ $\mu_1 = \mu_2 = \mu_3 = 0$	$\mathcal{S}$	*stable*

where $\mathcal{S}$ *stands for arbitrary left-invariant unit vector field of the form* $\xi = \cos t \cos s\, e_1 + \cos t \sin s\, e_2 + \sin t\, e_3$ *with fixed parameters* t *and* s.

Proof. If one takes the left-invariant variations, then (9) takes the form

$$W(h,h) = \left(\frac{\rho_j^2}{4} - 1\right)h_i^2 + \left(\frac{\rho_i^2}{4} - 1\right)h_j^2.$$

Hence if

$$\min(|\rho_i|, |\rho_j|) \geq \rho_m = 2 \quad (i \neq j \neq m \neq i),$$

then $\xi = e_m$ generates a **stable** totally geodesic submanifold. If $\rho_i < 2$ or $\rho_j < 2$, then choosing $h_j \neq 0$ or $h_i \neq 0$ we get $W(h,h) < 0$ which means that the submanifold $\xi(G)$ is **unstable**.

Below, we check all unimodular three-dimensional Lie groups with left-invariant metric and corresponding totally geodesic unit vector fields on left-invariant stability or instability.

- The group $SO(3)$.

  1. $\lambda_1 = \lambda_2 = \lambda_3 = 2$. Here ξ is arbitrary unit left-invariant and $\xi(G)$ is a **classically stable** totally geodesic submanifold in $T_1(G)$ by Theorem 2.2.

2. Put $\lambda_1 = \lambda_2 = 2 + \delta,\ \lambda_3 = 2$. Here $\xi = e_3$. Since

$$\rho_1 = 2(1+\delta) = \rho_2 = 2(1+\delta) > \rho_3 = 2,$$

$\xi(G)$ is a **left-invariant stable** totally geodesic submanifold in $T_1(G)$.

3. Put $\lambda_1 = \lambda_2 = 2 + \varepsilon,\ \lambda_3 = 2 + \varepsilon - \sqrt{\varepsilon(\varepsilon+4)} > 0$. Rotating the frame in $e_1 \wedge e_2$ plane, we can always put $\xi = e_1$.

   The connection coefficients are

$$\mu_1 = 1 + \frac{\varepsilon - \sqrt{\varepsilon(\varepsilon+4)}}{2},\ \mu_2 = 1 + \frac{\varepsilon - \sqrt{\varepsilon(\varepsilon+4)}}{2},\ \mu_3 = 1 + \frac{\varepsilon + \sqrt{\varepsilon(\varepsilon+4)}}{2}.$$

   The principal Ricci curvatures are

$$\rho_1 = 2, \quad \rho_2 = 2, \quad \rho_3 = \frac{1}{2}\left(2 + \varepsilon - \sqrt{\varepsilon(\varepsilon+4)}\right)^2.$$

   We have $\rho_1 = \rho_2 = 2 > \rho_3 > 0$ and $W(h,h) = (\rho_3^2/4 - 1)h_2^2 < 0$ for $h_2 \neq 0$. Hence, $\xi(G)$ is an **unstable** totally geodesic submanifold in $T_1(G)$.

4. Put $\lambda_1 = 2, \lambda_2 = \lambda_3 = 2 - \varepsilon,\ 0 < \varepsilon < 2$. Here $\xi = e_1$. The connection coefficients and the Ricci principal curvatures are

$$\mu_1 = 1 - \varepsilon,\ \mu_2 = 1,\ \mu_3 = 1; \quad \rho_1 = 2,\ \rho_2 = 2(1-\varepsilon),\ \rho_3 = 2(1-\varepsilon).$$

   We have $\rho_1 = 2 > \rho_2 = \rho_3 > -2$ and hence $\rho_2^2 = \rho_3^2 < 4$. Therefore, $\xi(G)$ is an **unstable** totally geodesic submanifold in $T_1(G)$.

5. Put $\lambda_1 = \varepsilon + \sqrt{4+\varepsilon^2},\ \lambda_2 = \sqrt{4+\varepsilon^2},\ \lambda_3 = \sqrt{4+\varepsilon^2}$. In this case $\xi = \cos t e_2 + \sin t e_3$. Rotating the frame, we may put $\xi = e_3$. Then

$$\mu_1 = \frac{\sqrt{\varepsilon^2+4} - \varepsilon}{2}, \quad \mu_2 = \mu_3 = \frac{\sqrt{\varepsilon^2+4} + \varepsilon}{2} = 1/\mu_1.$$

   The principal Ricci curvatures are

$$\rho_1 = 2 + \varepsilon(\sqrt{\varepsilon^2+4} + \varepsilon) > \rho_2 = \rho_3 = 2,$$

   and hence $\xi(G)$ is a **left-invariant stable** totally geodesic submanifold in $T_1(G)$.

6. $\lambda_1 > \lambda_2 > \lambda_3 > 0,\ \lambda_m^2 - (\lambda_i - \lambda_k)^2 = 4$. Denote $\lambda_2 - \lambda_3 = \delta > 0, \lambda_1 - \lambda_2 = \varepsilon > 0$. Then $\lambda_1 - \lambda_3 = \varepsilon + \delta$. Here we have 3 distinct cases.

   (i) $\lambda_1^2 = (\lambda_2 - \lambda_3)^2 + 4,\ \xi = e_1$. Then

$$\lambda_1 = \sqrt{4+\delta^2}, \quad \lambda_2 = \sqrt{4+\delta^2} - \varepsilon > 0, \quad \lambda_3 = \sqrt{4+\delta^2} - \varepsilon - \delta > 0.$$

The connection coefficients are

$$\mu_1 = \frac{\sqrt{\delta^2+4}-\delta}{2} - \varepsilon, \quad \mu_2 = \frac{\sqrt{\delta^2+4}-\delta}{2}, \quad \mu_3 = \frac{\sqrt{\delta^2+4}+\delta}{2}.$$

The principal Ricci curvatures are

$$\rho_1 = 2 > \rho_2 = 2-\varepsilon(\sqrt{\delta^2+4}+\delta) > \rho_3 = 2-(\varepsilon+\delta)(\sqrt{\delta^2+4}-\delta) > -2,$$

and we obtain an **unstable** totally geodesic submanifold in $T_1(G)$.

(ii) $\lambda_2^2 = (\lambda_1 - \lambda_3)^2 + 4$, $\xi = e_2$. Then

$$\lambda_1 = \sqrt{4+(\varepsilon+\delta)^2}+\varepsilon, \quad \lambda_2 = \sqrt{4+(\varepsilon+\delta)^2}, \quad \lambda_3 = \sqrt{4+(\varepsilon+\delta)^2}-\delta > 0.$$

The connection coefficients are

$$\mu_1 = \frac{\sqrt{(\varepsilon+\delta)^2+4}-(\varepsilon+\delta)}{2}, \quad \mu_2 = \frac{\sqrt{(\varepsilon+\delta)^2+4}+\varepsilon-\delta}{2},$$

$$\mu_3 = \frac{\sqrt{(\varepsilon+\delta)^2+4}+\varepsilon+\delta}{2}.$$

The principal Ricci curvatures are

$$\rho_1 = 2+\varepsilon(\sqrt{(\varepsilon+\delta)^2+4}+\varepsilon+\delta), \rho_2 = 2, \rho_3 = 2-\delta\big(\sqrt{(\varepsilon+\delta)^2+4}-(\varepsilon+\delta)\big).$$

Here $\rho_1 > \rho_2 = 2 > \rho_3 > -2$ and we obtain an **unstable** totally geodesic submanifold in $T_1(G)$.

(iii) $\lambda_3^2 = (\lambda_1 - \lambda_2)^2 + 4$, $\xi = e_3$. Then

$$\lambda_1 = \sqrt{4+\varepsilon^2}+\varepsilon+\delta, \quad \lambda_2 = \sqrt{4+\varepsilon^2}+\delta, \quad \lambda_3 = \sqrt{4+\varepsilon^2}.$$

The connection coefficients are

$$\mu_1 = \frac{\sqrt{\varepsilon^2+4}-\varepsilon}{2}, \quad \mu_2 = \frac{\sqrt{\varepsilon^2+|4}+\varepsilon}{2} = 1/\mu_1, \quad \mu_3 = \frac{\sqrt{\varepsilon^2+4}+\varepsilon}{2}+\delta.$$

The principal Ricci curvatures are

$$\rho_1 = 2+(\varepsilon+\delta)(\sqrt{\varepsilon^2+4}+\varepsilon) > \rho_2 = 2+\delta(\sqrt{\varepsilon^2+4}-\varepsilon) > \rho_3 = 2,$$

and we obtain a **left-invariant stable** totally geodesic submanifold in $T_1(G)$.

- The group $SL(2,R)$. Here we have $\xi = e_3$ or $\xi = e_1$.

1. In the case $\xi = e_3$, we have $\lambda_3^2 - (\lambda_1 - \lambda_2)^2 = 4$. Put $\lambda_1 - \lambda_2 = \varepsilon > 0$. Then $\lambda_3 = -\sqrt{4+\varepsilon^2}$, $\lambda_2 = a > 0$, $\lambda_1 = a + \varepsilon$. The connection coefficients are

$$\mu_1 = -\frac{\sqrt{\varepsilon^2+4}+\varepsilon}{2}, \quad \mu_2 = -\frac{\sqrt{\varepsilon^2+4}-\varepsilon}{2} = 1/\mu_1, \quad \mu_3 = a + \frac{\sqrt{\varepsilon^2+4}+\varepsilon}{2}.$$

The principal Ricci curvatures are

$$\rho_1 = -2 - a\left(\sqrt{\varepsilon^2+4} - \varepsilon\right), \rho_2 = -2 - (a+\varepsilon)\left(\sqrt{4+\varepsilon^2} + \varepsilon\right), \rho_3 = 2.$$

Observe that $\rho_3 = 2 > \rho_1 > \rho_2$, but $\rho_2 < \rho_1 < -2$. Therefore, $\rho_2^2 > \rho_1^2 > 4$ and hence $(\rho_2^2/4 - 1)h_3^2 + (\rho_3^2/4 - 1)h_2^2 > 0$. So we have $\xi(G)$ a **left-invariant stable** totally geodesic submanifold in $T_1(G)$.

2. In the case $\xi = e_1$, we have $\lambda_1^2 - (\lambda_2 - \lambda_3)^2 = 4$. Put $\lambda_3 = -a$ $(a > 0)$, $\lambda_2 = \lambda_3 + \varepsilon = \varepsilon - a > 0$, $\lambda_1 = \sqrt{4+\varepsilon^2}$. (Observe, that $\lambda_1 - \lambda_2 = \sqrt{\varepsilon^2+4} - \varepsilon + a > -\lambda_3 = a$.) Besides, $\lambda_1 \geq \lambda_2$. Therefore, $\sqrt{\varepsilon^2+4} \geq \varepsilon - a > 0$.
   The connection coefficients are

$$\mu_1 = -a - \frac{\sqrt{\varepsilon^2+4}-\varepsilon}{2}, \quad \mu_2 = \frac{\sqrt{\varepsilon^2+4}-\varepsilon}{2}, \quad \mu_3 = \frac{\sqrt{\varepsilon^2+4}+\varepsilon}{2} = 1/\mu_2.$$

The principal Ricci curvatures are

$$\rho_1 = 2, \quad \rho_2 = -2 - a\left(\sqrt{\varepsilon^2+4} + \varepsilon\right), \quad \rho_3 = -2 + (\varepsilon - a)\left(\sqrt{\varepsilon^2+4} - \varepsilon\right).$$

Observe that $\rho_2 < -2$ but $-2 < \rho_3 < 2$. Indeed, $\varepsilon - a \leq \sqrt{\varepsilon^2+4}$, and hence

$$(\varepsilon - a)\left(\sqrt{\varepsilon^2+4} - \varepsilon\right) \leq \sqrt{\varepsilon^2+4}\left(\sqrt{\varepsilon^2+4} - \varepsilon\right) = 4 - \varepsilon\left(\sqrt{\varepsilon^2+4} - \varepsilon\right) < 4.$$

Therefore the $\xi(G)$ is an **unstable** totally geodesic submanifold in $T_1(G)$.

- The group $E(2)$. The flat case was considered in Theorem 2.3. Consider the case $\lambda_1^2 - \lambda_2^2 = 4$, $\lambda_1 > 0, \lambda_2 > 0$, and $\xi = e_1$. Put $\lambda_1 = \sqrt{4+a^2}, \lambda_2 = a > 0$, $\lambda_3 = 0$. Then

$$\mu_1 = -\frac{\sqrt{4+a^2} - a}{2}, \quad \mu_2 = \frac{\sqrt{4+a^2} - a}{2}, \quad \mu_3 = \frac{\sqrt{4+a^2} + a}{2} = 1/\mu_2$$

and

$$\rho_1 = 2, \quad \rho_2 = -2, \quad \rho_3 = -2 + a(\sqrt{4+a^2} - a).$$

So we have

$$\rho_1 = 2 > \rho_3 > \rho_2 = -2, \quad \rho_3^2 < 4,$$

and hence $(\rho_3^2/4 - 1)h_2^2 < 0$ for $h_2 \neq 0$. Therefore, $\xi(G)$ is an **unstable** totally geodesic submanifold in $T_1(G)$.

- The group $E(1,1)$. Here again we have 2 options.
  1. Consider $\lambda_3^2 - \lambda_1^2 = 4, \lambda_1 > 0, \lambda_2 = 0, \lambda_3 < 0$. The field here is $\xi = e_3$. Put $\lambda_1 = a$ and $\lambda_3 = -\sqrt{a^2+4}$. Then

$$\mu_1 = -\frac{a+\sqrt{a^2+4}}{2}, \quad \mu_3 = \frac{a+\sqrt{a^2+4}}{2}, \quad \mu_2 = \frac{a-\sqrt{a^2+4}}{2} = 1/\mu_1,$$

  and the principal Ricci curvatures are

$$\rho_1 = -2, \quad \rho_3 = 2, \quad \rho_2 = -\frac{1}{2}(a+\sqrt{a^2+4})^2 = -2 - a(a+\sqrt{4+a^2}).$$

  Evidently, $\rho_3 = 2 > \rho_1 = -2 > \rho_2$ but $\rho_2^2 > \rho_1^2 = 4$. Therefore,

$$(\rho_2^2/4 - 1)h_1^2 \geq 0,$$

  and we have a **left-invariant stable** totally geodesic submanifold in $T_1(G)$.
  2. Consider $\lambda_1^2 - \lambda_3^2 = 4, \lambda_1 > 0, \lambda_2 = 0, \lambda_3 < 0$. In this case $\xi = e_1$. Put $\lambda_1 = \sqrt{a^2+4}$ and $\lambda_3 = -a < 0$. Then

$$\mu_1 = -\frac{a+\sqrt{a^2+4}}{2}, \quad \mu_2 = \frac{\sqrt{a^2+4}-a}{2}, \quad \mu_3 = \frac{\sqrt{a^2+4}+a}{2} = 1/\mu_2,$$

  and the principal Ricci curvatures are

$$\rho_1 = 2, \quad \rho_2 = -\frac{1}{2}(a+\sqrt{a^2+4})^2 = -2 - a(a+\sqrt{a^2+4}), \quad \rho_3 = -2.$$

  Observe that $\rho_1 = 2 > \rho_3 = -2 > \rho_2$ but $\rho_2^2 > \rho_3^2 = 4$. Therefore,

$$(\rho_2^2/4 - 1)h_1^2 \geq 0,$$

  and we have a **left-invariant stable** totally geodesic submanifold in $T_1(G)$.
- The group Nil^3. In this case $\lambda_1 = 2, \lambda_2 = 0, \lambda_3 = 0$ and the field $\xi = e_1$. It is easy to calculate

$$\mu_1 = -1, \quad \mu_2 = 1, \quad \mu_3 = 1 = 1/\mu_2,$$

$$\rho_1 = 2, \quad \rho_2 = -2, \quad \rho_3 = -2,$$

  and observe that $\rho_1 = 2 > \rho_2 = \rho_3 = -2$. Therefore, $\xi(G)$ is a **left-invariant stable** totally geodesic submanifold in $T_1(G)$.
- The flat torus T^3 was considered in Theorem 2.3. □

Remark 2. The results of Theorem 2.4 that concern instability correlate with instability results from [14], where the second variation of volume was calculated with respect to the field variations and the variation field was chosen with constant variation functions, i.e., left-invariant in our terminology.

Summarizing the results of Theorem 2.4, we can observe that $\xi(G)$ *is stable with respect to left-invariant variations with totally geodesic unit vector field if and only if* ξ *is the unit eigenvector of the Ricci operator which corresponds to minimal in absolute value principal Ricci curvature* $\rho = 2$.

3 Non-Unimodular Groups

If G is a three-dimensional non-unimodular Lie group with a left-invariant metric, then there is the left-invariant orthonormal frame (e_1, e_2, e_3) such that

$$[e_1, e_2] = \alpha\, e_2 + \beta\, e_3, \quad [e_1, e_3] = -\beta\, e_2 + \delta\, e_3, \quad [e_2, e_3] = 0,$$

where α, β, and δ ar all constant satisfying $\alpha > \delta, \alpha \geq -\delta$. Let us call this frame *canonical* one.

The non-unimodular group is not compact and does not admit a compact factor [15]. That is why one should consider formula (3) over each domain $F \subset G$ with compact closure. *We say that the* $\xi(G)$ *is an unstable minimal/totally geodesic unit vector field if there is a domain* $F \subset G$ *with a compact closure such that the second variation* $\delta^2 \mathrm{Vol}_\xi(F) < 0$.

The author described the groups which admit the totally geodesic left-invariant vector fields [21]. Here we complete the theorem with stability property as follows.

Theorem 3.1. *Let G be a three-dimensional non-unimodular Lie group with a left-invariant metric. Let* ξ *be a left-invariant unit vector field on G and* (e_1, e_2, e_3) *the canonical orthonormal frame of its Lie algebra. Suppose* $\xi(G) \subset T_1(G)$ *is totally geodesic. Then*

- $\beta = \delta = 0$ *and* $\xi = e_3$ *is a parallel unit vector field; the* $\xi(G)$ *is a* ***stable*** *totally geodesic submanifold in* $T_1(G)$*;*
- $\alpha\delta = -1, \beta = \pm 1$ *and* ξ *is of the form*

$$\xi = \frac{\beta}{\sqrt{1+\alpha^2}}\, e_2 + \frac{\alpha}{\sqrt{1+\alpha^2}}\, e_3;$$

the $\xi(G)$ *is an* ***unstable*** *totally geodesic submanifold in* $T_1(G)$.

Proof. As it was proved in [21], if $\beta = \delta = 0$, then $\xi = e_3$ is a field of unit normals of some totally geodesic 2-foliation on G and $A_\xi = -\nabla\xi = 0$. Hence, in (8) all the terms with ξ turn into zero. Equation (1) implies that $\xi(G)$ is horizontal while its field of normals is vertical. Therefore, $X_2 = K(\tilde{X}) = 0$ and $Z_1 = \pi_*(\tilde{N}) = 0$. Equation (8) implies

$$\widetilde{\mathrm{Ric}}(\tilde{N}) = 0.$$

Therefore, $W(h,h) \geq 0$ and hence $\xi(G)$ is stable.

Consider the case

$$\alpha\delta = -1, \quad \beta = \pm 1, \quad \xi = \frac{\beta}{\sqrt{1+\alpha^2}}\, e_2 + \frac{\alpha}{\sqrt{1+\alpha^2}}\, e_3.$$

Observe, that the conditions $\alpha > \delta, \alpha \geq -\delta$ and $\alpha\delta = -1$ imply $\alpha \geq 1$. For such a vector field, we have

$$x_1 = 0, \quad x_2 = \frac{\beta}{\sqrt{1+\alpha^2}}, \quad x_3 = \frac{\alpha}{\sqrt{1+\alpha^2}}. \tag{10}$$

The table of covariant derivatives is

∇	e_1	e_2	e_3
e_1	0	$\beta\, e_3$	$-\beta\, e_2$
e_2	$-\alpha\, e_2$	$\alpha\, e_1$	0
e_3	$\frac{1}{\alpha} e_3$	0	$-\frac{1}{\alpha} e_1$

Then

$$A_\xi = \begin{pmatrix} 0 & -\alpha\, x_2 & \frac{1}{\alpha} x_3 \\ \beta\, x_3 & 0 & 0 \\ -\beta\, x_2 & 0 & 0 \end{pmatrix}, \quad A_\xi^t = \begin{pmatrix} 0 & \beta\, x_3 & -\beta\, x_2 \\ -\alpha\, x_2 & 0 & 0 \\ \frac{1}{\alpha} x_3 & 0 & 0 \end{pmatrix}$$

and

$$A_\xi^t A_\xi = \begin{pmatrix} 1 & 0 & 0 \\ 0 & \frac{\alpha^2}{1+\alpha^2} & \frac{-\alpha\beta}{1+\alpha^2} \\ 0 & \frac{-\alpha\beta}{1+\alpha^2} & \frac{1}{1+\alpha^2} \end{pmatrix}.$$

Therefore, the singular values of A_ξ are 0 and 1. The corresponding singular frames are

$$s_0 = \xi, \quad s_1 = e_1, \quad s_2 = \frac{-\alpha\,\beta}{\sqrt{1+\alpha^2}}\,e_2 + \frac{1}{\sqrt{1+\alpha^2}}\,e_3,$$

$$f_1 = A_\xi(s_1) = \frac{\beta\alpha}{\sqrt{1+\alpha^2}}\,e_2 - \frac{1}{\sqrt{1+\alpha^2}}\,e_3, \quad f_2 = A_\xi(s_2) = e_1.$$

Hence, the tangent and normal orthonormal framing of $\xi(G)$ is given by (2) as follows

$$\tilde{e}_0 = \xi^h,$$

$$\tilde{e}_1 = \frac{\xi_*(s_1)}{|\xi_*(s_1)|} = \frac{1}{\sqrt{2}}e_1^h - \frac{1}{\sqrt{2}}\Big(\frac{\alpha\,\beta}{\sqrt{1+\alpha^2}}\,e_2 - \frac{1}{\sqrt{1+\alpha^2}}\,e_3\Big)^v,$$

$$\tilde{e}_2 = \frac{\xi_*(s_2)}{|\xi_*(s_2)|} = \frac{1}{\sqrt{2}}\Big(\frac{-\alpha\,\beta}{\sqrt{1+\alpha^2}}\,e_2 + \frac{1}{\sqrt{1+\alpha^2}}\,e_3\Big)^h - \frac{1}{\sqrt{2}}e_1^v,$$

$$\tilde{n}_1 = \frac{\nu(f_1)}{|\nu(f_1)|} = \frac{1}{\sqrt{2}}\,e_1^h + \frac{1}{\sqrt{2}}\Big(\frac{\alpha\,\beta}{\sqrt{1+\alpha^2}}\,e_2 - \frac{1}{\sqrt{1+\alpha^2}}\,e_3\Big)^v,$$

$$\tilde{n}_2 = \frac{\nu(f_2)}{|\nu(f_2)|} = \frac{1}{\sqrt{2}}\Big(\frac{-\alpha\,\beta}{\sqrt{1+\alpha^2}}\,e_2 + \frac{1}{\sqrt{1+\alpha^2}}\,e_3\Big)^h + \frac{1}{\sqrt{2}}\,e_1^v.$$

To calculate the partial Ricci curvature for $\xi(G)$ by (8), we need the components of the Riemannian tensor of G with respect to the canonical frame [21].

	e_1	e_2	e_3
$R(e_1, e_2)\bullet$	$\alpha^2 e_2 - \beta(\alpha - \delta)e_3$	$-\alpha^2 e_1$	$\beta(\alpha - \delta)e_1$
$R(e_1, e_3)\bullet$	$-\beta(\alpha - \delta)e_2 + \delta^2 e_3$	$\beta(\alpha - \delta)e_1$	$-\delta^2 e_1$
$R(e_2, e_3)\bullet$	0	$\alpha\,\delta\,e_3$	$-\alpha\,\delta\,e_2$

The derivatives of the curvature tensor need routine calculations which can be conducted with MAPLE.

	$(\nabla_\bullet R)(e_1, e_2)e_1$	$(\nabla_\bullet R)(e_1, e_2)e_2$	$(\nabla_\bullet R)(e_1, e_2)e_3$
e_1	$2\beta^2(\alpha - \delta)e_2 + \beta(\alpha^2 - \delta^2)e_3$	$-2\beta^2(\alpha - \delta)e_1$	$-\beta(\alpha^2 - \delta^2)e_1$
e_2	0	$\beta\alpha\,(\alpha - \delta)e_3$	$-\beta\alpha\,(\alpha - \delta)e_2$
e_3	0	$\alpha\,\delta\,(\alpha - \delta)e_3$	$-\alpha\,\delta\,(\alpha - \delta)e_2$

	$(\nabla_\bullet R)(e_1, e_3)e_1$	$(\nabla_\bullet R)(e_1, e_3)e_2$	$(\nabla_\bullet R)(e_1, e_3)e_3$
e_1	$\beta(\alpha^2 - \delta^2)e_2 - \beta^2(\alpha - \delta)e_3$	$-\beta(\alpha^2 - \delta^2)e_1$	$2\beta^2(\alpha - \delta)e_1$
e_2	0	$\alpha\,\delta\,(\alpha - \delta)e_3$	$-\alpha\,\delta\,(\alpha - \delta)e_2$
e_3	0	$-\beta\delta\,(\alpha - \delta)e_3$	$\beta\delta\,(\alpha - \delta)e_2$

	$(\nabla_{\bullet} R)(e_2, e_3)e_1$	$(\nabla_{\bullet} R)(e_2, e_3)e_2$	$(\nabla_{\bullet} R)(e_2, e_3)e_3$
e_1	0	0	0
e_2	$\alpha\beta(\alpha-\delta)e_2 + \alpha\delta(\alpha-\delta)e_3$	$-\alpha\beta(\alpha-\delta)e_1$	$-\alpha\delta(\alpha-\delta)e_1$
e_3	$\alpha\delta(\alpha-\delta)e_2 - \beta\delta(\alpha-\delta)e_3$	$-\alpha\delta(\alpha-\delta)e_1$	$\beta\delta(\alpha-\delta)e_1$

Take now the field of normal variation $\tilde{N} = h_1\tilde{n}_1 + h_2\tilde{n}_2$. To calculate $K(\tilde{e}_1, \tilde{N})$, put

$$X_1 = \frac{1}{\sqrt{2}}e_1, \qquad X_2 = -\frac{1}{\sqrt{2}}\left(\frac{\alpha\beta}{\sqrt{1+\alpha^2}}e_2 - \frac{1}{\sqrt{1+\alpha^2}}e_3\right),$$

$$Y_1 = \pi_*(\tilde{N}) = \frac{1}{\sqrt{2}}\left(h_1e_1 + h_2\left(-\frac{\alpha\beta}{\sqrt{1+\alpha^2}}e_2 + \frac{1}{\sqrt{1+\alpha^2}}e_3\right)\right),$$

$$Y_2 = K(\tilde{N}) = \frac{1}{\sqrt{2}}\left(h_2e_1 + h_1\left(\frac{\alpha\beta}{\sqrt{1+\alpha^2}}e_2 - \frac{1}{\sqrt{1+\alpha^2}}e_3\right)\right),$$

and apply (8). The MAPLE calculations yield

$$\begin{aligned}
&\langle R(X_1,Y_1)Y_1,X_1\rangle = -\frac{1}{4}\frac{\alpha^4+\alpha^2+1}{\alpha^2}h_2^2, \quad ||R(X_1,Y_1)\xi||^2 = 0,\\
&||R(\xi,Y_2)X_1 + R(\xi,X_2)Y_1||^2 = \frac{\alpha^6-\alpha^4+3\alpha^2+1}{\alpha^2(1+\alpha^2)}h_2^2,\\
&\|X_2\|^2\|Y_2\|^2 - \langle X_2,Y_2\rangle^2 = \tfrac{1}{4}h_2^2,\\
&\langle R(X_1,Y_1)Y_2,X_2\rangle = \frac{1}{4}\frac{\alpha^4+\alpha^2+1}{\alpha^2}h_2^2, \quad \langle R(\xi,X_2)X_1, R(\xi,Y_2)Y_1\rangle = 0,\\
&\langle(\nabla_{X_1}R)(\xi,Y_2)Y_1,X_1\rangle = -\frac{1}{4}\frac{\alpha^6-\alpha^4+5\alpha^2-1}{\alpha^2(1+\alpha^2)}h_2^2,\\
&\langle(\nabla_{Y_1}R)(\xi,X_2)X_1,Y_1\rangle = -\frac{1}{4}\frac{\alpha^6+10\alpha^4+4\alpha^2+7}{\alpha^2(1+\alpha^2)}h_2^2.
\end{aligned}$$

After substitution into (8) and the MAPLE algebraic transformations, we get

$$\tilde{K}(\tilde{e}_1,\tilde{N}) = \frac{1}{4}\frac{\alpha^6+10\alpha^4+4\alpha^2+7}{\alpha^2(1+\alpha^2)}h_2^2.$$

In a similar way,

$$\tilde{K}(\tilde{e}_2, \tilde{N}) = \frac{1}{4}\frac{5\alpha^8 - \alpha^6 + 3\alpha^4 + 13\alpha^2 - 8}{\alpha^2(1+\alpha^2)^2}h_1^2 + \frac{\alpha^8 + 2\alpha^4 + 1}{\alpha^2(1+\alpha^2)^2}h_2^2,$$

$$\tilde{K}(\tilde{e}_0, \tilde{N}) = \frac{1}{4}\frac{\alpha^4 + 14\alpha^2 - 11}{(1+\alpha^2)^2}h_1^2 - \frac{1}{4}\frac{3\alpha^4 + 2\alpha^2 - 9}{(1+\alpha^2)^2}h_2^2.$$

As a result, the partial Ricci curvature of $\xi(G)$ obtains the form

$$\widetilde{\mathrm{Ric}}(\tilde{N}) = \frac{1}{4}\frac{5\alpha^8 + 17\alpha^4 + 2\alpha^2 - 8}{\alpha^2(1+\alpha^2)^2}h_1^2 + \frac{1}{4}\frac{5\alpha^8 + 8\alpha^6 + 20\alpha^4 + 20\alpha^2 + 11}{\alpha^2(1+\alpha^2)^2}h_2^2.$$

In this case, we cannot consider left-invariant variations because of the boundary conditions. Nevertheless, one can consider a left-invariant variation over a subdomain $F_1 \subset F$ such that $mes(\bar{F} \setminus F_1) < \varepsilon$ for however small ε. If the second left-invariant variation over F_1 is negative and bounded away from zero, then by taking F_1 sufficiently large we always can make $\delta^2 \mathrm{Vol}_\xi(F) < 0$.

If the variation field $\tilde{N}$ is left-invariant, then

$$\sum_{i=0}^{2} ||\tilde{\nabla}^{\perp}_{\tilde{e}_i}\tilde{N}||^2 = \left(\sum_{i=0}^{2}(\tilde{\gamma}^2_{1i})^2\right)(h_1^2 + h_2^2),$$

where $\tilde{\gamma}^2_{1i} = \tilde{g}(\tilde{\nabla}_{\tilde{e}_i}\tilde{n}_1, \tilde{n}_2)$ are the coefficients of the $\xi(G)$ normal bundle connection with respect to the chosen frame. Calculating, we get

$$\tilde{\nabla}_{\tilde{e}_0}\tilde{n}_1 = \frac{3}{4}\frac{\sqrt{2}}{\sqrt{1+\alpha^2}}\left(\beta\alpha e_2 + e_3\right)^h + \frac{\sqrt{2}}{1+\alpha^2}e_1^v, \quad \tilde{\nabla}_{\tilde{e}_1}\tilde{n}_1 = 0,$$

$$\tilde{\nabla}_{\tilde{e}_2}\tilde{n}_1 = \frac{1}{2}\frac{1}{\sqrt{1+\alpha^2}}\left(\beta(2\alpha^2 - 1)e_2 - \alpha e_3\right)^h - \frac{1}{2}\frac{\alpha^4 + \alpha^2 - 2}{\alpha(1+\alpha^2)^2}e_1^v.$$

Now one can easily find

$$\tilde{\gamma}^2_{10} = \frac{\alpha^2 + 2}{1+\alpha^2}, \quad \tilde{\gamma}^2_{11} = 0, \quad \tilde{\gamma}^2_{12} = -\frac{1}{4}\frac{\sqrt{2}(3\alpha^2 - 2)}{\alpha}.$$

After substitution and MAPLE algebraic transformations, the left-invariant part of integrand in (3) takes the form

$$W(h,h) = -\frac{1}{8}\frac{\alpha^8 - 14\alpha^6 + 13\alpha^4 - 24\alpha^2 - 20}{\alpha^2(1+\alpha^2)^2}h_1^2 - \frac{1}{8}\frac{\alpha^8 + 2\alpha^6 + 19\alpha^4 + 12\alpha^2 + 18}{\alpha^2(1+\alpha^2)^2}h_2^2. \quad (11)$$

The factor at h_2 is always negative and hence the submanifold $\xi(G)$ is **unstable**. The proof is complete. □

Closing Observation

Analyzing Remarks 1 and 2, one can *conjecture* that if the horizontal and vertical projections of classical normal variation vector field are in $\xi^{\perp}$, then the classical stability or instability of minimal (or totally geodesic) submanifold $\xi(M) \subset T_1(M)$ is equivalent to stability or instability of the unit vector field in the meaning of [11].

Acknowledgements The author thanks Prof. Vladimir Rovenski (Haifa University, Israel) for hospitality during the 2-nd International Workshop on Geometry and Symbolic Computations (Haifa, May 15–20, 2013) and the MAPLE developers for the perfect tool in complicated calculations. The author also thanks Ye. Petrov (Kharkiv National University) for careful reading of preliminary version and helpful suggestions.

References

1. Borisenko A. and Yampolsky A., *Riemannian geometry of bundles*, Russian Math. Surveys, (1991), **46**(6), 55–106.
2. Boeckx E. and Vanhecke L., *Unit tangent sphere bundles with constant scalar curvature,* Czechoslovak Math. J. **51** (2001), 523–544.
3. Boeckx E. and Vanhecke L., *Harmonic and minimal radial vector fields*, Acta Math. Hungar. **90** (2001), 317–331.
4. Boeckx E. and Vanhecke L., *Harmonic and minimal vector fields on tangent and unit tangent bundles*, Differential Geom. Appl. **13** (2000), 77–93.
5. Boeckx E. and Vanhecke L., *Isoparametric functions and harmonic and minimal unit vector fields*, Contemp. Math. **288** (2001), 20–31.
6. Borisenko A. and Yampolsky A. The Sectional curvature of Sasaki metric of the T_1M^n. Ukr. geom. sbornik, **30**, (1987), 10–17 (Engl. transl.: Journ. Sov. Math., (1990), **51**(5), 2503–2508.
7. Dombrowski P.,*On the geometry of the tangent bundle*, J. Reine Angew. Mathematik, **210** (1962), 73–88.
8. Gil-Medrano O., González-Dávila J.C. and Vanhecke L., *Harmonic and minimal invariant unit vector fields on homogeneous Riemannian manifolds*, Houston J. Math. 27 (2001), 377–409.
9. Gil-Medrano O., *Relationship between volume and energy of unit vector fields*, Diff. Geom. Appl. **15** (2001), 137–152.
10. Gil-Medrano O., Llinares-Fuster E., Minimal unit vector fields, *Tôhoku Math. J.*, 54 (2002), 71–84.
11. Gil-Medrano O., Llinares-Fuster E., *Second variation of volume and energy of vector fields. Stability of Hopf vector fields*, Math. Ann. 320 (2001), 531–545.
12. Gluck H. and Ziller W., *On the volume of a unit vector field on the tree sphere*, Comment. Math. Helv. **61** (1986), 177–192.
13. González-Dávila J.C. and Vanhecke L., *Examples of minimal unit vector fields*, Ann Glob. Anal. Geom. **18** (2000), 385–404.
14. González-Dávila J.C. and Vanhecke L., *Energy and volume of unit vector fields on three-dimensional Riemannian manifolds*, Diff. Geom. Appl. **16** (2002), 225–244.
15. Milnor J., *Curvatures of left invariant metrics on Lie groups,* Adv. in Math, 21 (1976), 293–329.
16. Simons J., *Minimal Varieties in Riemannian Manifolds,* Annals of Math. Second Series **88/1** (Jul., 1968), 62–105.
17. Tsukada K. and Vanhecke L., *Invariant minimal unit vector fields on Lie groups*, Period. Math. Hungar. **40** (2000), 123–133.

18. Yampolsky A., *Totally geodesic property of Hopf vector field*, Acta Math. Hungar. **101/1-2** (2003), 93–112.
19. Yampolsky A., *Full description of totally geodesic unit vector field on Riemannian 2-manifold*, J. of Math. Physics, Analysys, Geometry **1/3** 2004, 355–365.
20. Yampolsky A., *On special types of minimal and totally geodesic unit vector fields*, Proceedings of the Seventh International Conference in Geometry, Integrability and Quantization June 2–10, 2005, Varna, Bulgaria, 292–306.
21. Yampolsky A., *Invariant totally geodesic unit vector fields on three-dimensional Lie groups*, J. of Math. Physics, Analysys, Geometry **3/2** (2007), 253–276.

Part II
Applications

Rotational Liquid Film Interacted with Ambient Gaseous Media

Gaissinski I., Levy Y., Rovenski V., and Sherbaum V.

Abstract Annular jets of an incompressible liquid moving in a gas at rest are of interest for applications. The experimental study of annular liquid jets shows existing *tulip* and *bubble* jet shapes and also predicts the existence of periodic shape. However, sufficient simplifications of mathematical models of the flow details were made: the effects of the forces of surface tension of the longitudinal motion and the variability of the tangential velocity component of the centrifugal forces in the field were neglected. In this work, the equations described the flow of rotational annular jets of viscous liquid in an undisturbed medium with allowance of the abovementioned effects. The basic model was obtained through the use of quasi-two-dimensional momentum balance equations in the metric space with the co- and contravariant basis vectors suitable for surfaces with complicated shape. The pressure difference outside and within the jet was obtained and analyzed. The results of calculations show the dependence of the jet shape on the relative contributions of the initial rotation rate, viscosity, surface tension, gravity forces, and pressure difference. An exact solution to the problem of the motion of a thin cylindrical shell due to different internal and external pressures is obtained. Analysis of nonlinear instabilities of the Rayleigh–Taylor type in meridional cross section was carried out. It is shown that the instabilities, which appear due to pressure drop, cannot be stabilized by rotation.

Gaissinski I. (✉) • Levy Y. • Sherbaum V.
Faculty Aerospace Engineering, Technion – Israel Institute of Technology, Haifa, Israel
e-mail: igorg@tx.technion.ac.il

Rovenski V.
Mathematical Department, University of Haifa, Haifa, Israel
e-mail: rovenski@math.haifa.ac.il

V. Rovenski and P. Walczak (eds.), *Geometry and its Applications*, Springer Proceedings in Mathematics & Statistics 72, DOI 10.1007/978-3-319-04675-4_9,

Keywords Rotational liquid film • Stress tensor • Surface tension • Two-dimensional model • Rayleigh–Taylor instability • Perturbation

Mathematics Subject Classifications (2010): 76D17, 76E05, 76E07, 34K2

Introduction

During the first stage of atomization process (atomization is the making of an aerosol, which is a colloid suspension of fine solid particles or liquid droplets in a gas), free rotating films of liquid are formed [11]. As a rule, they take cylindrical or conical forms. For these kinds of films two main mechanisms can be distinguished which disturb the film shape. The first of them is connected to instability and to the growth of flexural perturbations due to the dynamic action of the surrounding gas. The second mechanism of the film distortion of the perfect shape of the film is due to the liquid rotation around the film axis.

In a study [4] special attention is paid to the non-axisymmetric perturbation development due to the film rotation. Due to competition between centrifugal and surface forces, the film acquires a wavy shape even when the flow is stable. Shapes of swirling liquid annular jets were studied in by Epikhin [3]. He accounted for viscosity and friction between air and liquid. These results gave solutions for the effect of these factors on the steady film configuration and its initial velocity profile. The numerical and experimental investigation of stable wavy shapes in free films of ideal liquids was carried out in [5]. But the surface tension effect on the longitudinal motion of the film was neglected. The present work is dedicated to the study of rotating liquid films interacting with the ambient air at different pressures. Asymptotic analysis and numerical simulation were applied. The effect of the liquid viscosity was taken into account.

1 The Main Equations

Let us consider a free rotating liquid film following out of an annular nozzle with the assumption that the disintegration of the film takes place fairly far downstream. The motion may naturally be described on the basis of the quasi-two-dimensional equations of thin film dynamics for the momentum balance (Fig. 1); $d\theta^1 d\theta^2$ is the element of the film. The overall momentum within the element is expressed as $\rho h\sqrt{a}\mathbf{V}d\theta^1 d\theta^2$. Entov and Yarin [13] have obtained the quasi-two-dimensional equations in the frame of reference θ^i $(i = 1, 2)$ associated with the *middle surface* of a film (i.e., the conditional area located in the middle of the film thickness). They considered it as a two-dimensional continuum:

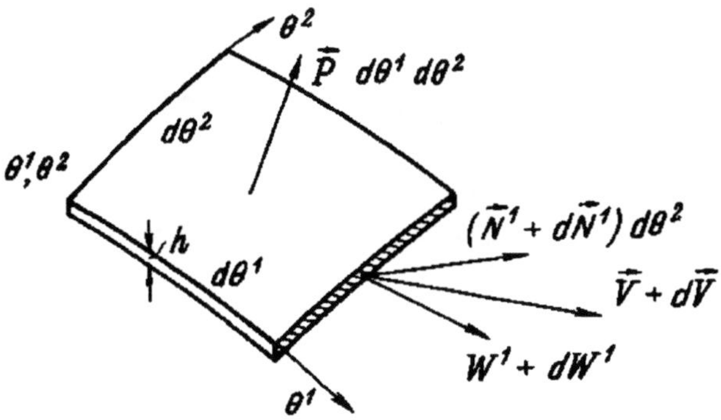

Fig. 1 Element of the film surface with vector components

$$\begin{aligned}
&\partial_t\left(g_1 h\right)+\partial_{\theta^1}\left(g_1 h W^1\right)+\partial_{\theta^2}\left(g_1 h W^2\right)=0,\\
&\partial_t\left(\rho \mathbf{V} g_1 h\right)+\partial_{\theta^1}\left(\rho \mathbf{V} W^1 g_1 h\right)+\partial_{\theta^2}\left(\rho \mathbf{V} W^2 g_1 h\right)=\partial_{\theta^1}\mathbf{N}^1+\partial_{\theta^2}\mathbf{N}^2+\mathbf{q} g_1,\\
&g_1=\sqrt{EG-F^2},\quad E=\mathbf{a}_1\cdot\mathbf{a}_1,\quad F=\mathbf{a}_1\cdot\mathbf{a}_2,\quad G=\mathbf{a}_2\cdot\mathbf{a}_2.
\end{aligned} \tag{1.1}$$

The following definitions are used: $\rho = \rho(\theta^1, \theta^2)$ is the liquid density; $\mathbf{V} = \mathbf{V}(\theta^1, \theta^2)$ is the velocity field; $h = h(\theta^1, \theta^2)$ is the film thickness; M is the middle surface, $r = r(\theta^1, \theta^2)$ of the film with the parametrization θ^α being the curvilinear coordinates (Fig. 1); $\mathbf{q}$ is the distributed external force per unit area of the film; $\mathbf{W}$ is the liquid velocity relative to the frame of reference associated with the median surface of the film (the reference frame velocity is $\mathbf{U}$ and $\mathbf{W} = \mathbf{V} - \mathbf{U}$); $N^\alpha = N^{\alpha\gamma}\mathbf{a}_\gamma$ is the internal surface forces per unit length of film cross section along a line $\theta^\alpha = \text{const}$; and $N^{\alpha\gamma}$ is the symmetrical contravariant tensor. The corresponding metrics is characterized by the covariant base vectors $\mathbf{a}_\alpha = \partial\mathbf{r}/\partial\theta^\alpha$, $d\mathbf{r} = \mathbf{a}_\alpha d\theta^\alpha$, and $\alpha = 1,\ 2$; $\mathbf{a}_3$ is the unit normal to the surface M; the contravariant base vectors $\mathbf{a}^\alpha$ and co- and contravariant components of the metric tensor $a_{\alpha\beta}$, $a^{\alpha\beta}$ satisfy the geometric formulas

$$a_{\alpha\beta}=\mathbf{a}_\alpha\mathbf{a}_\beta,\ \ a^{\alpha\beta}=\mathbf{a}^\alpha\mathbf{a}^\beta,\ \ \mathbf{a}^\alpha\mathbf{a}_\beta=\delta^\alpha_\beta,\ \ a_{\alpha\beta}a^{\beta\gamma}=\delta^\gamma_\alpha,\ \ \det a_{\alpha\beta}=a,$$

where δ^γ_α is the Kronecker delta function $g^{\alpha m}g_{m\beta} = \delta^\alpha_\beta$. The components of the tensor $N^{\alpha\gamma}$ are given in the form [13]

$$N^{\alpha\gamma}=\sqrt{g_1}\left(\sigma^{\alpha\gamma}h+2\sigma_* a^{\alpha\gamma}\right), \tag{1.2}$$

where σ_* is the surface tension coefficient and $\sigma^{\alpha\gamma}$ is the contravariant surface stress tensor.

The first equation of (1.1) is continuity, and the second one is the momentum equation. In the case of axisymmetric stable flow (Fig. 2) the motion may be described on the basis of the thin film dynamics (1.1) and (1.2) for quantities averaged over the thickness of the film. For the case of axisymmetric stable flow equations take the following form [13]:

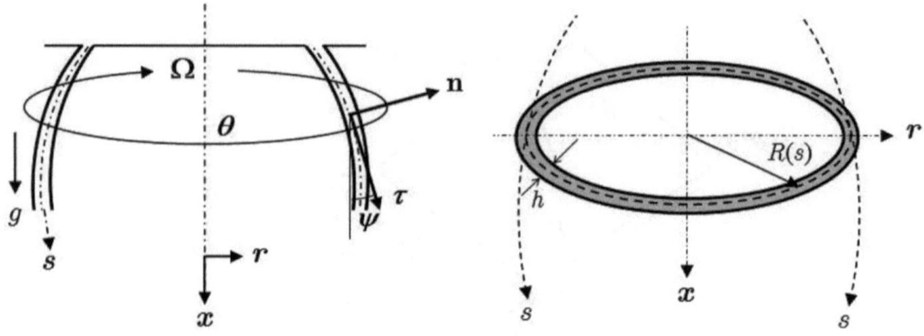

Fig. 2 Scheme of a liquid film with motion direction

$$
\begin{aligned}
&hRV_\tau = h_0 R_0 V_{\tau 0},\\
&\rho\left(hRV_\tau \frac{dV_\tau}{ds} - V_\theta^2 h \sin\psi\right) = \frac{d\sigma_{\tau\tau} Rh}{ds} - h\sigma_{\theta\theta}\sin\psi + g\rho Rh\cos\psi,\\
&\rho(hRV_\tau^2 \frac{d\psi}{ds} - V_\theta^2 h\cos\psi) = \sigma_{\tau\tau} Rh\frac{d\psi}{ds} - h\sigma_{\theta\theta}\cos\psi\\
&\quad + 2\sigma_*(R\frac{d\psi}{ds} - \cos\psi) - g\rho Rh\sin\psi - (p_i - p_o)\,R,\\
&\rho(hRV_\tau \frac{dV_\theta}{ds} - V_\theta V_\tau h\sin\psi) = \frac{d\sigma_{\theta\tau} Rh}{ds} + \sigma_{\tau\theta} h\sin\psi.
\end{aligned}
\tag{1.3}
$$

The terms which take into account pressure drop are included additionally by comparison with studies [13]. The first equation of the system (1.1) gives the equation of continuity; the remaining equations are projections of the momentum equation on the directions of the tangent τ and the normal $\mathbf{n}$ to the middle surface of the film and on the direction of the variation of the angular (azimuthal) coordinate θ on the middle surface.

The following notation is introduced: R is the radius of the middle surface of the film (Fig. 2); h is its thickness; V_τ and V_θ are the longitudinal and rotational components of the fluid velocity vector; the index 0 denotes the values of the quantities at the nozzle exit; ρ is the liquid density; s is the coordinate reckoned along the generator of the middle surface of the film; ψ is the angle between the tangent τ and the axis of symmetry x of the film; p_o and p_i are the outer and inner ambient media static pressures; $d\psi/ds$ is the curvature of the generator of the middle surface of the film; g is the acceleration due to the force of gravity g which is directed along the axis x; $\sigma_{\tau\tau}$, $\sigma_{\tau\theta}$, and $\sigma_{\theta\theta}$ are the components of the stress tensor in the coordinate system associated with the middle surface of the film, which are determined for a Newtonian liquid (with allowance for the condition that there are no stresses on the film surfaces) by the relationships

$$\sigma_{\tau\tau} = 2\mu\Big(2\frac{dV_\tau}{ds} + \frac{V_\tau \sin\psi}{R}\Big), \quad \sigma_{\tau\theta} = \mu\Big(\frac{dV_\theta}{ds} - \frac{V_\theta \sin\psi}{R}\Big),$$
$$\sigma_{\theta\theta} = 2\mu\Big(\frac{dV_\tau}{ds} + \frac{2V_\tau \sin\psi}{R}\Big), \tag{1.4}$$

where μ is the liquid viscosity. Choosing as scales for R and s the radius R_0, for V_τ and V_θ the velocity $V_{\tau 0}$, for h the thickness h_0, and for the stresses parameter $\mu V_{\tau 0}/R_0$, we obtain, using (1.3) and (1.4), the following dimensionless system:

$$\frac{dV_\tau}{ds} = \frac{V_\theta^2}{V_\tau R}\sin\psi + Re^{-1}\left[\frac{d}{ds}\left(\frac{\sigma_{\tau\tau}}{V_\tau}\right) - \frac{\sigma_{\theta\theta}}{V_\tau R}\sin\psi\right] + Fr^{-1}\frac{\cos\psi}{V_\tau},$$
$$\frac{d\psi}{ds} = \left(V_\tau - We^{-1}R - Re^{-1}\frac{\sigma_{\tau\tau}}{V_\tau}\right)^{-1}\left[\left(\frac{V_\theta^2}{V_\tau R} - We^{-1} - Re^{-1}\frac{\sigma_{\theta\theta}}{V_\tau R}\right)\cos\psi\right.$$
$$\left. - Fr^{-1}\frac{\sin\psi}{V_\tau} - \frac{Eu}{}We R\right],$$
$$\frac{dV_\theta}{ds} + \frac{V_\theta}{R}\sin\psi = Re^{-1}\left[\frac{d}{ds}\left(\frac{\sigma_{\tau\theta}}{V_\tau}\right) - \frac{\sigma_{\tau\theta}}{V_\tau R}\sin\psi\right],$$
$$\frac{dR}{ds} = \sin\psi, \qquad \frac{dx}{ds} = \cos\psi, \tag{1.5}$$

where dimensionless stress tensor components are

$$\sigma_{\tau\tau} = 2\left(2\frac{dV_\tau}{ds} + \frac{V_\tau \sin\psi}{R}\right), \quad \sigma_{\tau\theta} = \frac{dV_\theta}{ds} - \frac{V_\theta \sin\psi}{R},$$
$$\sigma_{\theta\theta} = 2\left(\frac{dV_\tau}{ds} + \frac{2V_\tau \sin\psi}{R}\right). \tag{1.6}$$

The last two equations in (1.5) express obvious geometric relationships. The dimensionless values

$$We = \frac{\rho h_0 V_{\tau 0}^2}{2\sigma_*}, \quad Re = \frac{\rho R_0 V_{\tau 0}}{\mu}, \quad Fr = \frac{V_{\tau 0}^2}{g R_0}, \quad Eu = \frac{(p_i - p_o)\, R_0}{2\sigma_*}$$

are the Weber, Reynolds, Froude, and Euler numbers, correspondingly. The system (1.5) requires formulation of conditions on the near and far ends of the film. On the nozzle exit we have

$$x|_{s=0} = 0, \quad R|_{s=0} = 1, \quad V_\tau|_{s=0} = 1, \quad V_\theta|_{s=0} = \Omega R_0/V_{\tau 0}$$
$$= V_{\theta 0}/V_{\tau 0} \equiv V_\Omega, \quad \psi|_{s=0} = \psi_0, \tag{1.7}$$

where Ω is the angular velocity (Fig. 2).

The effect of the boundary conditions at the far end of an annular jet with $x = L$ appears to a great extent only in a narrow boundary layer and for fairly high does not propagate upstream. In this context it is enough to solve Cauchy's problem (1.5)–(1.7).

2 Asymptotic Analysis

We consider the Newtonian liquid films in the equality conditions of the outer and inner static pressures ($Eu = 0$); gravity effect is neglected ($Fr \to \infty$). It follows from (1.5) and (1.7) that at a definite value of the rotation velocity at the nozzle exit $V_\Omega = We^{-1/2}$ and with $\psi_0 = 0$ for ideal liquid (inverse Reynolds number $\varepsilon = Re^{-1} \to 0$), there are no oscillations, and $R \equiv 1$; consequently, the middle surface of the film has a cylindrical shape (see Appendix).

For the given initial rotation velocity and the film exit angle such that

$$V_\Omega = We^{-1/2} + \beta_\Omega, \quad \beta_\Omega << We^{-1/2}, \quad \psi_0 << 1, \tag{2.1}$$

small oscillations in the main parameters of the film must take place along x-axis, and the presence of viscosity cannot alter this picture qualitatively. Consequently, assuming that the inequalities (2.1) are fulfilled, we represent the unknown values in the form

$$V_\tau = 1 + \alpha(s), \quad V_\theta = We^{-1/2} + \beta(s), \quad R = 1 + \gamma(s), \tag{2.2}$$

where α and γ are small in comparison with unity and β is small in comparison with $We^{-1/2}$.

The representation (2.2) for R is not subject to doubt when $\psi_0 = 0$; in the case $0 < \psi_0 << 1$ we may also expect a solution oscillating periodically near $R = 1$ which will be constructed in what follows. In linear approximation $s \approx x$ and

$$\psi \approx \sin\psi \approx \tan\psi = dR/dx = d\gamma/dx << 1. \tag{2.3}$$

Substituting (2.2) and (2.3) into (1.5) and (1.6), we obtain after linearization with respect to α, β, and γ

$$\begin{aligned}
\frac{d\alpha}{dx} &= We^{-1}\frac{d\gamma}{dx} + \varepsilon\left(4\frac{d^2\alpha}{dx^2} + 2\frac{d^2\gamma}{dx^2}\right),\\
\frac{d^2\gamma}{dx^2} &= \left(1 - We^{-1}\right)^{-1}\left[We^{-1}\left(2\beta\, We^{1/2} - \gamma - \alpha\right) - \varepsilon\left(2\frac{d\alpha}{dx} + 4\frac{d\gamma}{dx}\right)\right],\\
\frac{d\beta}{dx} &= -\,We^{-1/2}\frac{d\gamma}{dx} + \varepsilon\left(\frac{d^2\beta}{dx^2} - We^{-1/2}\frac{d^2\gamma}{dx^2}\right).
\end{aligned} \tag{2.4}$$

Using (2.1) and (2.3) for (1.7), we find that the solutions of the system (2.4) must satisfy the conditions

$$\alpha|_{x=0} = \gamma|_{x=0} = 0, \quad \beta|_{x=0} = \beta_\Omega, \quad d\gamma/dx\,|_{x=0} = \psi_0. \tag{2.5}$$

Integrating the first and the third equations of the system (2.4), we obtain

$$\alpha = C_1 + We^{-1}\gamma + \varepsilon\left(4\frac{d\alpha}{dx} + 2\frac{d\gamma}{dx}\right),$$
$$\beta = D_1 - We^{-1}\gamma + \varepsilon\left(\frac{d\beta}{dx} - We^{-1}\frac{d\gamma}{dx}\right), \tag{2.6}$$

where C_1 and D_1 are indeterminate constants. Substituting (2.6) into the second equation of the system (2.4) and neglecting values $\mathcal{O}\left(\varepsilon^2\right)$ we get the following equation:

$$\frac{d^2\gamma}{dx^2} + \gamma\frac{3 + We^{-1}}{We - 1} = \frac{2\,We^{1/2}D_1 - C_1}{We - 1} - 4\,\varepsilon\frac{We + 1}{We - 1}\cdot\frac{d\gamma}{dx}. \tag{2.7}$$

Let us find the solution of this equation in the case of low viscosity, $Re >> 1$, by means of the asymptotic multi-scale method [8, 12]. Note that for typical values of the parameters, $\rho = 10^3\,\mathrm{kg/m^3}$, $R_0 = 10^{-2}\,\mathrm{m}$, $V_{\tau 0} = 1.0\,\mathrm{m/s}$, the value of ε is low, due to $Re \approx 10^4$. Since the important effect of viscosity can appear only at far downstream the film exit, we introduce the *slow* variable $X = \varepsilon x = \mathcal{O}(1)$, which is locally independent from x. Representing the solution in the form of the asymptotic series $\gamma = \gamma_0 + \varepsilon\gamma_1$ and neglecting terms of order higher than Re^{-1}, we obtain, from (2.7),

$$\frac{\partial^2}{\partial x^2}(\gamma_0 + \varepsilon\gamma_1) + 2\varepsilon\frac{\partial^2\gamma_0}{\partial x\partial X} + \frac{3 + We^{-1}}{We - 1}(\gamma_0 + \varepsilon\gamma_1) = \frac{2\,We^{1/2}D_1 - C_1}{We - 1}$$
$$- 4\varepsilon\frac{(We + 1)^2}{We\,(We-)}\frac{\partial\gamma_0}{\partial x}. \tag{2.8}$$

Separating out the dominant terms from (2.8), we obtain the equation

$$\frac{\partial^2\gamma_0}{\partial x^2} + \gamma_0\frac{3 + We^{-1}}{We - 1} = \frac{2\,We^{1/2}D_1 - C_1}{We - 1},$$

whose solution is

$$\gamma_0 = A(X)\,e^{imx} + B(X)\,e^{-imx} + \gamma_r, \quad m = \left[\frac{3\,We + 1}{We(\,We - 1)}\right]^{1/2},$$
$$\gamma_r = \frac{2\,We^{1/2}D_1 - C_1}{3 + We^{-1}}. \tag{2.9}$$

It is assumed that in practice $We > 1$. In order to determine the unknown functions A and B in (2.9), we will consider the terms of the order ε in (2.8):

$$\frac{\partial^2 \gamma_1}{\partial x^2} + m^2 \gamma_1 = -4\varepsilon \frac{(We+1)^2}{We\,(We-1)} \frac{\partial \gamma_0}{\partial x} - 2\frac{\partial^2 \gamma_0}{\partial x \partial X}. \tag{2.10}$$

Substituting (2.9) in the r.h.s. of (2.10) and requiring absence solutions of type $\gamma_1 \sim x \exp(\pm imx)$, which are inadmissible in an asymptotic series, we obtain

$$A = A_0 e^{-PX}, \quad B = B_0 e^{-PX}, \quad P = \frac{2\,(We+1)^2}{We\,(We-1)}, \tag{2.11}$$

where A_0 and B_0 are arbitrary constants.

Cutting off the asymptotic series, we obtain by means of (2.9) and (2.11)

$$\gamma_0 = A_0 e^{-PX+imx} + B_0 \exp e^{-PX-imx} + \gamma_r. \tag{2.12}$$

In accordance with conditions for γ in (2.5) and (2.12), we obtain

$$A_0 = -\frac{1}{2}\left[i\,(\psi_0/m) + \gamma_r\right], \quad B_0 = \frac{1}{2}\left[i\,(\psi_0/m) - \gamma_r\right].$$

These latter equations give, with allowance for (2.12),

$$\gamma = \gamma_r \left[1 - e^{-PX} \cos mx\right] + \frac{\psi_0}{m} e^{-PX} \sin mx. \tag{2.13}$$

Using (2.13) and omitting in (2.6) the important terms of order ε, we obtain

$$\begin{aligned} \alpha &= C_1 + \gamma_r\, We^{-1}\left[1 - e^{-PX}\cos mx\right] + We^{-1}\tfrac{\psi_0}{m} e^{-PX} \sin mx, \\ \beta &= D_1 - \gamma_r\, We^{-1/2}\left[1 - e^{-PX}\cos mx\right] - We^{-1}\tfrac{\psi_0}{m} e^{-PX} \sin mx. \end{aligned} \tag{2.14}$$

By means of the boundary conditions (2.5) for α and β we find from (2.14) that $C_1 = 0,\ D_1 = \beta_\Omega$. Using Eqs. (2.2), (2.9), (2.11), (2.13), and (2.14), we find

$$\begin{aligned} R &= 1 + \tfrac{2}{3}\beta_\Omega \tfrac{We^{3/2}}{We+\frac{1}{3}} \left(1 - e^{-PX}\cos mx\right) + \tfrac{\psi_0}{m}\, e^{-PX} \sin mx, \\ V_\tau &= 1 + \tfrac{2}{3}\beta_\Omega \tfrac{We^{3/2}}{We+\frac{1}{3}} \left(1 - e^{-PX}\cos mx\right) + \tfrac{\psi_0}{m}\, We^{-1}\, e^{-PX} \sin mx, \\ V_\theta &= We^{-1/2} + \tfrac{1}{3}\beta_\Omega \tfrac{We+1}{We+\frac{1}{3}} + \tfrac{2}{3}\beta_\Omega \tfrac{We^{3/2}}{We+\frac{1}{3}}\, e^{-PX}\cos mx - \tfrac{\psi_0}{m}\, We^{-1/2}\, e^{-PX} \sin mx, \\ \psi &= \tfrac{2}{3}\,\beta_\Omega \tfrac{We^{3/2}}{We+\frac{1}{3}} m\, e^{-PX} \sin mx + \psi_0\, e^{-PX}\cos mx, \\ X &= \varepsilon x, \quad m = \left[\tfrac{3\,We+1}{We(We-1)}\right]^{1/2}, \quad P = \tfrac{2(We+1)^2}{We(We-1)}, \qquad We > 1. \end{aligned} \tag{2.15}$$

The thickness of the film is calculated by using the continuity equation as $h = 1/RV_\tau$. We also note that the general solution of the system (2.4) has the form

$$\begin{aligned}
\gamma &= e^{-PX}\left(F_1 \sin mx + F_2 \cos mx\right) + \gamma_r + \mathcal{O}(\varepsilon),\\
\alpha &= C_2\, e^{-\frac{L-x}{4\varepsilon}} + We^{-1}\, e^{-PX}\left(F_1 \sin mx + F_2 \cos mx\right) + \gamma_r\, We^{-1} + C_1 + \mathcal{O}(\varepsilon),\\
\beta &= D_2\, e^{-\frac{L-x}{4\varepsilon}} - We^{-1/2}\, e^{-PX}\left(F_1 \sin mx + F_2 \cos mx\right) - \gamma_r We^{-1/2} + D_1 + \mathcal{O}(\varepsilon),
\end{aligned} \tag{2.16}$$

where the new indeterminate constants $F_1,\ F_2,\ C_2$, and D_2 appear. It is easily seen that there is a boundary layer of thickness $\mathcal{O}(\varepsilon)$ far downstream nozzle exit, $x = L$, where perturbations of the velocity components α and β are finely adjusted to the boundary conditions $\alpha|_{x=L} = \alpha_L,\ \beta|_{x=L} = \beta_L$, where α_L and β_L are prescribed values.

Outside this boundary layer the first terms in the expressions for α and β in (2.16) are unimportant like the terms of order $\mathcal{O}(\varepsilon)$, and the four constants $F_1,\ F_2,\ C_2$, and D_2 are determined by the boundary conditions (2.5). The values C_1 and D_1 coincide with those given above, and moreover we have

$$F_1 = \psi_0/m, \qquad F_2 = -\frac{2}{3}\frac{\beta_\Omega We^{3/2}}{We + \frac{1}{3}}.$$

Correspondingly, (2.16) gives rise to the asymptotic solution (2.15) which holds everywhere outside a narrow layer of thickness $\mathcal{O}(\varepsilon)$ in the vicinity $x = L$. The boundary conditions at $x = L$ determine the constants C_2 and D_2. The solution constructed for the problem (2.15) describes the main effect due to the influence of low viscosity: at a large downstream distance the low viscosity makes a contribution comparable with the oscillation amplitudes of s, which decreases as a result.

It is easy to be satisfied by means (2.15) that in the considered approximation (when the interaction between liquid film and ambient media is assumed to be neglected) the kinetic energy

$$E = \frac{1}{2}\rho\left(V_\tau^2 + V_\theta^2\right)$$

of the liquid (exactly like the momentum projection on the x-axis, $\rho R V_\theta$) is conserved and only an energy transfer from the rotational to the longitudinal motion takes place, and vice versa, as it happens in the absence of viscous stresses.

It follows from (2.15) that with increase X the oscillations damp, and the film has parameters which differ from those on the nozzle exit. Thus, when $\beta_\Omega > 0$, the film expands, and the longitudinal velocity component increases, while the rotational one decreases.

3 Laminar Jet with Different Outer and Inner Pressures of Ambient Media

Laminar flows of rotating annular jets may be described by the Navier–Stokes equations in orthogonal coordinate system $\{\mathbf{s}, \mathbf{n}, \theta\}$ attached to the middle surface of the jet as it was shown in Fig. 2. We introduce an additional dimensionless term, χ, considering the pressure drop between outer, "o," and inner, "i," media:

$$\chi = \frac{p_{\text{o,st}} - p_{\text{i,st}}}{\rho V_{\tau 0}^2},$$

where $p_{\text{o,st}}$ *and* $p_{\text{i,st}}$ are the outer and inner static pressures, accordingly. We introduce the variable $N = n\varepsilon_0^{-1}$ and present the solution in the form of a power series expansion with the parameter $\varepsilon_0 = Re^{-1/2}$, assuming the normal velocity and thickness of the jet to be small values of first order. In the first approximation, we obtain a system of equations describing the rotating liquid jet flow in the form

$$\begin{aligned}
&\frac{\partial}{\partial s}(RV_\tau) + \frac{\partial}{\partial N}(RV_n) = 0,\\
&V_\tau \frac{\partial V_\tau}{\partial s} + V_n \frac{\partial V_\tau}{\partial N} = Fr^{-1}\cos\psi + \omega^2 R \sin\psi + \frac{\partial^2 V_\tau}{\partial N^2} - \frac{V_\tau^2}{r_0(s)}\\
&= -\frac{\partial p}{\partial N} - Fr^{-1}\sin\psi + \omega^2 R\cos\psi,\\
&V_\tau \frac{\partial \omega}{\partial s} + V_n \frac{\partial \omega}{\partial N} = -2\frac{V_\tau \omega}{R(s)}\sin\psi + \frac{\partial^2 \omega}{\partial N^2}.
\end{aligned} \tag{3.1}$$

In the system (3.1), we define $r_0(s) \equiv k^{-1}(s)$ as the curvature radius of the jet surface $R(s)$ and ω as the angular velocity. The equations of the middle surface and bounding surfaces of the jet have, respectively, the form

$$\frac{dR}{ds} = \sin\psi, \quad \frac{dx}{ds} = \cos\psi, \quad N = 0, \tag{3.2}$$

$$V_{\tau,l}\frac{d\delta_l}{ds} = 2V_{n,l}, \quad N = \frac{1}{2}(-1)^{l+1}\delta(s), \quad l = \begin{cases} 1, \text{ for outer surface "o"}, \\ 2, \text{ for inner surface "i"}. \end{cases} \tag{3.3}$$

The boundary conditions on the interfaces express the absence of tangential stresses and the discontinuity of the normal stresses and, in the considered approximation, have the form

$$\begin{aligned}
&\frac{\partial V_\tau}{\partial N} = 0, \;\; p_l = p_{l,\text{st}} + 2(-1)^{l+1}\sigma_s R_s^{-1}, \;\; N = \frac{1}{2}(-1)^{l+1}\delta(s), \;\; \frac{2}{R_s}\\
&= \frac{1}{R(s)}\cos\psi - \frac{d\psi}{ds}.
\end{aligned} \tag{3.4}$$

Here, R_s is the effective curvature radius of the middle surface at the considered point; δ is the liquid film thickness; σ_o and σ_i are the liquid surface tension coefficients at the outer and inner surfaces of the liquid film, $\sigma_s = \frac{1}{2}(\sigma_i + \sigma_o)$. If the tangential and angular velocity profiles at the nozzle exit, $s = 0$, are uniform and have minor differences from the profiles for $s > 0$, $V_\tau = V_\tau(s)$, $\omega = \omega(s)$, then the first, third, and fourth equations of the system (3.1) have, with allowance for (3.3), the form

$$RV_\tau\delta = Q_0 = \text{const},\ \ V_\tau\frac{dV_\tau}{ds} = Fr^{-1}\cos\psi + \omega^2 R\sin\psi,\ \ V_\tau\frac{d\omega}{ds} = -2\frac{\omega V_\tau}{R}\sin\psi. \tag{3.5}$$

Let us consider the second equation of (3.1). For fixed value of the coordinate s, it is an ordinary equation of the first order in P:

$$\frac{dp}{dN} = -Fr^{-1}\sin\psi + V_\tau^2/r_0(s) + \omega^2 R(s)\cos\psi.$$

Integrating it across the film and using the boundary conditions (3.4) we find

$$p_o - p_i = \left[Fr^{-1}\sin\psi - V_\tau^2/r_0(s) - \omega^2 R(s)\cos\psi\right]\delta(s).$$

Expressing by means of the jet shape curvature and using (3.4), we obtain the following equation:

$$\frac{d\psi}{ds} = \frac{\cos\psi - R\,Eu + \left(Fr_0^{-1}\sin\psi - We_0^{-1}\omega^2 R\cos\psi\right)/V_\tau}{R - We_0^{-1}V_\tau},$$

$$We_0^{-1} = \frac{1}{2}Q_0\,We_s,\quad Fr_0^{-1} = \frac{1}{2}Q_0\,We_s Fr^{-1},\quad Eu = \frac{1}{2}We_s\chi, \tag{3.6}$$

where $We_s = 2\,We_i\,We_o/(We_i + We_o)$ is the average Weber at the middle surface of the liquid film. Here, the dimensionless parameters of the problem are replaced by modified dimensionless numbers in accordance with formulas of [2], where $Q_0 = 1$ was taken.

Thus, the problem of the flow of incompressible liquid jet in an ideal undisturbed medium reduces to Cauchy's problem for the system of ODE ((3.2), (3.5), and (3.6)) with initial conditions

$$V_\tau(0) = 1,\quad \omega(0) = \Omega,\quad R(0) = 1,\quad \psi(0) = \psi_0. \tag{3.7}$$

Equations (3.5) in the case $Fr_0^{-1} = 0$ can be integrated. With allowance for the initial conditions (3.7), we find

$$\omega(s) = \Omega/R^2(s),\qquad V_\tau(s) = \sqrt{1 + \Omega^2\left[1 - 1/R(s)\right]}. \tag{3.8}$$

We use the expression (3.8) to calculate the surface radius R^* in the *critical section* (where the tangential velocity component vanishes) that can take a place in the limit of high initial rotation velocity, $\Omega >> 1$:

$$R^* = \Omega/\sqrt{1+\Omega^2}. \tag{3.9}$$

It follows from (3.9) that for strong initial rotation, $\Omega >> 1$, all $R^*(\Omega)$ nears unity, and the *critical sections* are asymptotically shifted to the start of the flow.

4 Numerical Simulations

Now we study the evolution of rotating annular jet (film) of an ideal liquid. This problem was formulated in Sect. 1 (system (1.5)) for $\varepsilon = Re^{-1} \to 0$ with initial conditions (1.7). After transformation of this system to a form which is solved for derivatives, Cauchy's problem indicated is integrated numerically by the Runge–Kutta method. The Runge and Kutta method showed that by combining the results of two additional Euler steps, the error can be reduced to $\mathcal{O}(h^5)$.

These algorithms can be extended to arbitrarily large first-order systems of ODE:

$$\begin{aligned} &\tfrac{dy_i}{ds} = f_i\left(\varepsilon,\ s,\ y_1(s),\dots,y_M(s)\right), \quad i=1,\dots,M, \\ &y_i(s_0) = y_{i,0}. \end{aligned} \tag{4.1}$$

The Runge–Kutta fourth-order method for this problem is given by the following equations:

$$\begin{aligned} s_{n+1} &= s_n + h, \\ y_{i,n+1} &= y_{i,n} + \tfrac{1}{6}\left(k_i^{(1)} + 2k_i^{(2)} + 2k_i^{(3)} + k_i^{(4)}\right), \end{aligned} \tag{4.2}$$

where $h = (s_N - s_0)/N$ is the step, $n = 0,\dots,N$

$$\begin{aligned} k_i^{(1)} &= hf_i\left(\varepsilon,\ s_n,\ y_{1,n},\dots,y_{M,n}\right), \\ k_i^{(2)} &= hf_i\left(\varepsilon,\ s_n + \frac{1}{2}h,\ y_{1,n} + \frac{1}{2}k_1^{(1)},\dots,y_{M,n} + \frac{1}{2}k_M^{(1)}\right), \\ k_i^{(3)} &= hf_i\left(\varepsilon,\ s_n + \frac{1}{2}h,\ y_{1,n} + \frac{1}{2}k_1^{(2)},\dots,y_{M,n} + \frac{1}{2}k_M^{(2)}\right), \\ k_i^{(4)} &= hf_i\left(\varepsilon,\ s_n + h,\ y_{1,n} + k_1^{(3)},\dots,y_{M,n} + k_M^{(3)}\right). \end{aligned} \tag{4.3}$$

The calculations for the interval $[s_N - s_0]$ are considered complete if

$$\max_{i\in[1,M]}\left\{\left|y_{i,N}^{(h/2)} - y_{i,N}^{(h)}\right| / \max_{i\in[1,M]}\left(\left|y_{i,N}^{(h)}\right|, \left|y_{i,N}^{(h/2)}\right|\right)\right\} < \varepsilon_0,$$

$$y_{i,N}^{(h)} = y_i(s_N)\big|_{h=(s_N-s_0)/N}, \qquad y_{i,N}^{(h/2)} = y_i(s_N)\big|_{h/2=(s_N-s_0)/2N}, \tag{4.4}$$

where ε_0 is the calculation error.

To calculate an annular jet of viscous liquid one of the perturbation theory methods may be used, namely, the *method of successive approximation* in [8,12]. The first step consists of calculating the velocity distribution in an ideal liquid film, in a gravitational field: Cauchy's problem (1.5)–(1.7) for $\varepsilon = 0$. In the next step the velocity distributions and the derivatives dV_τ/ds, dV_θ/ds, $d\psi/ds$, dR/ds, and dx/ds may be found analytically using the solutions $V_\tau^{(0)}$, $V_\theta^{(0)}$, $\psi^{(0)}$, $R^{(0)}, and\ x^{(0)}$ of the system (1.5) obtained for $\varepsilon \to 0$. Then, using the already obtained data, the stresses $\sigma_{\tau\tau}$, $\sigma_{\tau\theta}$, $\sigma_{\theta\theta}$ and derivatives $d(\sigma_{\tau\tau}/V_\tau)/ds$, $d(\sigma_{\tau\theta}/V_\tau)/ds$ can be determined to solve the system (1.6), which should be integrated together with (1.5) for the general case $\varepsilon \neq 0$.

The solution to (4.1) will be found using iteration procedure below:

$$\begin{aligned}
dy_i^{(0)}/ds &= f_i\,(0,\ s,\ y_1(s), \ldots, y_M(s))\,,\\
dy_i^{(1)}/ds &= f_i\big(\varepsilon,\ s,\ y_1^{(0)}(s), \ldots, y_M^{(0)}(s)\big),\\
dy_i^{(2)}/ds &= f_i\big(\varepsilon,\ s,\ y_1^{(1)}(s), \ldots, y_M^{(1)}(s)\big),\\
&\ldots\ldots\ldots\ldots\ldots\ldots\ldots\ldots\ldots\ldots\\
dy_i^{(q)}/ds &= f_i\big(\varepsilon,\ s,\ y_1^{(q-1)}(s), \ldots, y_M^{(q-1)}(s)\big).
\end{aligned}$$

Follow the estimation

$$\begin{aligned}
&\left\|y_i^{(q)} - y_i^{(q-1)}\right\|_{L_2[0,S]} \leq \Big\| f_i\big(\varepsilon, s, y_1^{(q-1)}(s), \ldots, y_M^{(q-1)}(s)\big)\\
&\quad - f_i\big(\varepsilon, s, y_1^{(q-2)}(s), \ldots, y_M^{(q-2)}(s)\big)\Big\|_{L_2[0,S]}\\
&\leq \int_0^S \Big| f_i\big(\varepsilon,\ s,\ y_1^{(q-1)}(s), \ldots, y_M^{(q-1)}(s)\big) - f_i\big(\varepsilon,\ s,\ y_1^{(q-2)}(s), \ldots, y_M^{(q-2)}(s)\big)\Big|\, ds\\
&\leq |\varepsilon|^q\ \|f_i\|_{L_2[0,S]} \leq |\varepsilon|^q K_i \underset{q\to\infty}{\longrightarrow} 0
\end{aligned}$$

for all functions f_i limited on compact $[0, S]$. As a rule, for $\varepsilon << 1$, it is enough to limit of only the first iteration, $q = 1$. The implementation of this method leads to the following ODE system:

$$\begin{aligned}
\frac{dV_\tau^{(0)}}{ds} &= \frac{V_\theta^{(0)}V_\theta^{(0)}}{V_\tau^{(0)}R^{(0)}}\sin\psi^{(0)} + \frac{Fr^{-1}}{V_\tau^{(0)}}\cos\psi^{(0)},\\
\frac{d\psi^{(0)}}{ds} &= \frac{\cos\psi^{(0)}}{V_\tau^{(0)} - We^{-1}R^{(0)}}\left(\frac{V_\theta^{(0)}V_\theta^{(0)}}{V_\tau^{(0)}R^{(0)}} - We^{-1} - \frac{Fr^{-1}}{V_\tau^{(0)}}\tan\psi^{(0)} - \frac{Eu}{We}\frac{R^{(0)}}{\cos\psi^{(0)}}\right),\\
\frac{dV_\theta^{(0)}}{ds} &= -\frac{V_\theta^{(0)}}{R^{(0)}}\sin\psi^{(0)}.
\end{aligned}$$

Calculating the terms in (4.1) and multiplying $\varepsilon = Re^{-1}$ by the second derivations $d^2V_\tau^{(0)}/ds^2$ and $d^2V_\theta^{(0)}/ds^2$, one obtains

$$\begin{aligned}
&Re^{-1}\left[\tfrac{d}{ds}\left(\tfrac{\sigma_{\tau\tau}}{V_\tau}\right) - \tfrac{\sigma_{\theta\theta}}{V_\tau R}\sin\psi\right] \approx \varepsilon\left[\tfrac{d}{ds}\left(\tfrac{\sigma_{\tau\tau}}{V_\tau}\right) - \tfrac{\sigma_{\theta\theta}}{V_\tau R}\sin\psi\right]\Big|_{V_\tau=V_\tau^{(0)},V_\theta=V_\theta^{(0)},R=R^{(0)},\psi=\psi^{(0)}}\\
&\quad = \varepsilon\tfrac{4}{V_\tau^{(0)}}\left[\tfrac{d^2V_\tau^{(0)}}{ds^2} - \tfrac{dV_\tau^{(0)}}{ds}\left(\tfrac{1}{V_\tau^{(0)}}\tfrac{dV_\tau^{(0)}}{ds} + \tfrac{1}{2}\tfrac{\sin\psi^{(0)}}{R^{(0)}}\right) + \tfrac{V_\tau^{(0)}}{2R^{(0)}}\left(\cos\psi^{(0)}\tfrac{d\psi^{(0)}}{ds} - \tfrac{3\sin^2\psi^{(0)}}{R^{(0)}}\right)\right]\\
&Re^{-1}\tfrac{\sigma_{\tau\tau}}{V_\tau} \approx \tfrac{\sigma_{\tau\tau}}{V_\tau}Re^{-1}\Big|_{V_\tau=V_\tau^{(0)},V_\theta=V_\theta^{(0)},R=R^{(0)},\psi=\psi^{(0)}} = 2\varepsilon\left(2\tfrac{1}{V_\tau^{(0)}}\tfrac{dV_\tau^{(0)}}{ds} + \tfrac{\sin\psi^{(0)}}{R^{(0)}}\right),\\
&Re^{-1}\tfrac{\sigma_{\theta\theta}}{V_\tau R} \approx \tfrac{\sigma_{\theta\theta}}{V_\tau R}Re^{-1}\Big|_{V_\tau=V_\tau^{(0)},V_\theta=V_\theta^{(0)},R=R^{(0)},\psi=\psi^{(0)}} = \tfrac{2\varepsilon}{R}\left(\tfrac{1}{V_\tau^{(0)}}\tfrac{dV_\tau^{(0)}}{ds} + \tfrac{2\sin\psi^{(0)}}{R^{(0)}}\right),
\end{aligned}\tag{4.5}$$

$$\begin{aligned}
&Re^{-1}\left[\tfrac{d}{ds}\left(\tfrac{\sigma_{\tau\theta}}{V_\tau}\right) - \tfrac{\sigma_{\tau\theta}}{V_\tau R}\sin\psi\right] \approx \varepsilon\left[\tfrac{d}{ds}\left(\tfrac{\sigma_{\tau\theta}}{V_\tau}\right) - \tfrac{\sigma_{\tau\theta}}{V_\tau R}\sin\psi^{(0)}\right]\Big|_{V_\tau=V_\tau^{(0)},V_\theta=V_\theta^{(0)},R=R^{(0)},\psi=\psi^{(0)}}\\
&\quad = \varepsilon\left[\tfrac{1}{V_\tau^{(0)}}\tfrac{d^2V_\theta^{(0)}}{ds^2} - \tfrac{V_\theta^{(0)}}{V_\tau^{(0)}}\tfrac{\cos\psi^{(0)}}{R^{(0)}}\tfrac{d\psi^{(0)}}{ds} - \left(\tfrac{1}{V_\tau^{(0)}}\tfrac{dV_\tau^{(0)}}{ds} + \tfrac{2\sin\psi^{(0)}}{R^{(0)}}\right)\left(\tfrac{1}{V_\tau^{(0)}}\tfrac{dV_\theta^{(0)}}{ds} - \tfrac{V_\theta^{(0)}}{V_\tau^{(0)}}\tfrac{\sin\psi^{(0)}}{R^{(0)}}\right)\right]
\end{aligned}\tag{4.6}$$

$$\begin{aligned}
\tfrac{d^2V_\tau^{(0)}}{ds^2} &= \tfrac{d}{ds}\left[\tfrac{\left(V_\theta^{(0)}\right)^2}{V_\tau^{(0)}R^{(0)}}\sin\psi^{(0)} + Fr^{-1}\tfrac{\cos\psi^{(0)}}{V_\tau^{(0)}}\right]\\
&= \tfrac{V_\theta^{(0)}}{V_\tau^{(0)}}\left[\tfrac{\sin\psi^{(0)}}{R^{(0)}}\left(2\tfrac{dV_\theta^{(0)}}{ds} - \tfrac{V_\theta^{(0)}}{V_\tau^{(0)}}\tfrac{dV_\tau^{(0)}}{ds} - \tfrac{V_\theta^{(0)}}{R^{(0)}}\sin\psi^{(0)}\right) - \tfrac{Fr^{-1}}{V_\tau^{(0)}V_\theta^{(0)}}\tfrac{dV_\tau^{(0)}}{ds}\right.\\
&\qquad\left.+\left(\tfrac{V_\theta^{(0)}}{R^{(0)}}\cos\psi^{(0)} - Fr^{-1}\tfrac{\sin\psi^{(0)}}{V_\theta^{(0)}}\right)\tfrac{d\psi^{(0)}}{ds}\right],\\
\tfrac{d^2V_\theta^{(0)}}{ds^2} &= -\tfrac{d}{ds}\left[\tfrac{V_\theta^{(0)}}{R^{(0)}}\sin\psi^{(0)}\right] = -\tfrac{\sin\psi^{(0)}}{R^{(0)}}\left(\tfrac{dV_\theta^{(0)}}{ds} - \tfrac{V_\theta^{(0)}}{R^{(0)}}\sin\psi^{(0)}\right) - \tfrac{V_\theta^{(0)}}{R^{(0)}}\cos\psi^{(0)}\tfrac{d\psi^{(0)}}{ds}.
\end{aligned}\tag{4.7}$$

Substituting the results (4.5) into the base system (1.5)–(1.6), we obtain finally

$$\begin{aligned}
\tfrac{dV_\tau}{ds} &= F_1(s, V_\tau, V_\theta, \psi, R) + \varepsilon G_1(s, V_\tau, V_\theta, \psi, R),\\
\tfrac{d\psi}{ds} &= \left\{\left[\tfrac{V_\theta^2}{V_\tau R} - We^{-1} - \varepsilon\tfrac{2}{R}\left(\tfrac{F_1(.)}{V_\tau} + \tfrac{2\sin\psi}{R}\right)\right]\cos\psi - \tfrac{Fr^{-1}}{V_\tau}\sin\psi - \tfrac{Eu}{We}R\right\}\\
&\quad\times\left[V_\tau - We^{-1}R - 2\varepsilon\left(2\tfrac{F_1(.)}{V_\tau} + \tfrac{\sin\psi}{R}\right)\right]^{-1}\\
\tfrac{dV_\theta}{ds} &= F_2(s, V_\tau, V_\theta, \psi, R) + \varepsilon G_2(s, V_\tau, V_\theta, \psi, R),\\
\tfrac{dR}{ds} &= \sin\psi, \qquad \tfrac{dx}{ds} = \cos\psi,
\end{aligned}\tag{4.8a}$$

where

$$
\begin{aligned}
F_1(.) &= \tfrac{V_\theta^2}{V_\tau}\tfrac{1}{R}\sin\psi + Fr^{-1}\tfrac{\cos\psi}{V_\tau},\ F_2(.) = -\tfrac{V_\theta}{R}\sin\psi,\\
G_1(.) &= 4\tfrac{V_\theta}{V_\tau^2}\left\{\tfrac{\sin\psi}{R}\left(2F_2(.) - \tfrac{V_\theta}{V_\tau}F_1(.) - \tfrac{V_\theta}{R}\sin\psi\right) - \tfrac{V_\tau}{V_\theta}F_1(.)\left(\tfrac{F_1(.)}{V_\tau} + \tfrac{1}{2}\tfrac{\sin\psi}{R}\right)\right.\\
&\quad -\tfrac{Fr^{-1}}{V_\tau}\tfrac{F_1(.)}{V_\theta} + \tfrac{\cos^2\psi}{V_\tau - We^{-1}R}\left(\tfrac{V_\theta^2}{V_\tau}\tfrac{1}{R} - We^{-1} - Fr^{-1}\tfrac{\tan\psi}{V_\tau} - \tfrac{Eu}{We}\tfrac{R}{\cos\psi}\right)\\
&\quad \left.\times\left(\tfrac{V_\theta}{R} - Fr^{-1}\tfrac{\tan\psi}{V_\theta} + \tfrac{1}{2R}\tfrac{V_\tau^2}{V_\theta}\right) - \tfrac{3}{2}\tfrac{V_\tau^2}{V_\theta}\tfrac{\sin^2\psi}{R^2}\right\},\\
G_2(.) &= -\left[2\tfrac{V_\theta}{V_\tau R}\tfrac{\cos^2\psi}{V_\tau - We^{-1}R}\left(\tfrac{V_\theta^2}{V_\tau}\tfrac{1}{R} - We^{-1} - Fr^{-1}\tfrac{\tan\psi}{V_\tau} - \tfrac{Eu}{We}\tfrac{R}{\cos\psi}\right)\right.\\
&\quad \left.+\tfrac{1}{V_\tau}\tfrac{\sin\psi}{R}\left(F_2(.) - \tfrac{V_\theta}{R}\sin\psi\right) + \left(\tfrac{F_1(.)}{V_\tau} + 2\tfrac{\sin\psi}{R}\right)\left(\tfrac{F_2(.)}{V_\tau} - \tfrac{V_\theta}{V_\tau}\tfrac{\sin\psi}{R}\right)\right].
\end{aligned}
\tag{4.8b}
$$

Solving the system (4.8a,b) with initial conditions (1.7) numerically using Runge–Kutta algorithm (4.2)–(4.4), one obtains a new velocity distribution $V_\tau,\ V_\theta$ and the liquid film shape $R(x)$ with allowance for viscosity.

For the case without rotation, when $V_\theta|_{s=0} = 0$, the system (4.8a,b) is sufficiently simplified:

$$
\begin{aligned}
\tfrac{dV_\tau}{ds} &= F_1(s, V_\tau, \psi, R) + \varepsilon G_1(s, V_\tau, \psi, R),\\
\tfrac{d\psi}{ds} &= -\left\{\left[We^{-1} + \varepsilon\tfrac{2}{R}\left(\tfrac{F_1(.)}{V_\tau} + \tfrac{2\sin\psi}{R}\right)\right]\cos\psi + Fr^{-1}\tfrac{\sin\psi}{V_\tau} + \tfrac{Eu}{We}R\right\}\\
&\quad \times\left[V_\tau - We^{-1}R - 2\varepsilon\left(2\tfrac{F_1(.)}{V_\tau} + \tfrac{\sin\psi}{R}\right)\right]^{-1},\\
\tfrac{dR}{ds} &= \sin\psi,\quad \tfrac{dx}{ds} = \cos\psi,
\end{aligned}
$$

where

$$
\begin{aligned}
F_1(.) &= \tfrac{Fr^{-1}}{V_\tau}\cos\psi,\\
G_1(.) &= -4\left\{\tfrac{F_1(.)}{V_\tau}\left(\tfrac{F_1(.)}{V_\tau} + \tfrac{1}{2}\tfrac{\sin\psi}{R}\right) + \tfrac{Fr^{-1}}{V_\tau}\tfrac{F_1(.)}{V_\tau^2} + \tfrac{3}{2}\tfrac{\sin^2\psi}{R^2}\right.\\
&\quad \left.+\tfrac{\cos^2\psi}{V_\tau - We^{-1}R}\left(We^{-1} + Fr^{-1}\tfrac{\tan\psi}{V_\tau} + \tfrac{Eu}{We}\tfrac{R}{\cos\psi}\right)\left(\tfrac{1}{2R} - Fr^{-1}\tfrac{\tan\psi}{V_\tau^2}\right)\right\}.
\end{aligned}
$$

The numerical simulations were provided using MAPLE. The main results are present in Fig. 3 for the case with $p_o = p_i$ $(Eu = 0)$, $We = 13.74$, $Fr = 5.01 \times 10^4$ for two different Reynolds numbers, $Re = 2{,}000$ (Fig. 3a) and $Re = 100$ (Fig. 3b).

To show that the values $R(x)$ are not zero for $x > 5$ in Fig. 3 (line (a)) we presented their numerical values (see Table 1). It can be seen the value $R(x) > 10^{-3}$ even for $x > 12$.

To show agreement between results obtained from asymptotic analysis to numerical ones, let us consider the ideal liquid film ($\varepsilon = 0$) for the case $We = 7.14$, $Fr = \infty$, $V_\tau|_{s=0} = 1.0$, $\psi|_{s=0} = 0$, $V_\theta|_{s=0} = 1.2\,We^{-1/2} = 0.44$ on the interval $x \in [0, 25]$. This case corresponds to situation when the middle surface of the film has to oscillate in a stable flow according to (2.15) obtained in Sect. 2. The results are shown in Fig. 4a.

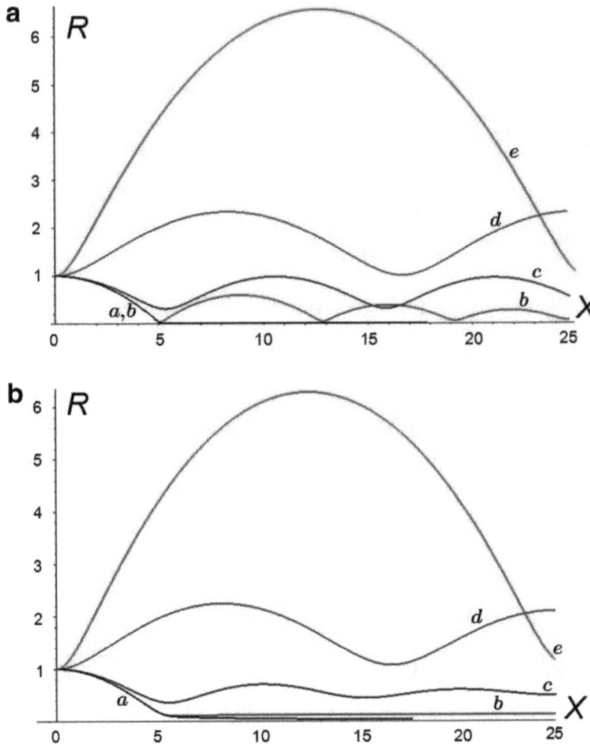

Fig. 3 Shape of the liquid film for the conditions: $We = 13.74$, $Fr = 5.01 \times 10^4$; $Re = 2,000$ case (**a**), Re = 100 case (**b**). Boundary conditions: $x_{|s=0} = 0$, $R_{|s=0} = 1$, $V_{\tau|s=0} = 1$, $\psi_{|s=0} = 0$, $V_{\theta|s=0} = 0.001, 0.01, 0.1, 0.5, 1.0$; the lines correspond to $V_\theta|_{s=0} = (a) \to 0$, $(b) \to 0.01$, $(c) \to 0.1$, $(d) \to 0.5$, $(e) \to 1.0$

Table 1 Numerical form of the shape $R(x)$

s	1	2	3	4	5	6	7
$x(s)$	0.99	1.99	2.97	3.94	4.88	5.87	6.87
$R(s)$	0.96084	0.84422	0.65247	0.38957	0.06469	0.00462	0.00395
s	8	9	10	11	12	13	14
s	7.87	8.87	9.87	10.87	11.87	12.87	13.87
$x(s)$	0.00345	0.00307	0.00276	0.00251	0.00229	0.00212	0.00195

It can be seen that for the case with ideal liquid we obtain undamped oscillations (Fig. 4a, red line) while accounting for viscosity; $\varepsilon = Re^{-1} = 0.01$ leads to natural damping.

Let us compare our modeling with numerical results [3] (Fig. 4b) with experiment [9] provided for an annular liquid jet with parameters $Re = 100$, $We = 110$, $Fr = \infty$, $Eu = 0$ and the initial angular velocities $V_\theta|_{s=0} = 2.0, 4.0$. Comparison shows good agreement of our model (solid lines) with experimental data (dash lines).

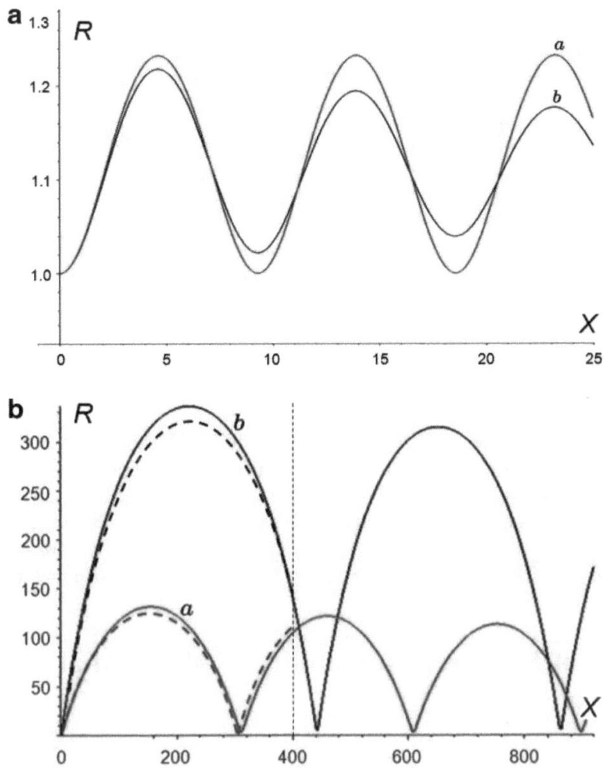

Fig. 4 (**a**) Shape of the liquid film for the conditions: $We = 7.143$, $Fr = \infty$, $\varepsilon = 0, 0.01$; boundary conditions: $x|_{s=0} = 0$, $R|_{s=0} = 1$, $V_\tau|_{s=0} = 1$, $V_\theta|_{s=0} = 1.2\,We^{-1/2} = 0.44$, $\psi|_{s=0} = 0$; line a corresponds to $\varepsilon = Re^{-1} = 0$, line b—$\varepsilon = Re^{-1} = 0.01$. (**b**) Shape of the liquid film for the conditions: $We = 110$, $Fr = \infty$, $Re = 100$, $Eu = 0$, boundary conditions: $x|_{s=0} = 0$, $R|_{s=0} = 1$, $V_\tau|_{s=0} = 1$, $V_\theta|_{s=0} = 2.0, 4.0$, $\psi|_{s=0} = 0$; the lines a and b correspond to $V_\theta|_{s=0} = 2.0$ and 4.0, accordingly; dash lines for $0 \le x \le 400$—experiment [3,9]

Our next step is to compare lines a and b with numerical lines $a*$ and $b*$ obtained for the oil annular rotational jet (for results, see Fig. 5a) with the following parameters and properties: flow rate $\dot{m} = 0.01777$ kg/s, initial geometry sizes $R_0 = 0.002$ m, $\delta_0 = 1.524 \cdot 10^{-4}$ m, fluid density $\rho = 765$ kg/m^3, viscosity $\mu = 9.2 \cdot 10^{-4}$ kg/m · s, and surface tension $\sigma_* = 0.025$ N/m [1], for the case when the inner pressure differs from outer one, $\Delta p \neq 0$, and pressure difference between inner and outer pressures is $\Delta p = 21, 138$ kPa. These conditions correspond to the Euler numbers $2.6 \cdot 10^3$, $1.7 \cdot 10^4$; $We = 347.0$, $Fr = 7586$, and $Re = 20{,}000$ correspond to the conditions above.

Comparing our simulations with experiment [10] for a water annular jet without rotation, the nozzle diameter $D_0 = 1.0$ mm and the pressure drop $\Delta p = 2.4$ Pa show good agreement for the two variants of inlet velocities, $V_{in} = 0.8$ and 1.6 m/s (Fig. 5b). Solid lines correspond to our calculations, and symbols ■ and • to experimental data for $V_{in} = 0.8$, 1.6 m/s, respectively.

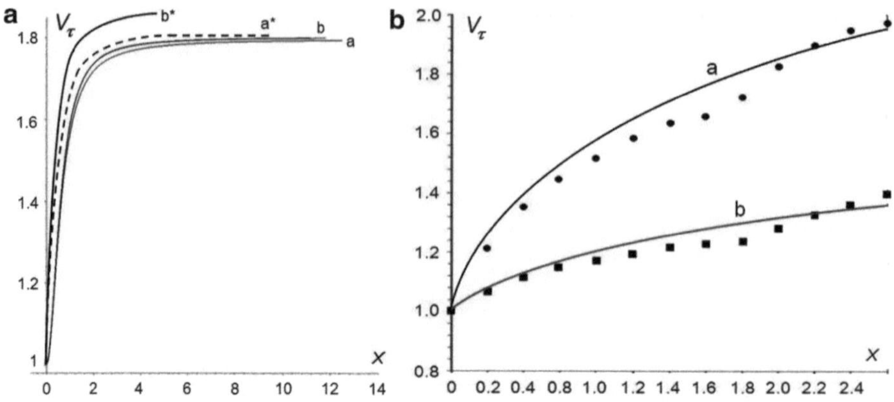

Fig. 5 (**a**) Longitudinal velocity V_τ evolution along the axial coordinate x for the conditions: $We = 347.0,\ Fr = 7586,\ Re = 20{,}000, \mathrm{Eu} = 2.6 \cdot 10^3,\ 1.7 \cdot 10^4$; boundary conditions: $x|_{s=0} = 0,\ R|_{s=0} = 1,\ V_\tau|_{s=0} = 1.5,\ V_\theta|_{s=0} = 0.5,\ \psi|_{s=0} = 0$; a and $a*$ $\Delta p = 21$ kPa, b and $b*$ $\Delta p = 138$ kPa. (**b**) Longitudinal velocity on the axial coordinate x for the boundary conditions: $x|_{s=0} = 0,\ R|_{s=0} = 1,\ V_\tau|_{s=0} = 1,\ V_\theta|_{s=0} = 0,\ \psi|_{s=0} = 0$; (*a*) $We = 2.2,\ Fr = 1.3,\ V_{\mathrm{in}} = 0.8$m/s, $\mathrm{Eu} = 0.00868$; (*b*) $We = 8.8,\ Fr = 5.2,\ V_{\mathrm{in}} = 1.6$m/s, $\mathrm{Eu} = 0.00868$

5 Analysis of Nonlinear Instability in Meridional Cross Section

Instabilities leading to oscillations in some particular properties of a system are intimately related to pattern formation. Nontrivial patterns form spontaneously as a result of the occurrence and propagation of the fronts of instability. These spontaneous patterns can be turned into functional structures at the corresponding length scale when the pattern-forming processes are properly designed and controlled. In this work, we propose a scenario for mass transfer instability in one-dimensional flow of a one-component fluid near its discontinuous liquid–gas phase transition. Instability leading to density oscillations occurs when the system fails to support steady-state flow due to the absence of mechanically stable uniform state as a consequence of a discontinuous transition. The main hydrodynamic instabilities are the Rayleigh–Taylor and Kelvin–Helmholtz [6, 8]. The instability approaches may be conditionally divided into two parts, linear and nonlinear. The linear approach is well known and simple for study, but it shows only which oscillating modes are stable and which are unstable. Nonlinear approach shows the singularities of the surface between jet and surrounding media that appear as a result of the spatial instability development. In this paragraph we will consider nonlinear approach to study instability in meridional cross section of the jet.

The hydrodynamic instability has 3D character, but for qualitative analysis we limit our research to study the Rayleigh–Taylor instability in meridional plane $R,\ \theta$ assuming the flow to be cylindrically axisymmetric. Then the motion equation may be written as [7]

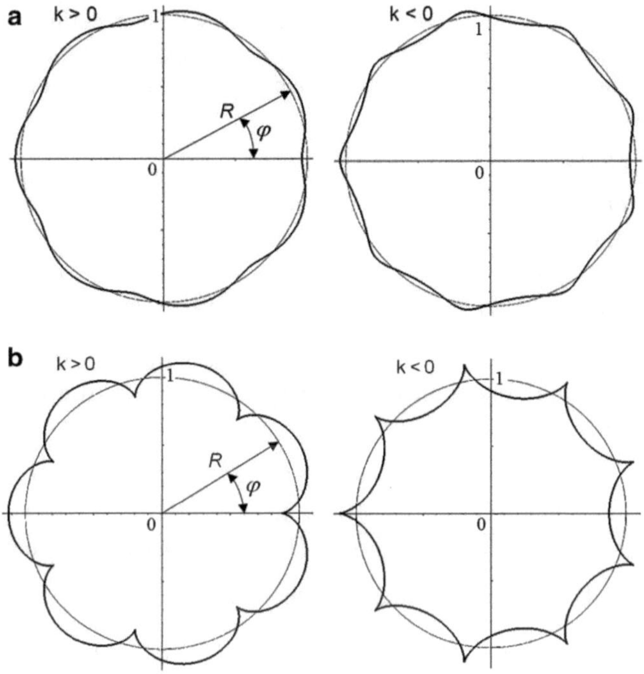

Fig. 6 (**a**) k-petal rose: $Z(\phi,t) = \exp(i\phi) - \lambda k^{-1}\exp(ik\phi)$ with $k = \pm 8$, $|\lambda| < 1$. (**b**) Self-intersecting k-petal rose: $Z(\phi,t) = \exp(i\phi) - \lambda k^{-1}\exp(ik\phi)$ with $k = \pm 8$, $|\lambda| = 1$

$$\rho\frac{\partial^2 \mathbf{r}}{\partial t^2} = \Delta p(r,\phi,t)\left(\mathbf{e}_z \times \frac{\partial \mathbf{r}}{\partial \phi}\right), \tag{5.1}$$

where ϕ is Lagrange coordinate, i.e., $dm = \rho d\phi$, and $\Delta p = p_i - p_o$ is assumed to be constant. Substituting $Z = x + iy = r\exp(i\phi)$ transforms (5.1) to the equation in polar coordinates

$$\rho\frac{\partial^2 Z}{\partial t^2} = i\Delta p\frac{\partial Z}{\partial \phi}. \tag{5.2}$$

Assuming that at the nozzle exit cross section the cylindrical film rotates with angular velocity $\Omega = V_{\theta,0}/R_0$ and has zero radial velocity, the initial conditions are given as follows:

$$Z(\phi,0) = \exp(i\phi) - \frac{\lambda_0}{k}\exp(ik\phi), \qquad \frac{\partial Z(\phi,0)}{\partial t} = i\Omega Z(\phi,0), \tag{5.3}$$

where $|\lambda_0| < 1$ and k is a wave mode. Note that the second initial condition is sufficiently different from the similar one in [6]. Let us find the solution to the problems (5.2) and (5.3) in the form (Fig. 6a)

$$Z(\phi,t) = R(t)\left[\exp(i\phi) - \lambda(t)k^{-1}\exp(ik\phi)\right]. \tag{5.4}$$

Note that at $|\lambda(t)| \geq 1$ the curve $Z(\phi, \tau)$ becomes self-intersecting (k-petal rose). Fig. 6b illustrates k-petal rose (for $\lambda = 1$). It would appear natural to assume that the film lives up to an instant of time t_* such that $|\lambda(t_*)| = 1$. Substituting (5.4) into (5.2) gives the system

$$\frac{d^2\lambda}{d\tau^2} + \frac{2}{R}\frac{dR}{d\tau}\frac{d\lambda}{d\tau} = a\lambda(\tau)(1-k), \qquad \frac{d^2R}{d\tau^2} = -aR(t), \qquad a = \frac{1}{2\pi}Eu/We \tag{5.5}$$

with initial conditions

$$R(0) = 1, \quad \frac{dR}{d\tau}\Big|_{\tau=0} = V_\theta\Big|_{\tau=0} = V_\Omega, \qquad \lambda(0) = \lambda_0, \quad \frac{d\lambda}{d\tau}\Big|_{\tau=0} = 0. \tag{5.6}$$

Here $\tau = t/t_0$ and $t_0 = R_0/V_{\tau 0}$. Obviously, the system (5.5) splits onto two equations. The second one with its boundary conditions (5.6) accepts analytical solution

$$R(\tau) = \begin{cases} \cos\sqrt{a}\tau + \frac{V_\Omega}{\sqrt{a}}\sin\sqrt{a}\tau, & \Delta p > 0 \\ V_\Omega\tau + 1, & \Delta p = 0 \\ \frac{1}{2}\left[\left(1 + (V_\Omega/\sqrt{|a|})\right)\exp\left(\sqrt{|a|}\tau\right) + \left(1 - (V_\Omega/\sqrt{|a|})\right)\exp\left(-\sqrt{|a|}\tau\right)\right], & \Delta p < 0. \end{cases}$$

Substituting (5.4) into the first equation of the system (5.5) gives

$$\begin{aligned} &\frac{d^2\lambda}{d\tau^2} + F(\tau)\frac{d\lambda}{d\tau} = \operatorname{sign}(\Delta p)\,|a|(1-k)\lambda(\tau), \\ &F(\tau) = 2\sqrt{|a|} \times \begin{cases} \frac{b - \tan\sqrt{a}\,\tau}{1 + b\tan\sqrt{a}\,\tau}, & \Delta p > 0 \\ \frac{b + \tanh\sqrt{|a|}\,\tau}{1 + b\tanh\sqrt{|a|}\,\tau}, & \Delta p < 0, \end{cases} \\ &\lambda(0) = \lambda_0, \qquad \frac{d\lambda}{d\tau}\Big|_{\tau=0} = 0, \end{aligned} \tag{5.7}$$

where $b = V_\Omega/\sqrt{|a|} = V_\Omega\sqrt{2\pi\, We/Eu}$. In the case when pressure drop is absent (5.7) has a trivial solution, $\lambda = \lambda_0$. In general case the problem (5.7) accepts analytical solution:

$$\lambda(\tau) = \frac{\lambda_0}{\sqrt{|k|}} \begin{cases} \frac{1}{\cos\sqrt{a}\,\tau + b\sin\sqrt{a}\,\tau} \begin{cases} \cos\sqrt{ak}\,\tau\left(\sqrt{k} + b\tan\sqrt{ak}\,\tau\right), & k > 0 \\ \cosh\sqrt{a|k|}\,\tau\left(\sqrt{|k|} + b\tanh\sqrt{a|k|}\,\tau\right), & k < 0 \end{cases} & \Delta p > 0 \\ \sqrt{|k|}, & \Delta p = 0 \\ \frac{1}{\cosh\sqrt{|a|}\,\tau + b\sinh\sqrt{|a|}\,\tau} \begin{cases} \cosh\sqrt{|ak|}\,\tau\left(\sqrt{|k|} + b\tanh\sqrt{|a|k}\,\tau\right), & k > 0 \\ \cos\sqrt{|ak|}\,\tau\left(\sqrt{|k|} + b\tan\sqrt{|ak|}\,\tau\right), & k < 0 \end{cases} & \Delta p < 0 \end{cases} \tag{5.8}$$

For nonrotating jet, $V_\Omega = 0$, the solution (5.8) is transformed to the simple form

$$\lambda(\tau) = \lambda_0 \begin{cases} \left.\begin{array}{l} \cos\sqrt{ak}\,\tau/\cos\sqrt{a}\,\tau, \ k > 0 \\ \cosh\sqrt{a|k|}\tau/\cos\sqrt{a}\,\tau, \ k < 0 \end{array}\right\} & \Delta p > 0, \\ 1, & \Delta p = 0, \\ \left.\begin{array}{l} \cosh\sqrt{|a|k}\,\tau/\cosh\sqrt{|a|}\,\tau, \ k > 0 \\ \cos\sqrt{|ak|}\,\tau/\cosh\sqrt{|a|}\,\tau, \ k < 0 \end{array}\right\} & \Delta p < 0. \end{cases} \tag{5.9}$$

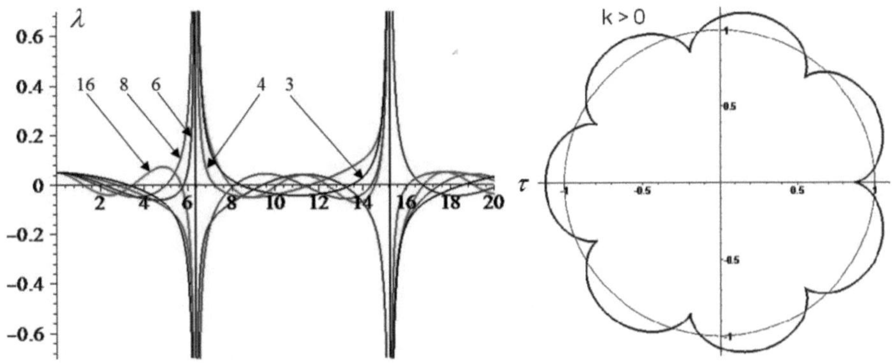

Fig. 7 Dependencies $\lambda(\tau)$ for $a = 0.125$ and $b = 0.8$; modes $k = 3,\ 4,\ 6,\ 8,\ 16$

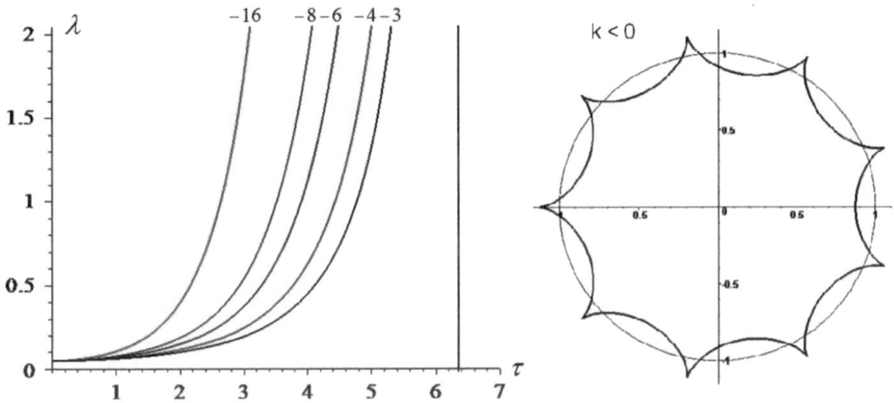

Fig. 8 Function $\lambda(\tau)$ for $a = 0.125$ and $b = 0.8$; modes $k = -3,\ -4,\ -6,\ -8,\ -16$

Let us study the modes $k = \pm2, \pm3, \pm4, \pm5, \pm6, \ldots$ in a wide range temporary interval, $\tau > 100$, for low initial value of parameter λ_0, $\lambda_0 = 0.05$. Figures 7–10 show dependence λ on dimensionless time τ; the curves correspond to the wave numbers $k = \pm3,\ \pm4,\ \pm6,\ \pm8,\ \pm16$.

It can be seen that for positive value a, which corresponds to positive pressure drop, $\Delta p > 0$ and for wide range of parameter $b = 0.8 \div 25$, all positive modes, $k > 1$, are unstable (Fig. 7). This result can be obtained analytically. In fact, see (5.8)

$$\lambda(\tau) = \lambda_0 \frac{1 + (b/\sqrt{k}) \tan \sqrt{ak}\,\tau}{1 + b \tan \sqrt{a}\,\tau} \cdot \frac{\cos \sqrt{ak}\,\tau}{\cos \sqrt{a}\,\tau}, \quad \Delta p > 0, \quad k > 0.$$

Thus, $\lambda(\tau) \to \infty$, when $\tau = \pi n / 2\sqrt{ak}$, $n = 1,\ 2, \ldots$, i.e., we obtain periodic singular unstable points that are consistent with the results shown in Fig. 7.

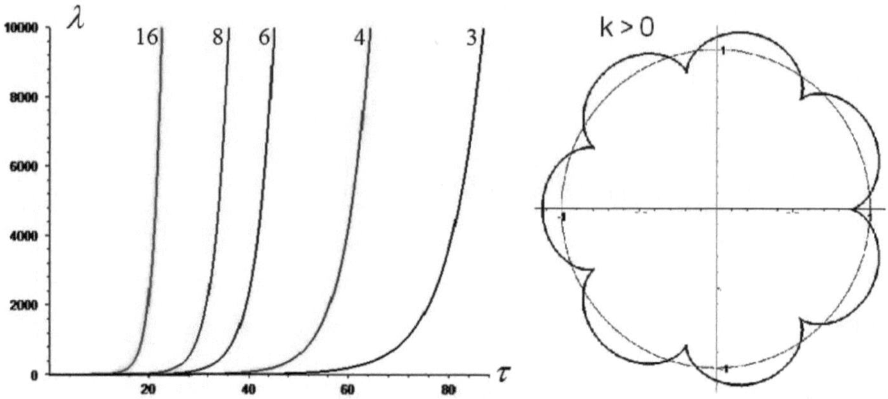

Fig. 9 Function $\lambda(\tau)$ for $a = -0.04$ and $b = 25.0$; modes $k = 3,\ 4,\ 6,\ 8,\ 16$

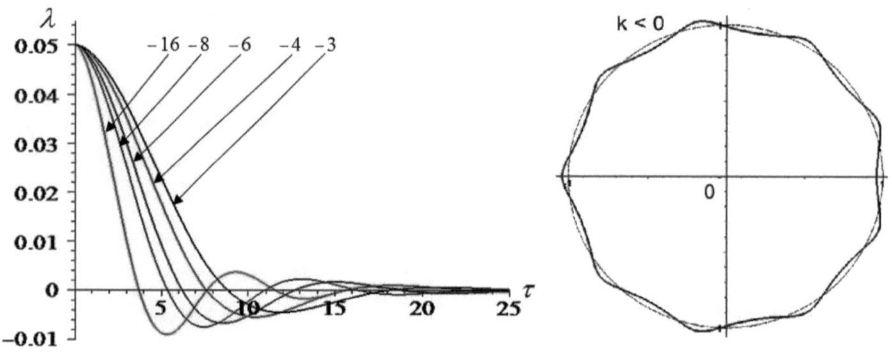

Fig. 10 Function $\lambda(\tau)$ for $a = -0.04$ and $b = 25.0$; modes $k = -3,\ -4,\ -6,\ -8,\ -16$

All modes with negative k are also unstable (Fig. 8). Really,

$$\lambda(\tau) = \lambda_0 \frac{1 + (b/\sqrt{|k|}) \tanh \sqrt{a|k|}\,\tau}{1 + b \tan \sqrt{a}\,\tau} \cdot \frac{\cosh \sqrt{a|k|}\,\tau}{\cos \sqrt{a}\,\tau}, \quad \Delta p > 0, \quad k < 0;$$

hence $\lambda(\tau) \to \infty$, when $\tau = \pi n / 2\sqrt{ak}$, $n = 1, 2, \ldots$, and at $\tau \to \infty$, i.e., we obtain periodic singular unstable points and asymptotic instability. For negative pressure drop corresponding to negative parameter a, all negative modes decreased (Fig. 10), while all positive modes increased (Fig. 9). The results shown in Figs. 9 and 10 may be obtained analytically. Indeed, from (5.8), we have

$$\lambda(\tau) = \lambda_0 \frac{1 + (b/\sqrt{|k|}) \tanh \sqrt{|a|k}\,\tau}{1 + b \tanh \sqrt{|a|}\,\tau} \cdot \frac{\cosh \sqrt{|ak|}\,\tau}{\cosh \sqrt{|a|}\,\tau}, \quad \Delta p < 0, \quad k > 0.$$

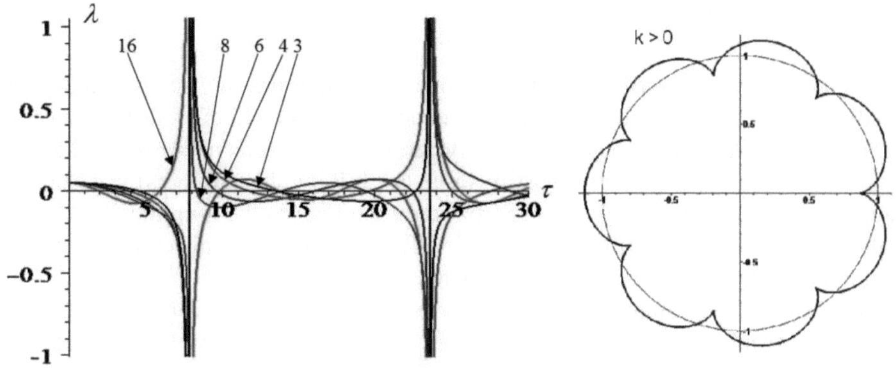

Fig. 11 Function $\lambda(\tau)$ for $a = 0.04$ and $b = 0$; modes $k = 3,\ 4,\ 6,\ 8,\ 16$

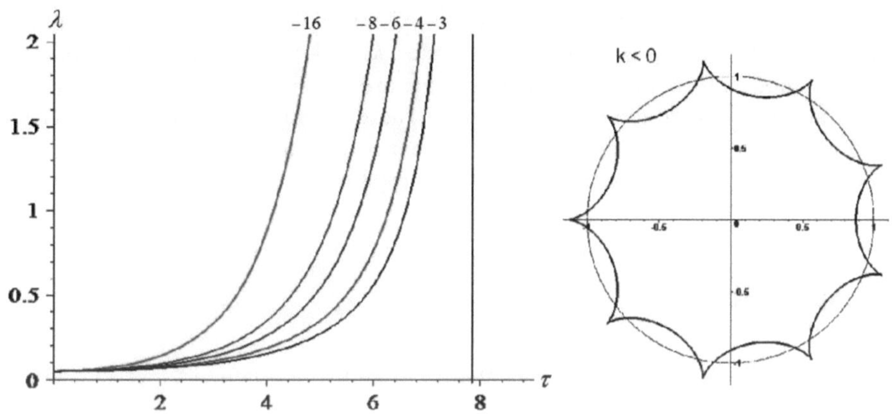

Fig. 12 Function $\lambda(\tau)$ for $a = 0.04$ and $b = 0$; modes $k = -3,\ -4,\ -6,\ -8,\ -16$

Hence, $\lambda(\tau) \to \infty$, at $\tau \to \infty$, for all $k < 0$, when $\Delta p > 0$, for all $k > 0$; and for all $k > 1$, when $\Delta p < 0$, i.e., there is asymptotic instability that is consistent with Figs. 8 and 9. For negative pressure drop, $\Delta p < 0$, one obtains $\lambda(\tau) \to 0$, at $\tau \to \infty$, for all negative modes, $k < 0$, i.e., there is asymptotic stability (compare with Fig. 10). The lines $\lambda(\tau)$ are gathering with increasing $|a|$ as it plays a role of dimensionless frequency in (5.8) and (5.9). In the case of nonrotating jet, when parameter $b = 0$, and at positive pressure drop ($a > 0$), both the positive and negative modes become unstable (Figs. 11 and 12) by comparison with the rotational jet (Fig. 7).

For negative pressure drop, all positive modes are unstable (Fig. 9), and negative modes are stable (similar to Fig. 10, i.e., in this case rotation does depend on the jet stability). The rotating velocity affects jet stability so that only two of considered positive modes, $k = 4$ and 16, become stable. The curves in Figs. 7, 8, 11, and 12 are periodic, so their behavior is shown in the range of one–two periods.

6 Conclusion

The equations that described the flow of rotating annular jets of viscous liquid in an undisturbed ideal medium are obtained and analyzed. The effect of surface tension gravity forces and inner–outer pressure difference on jet behavior was considered. Asymptotic analysis for low viscosity value (high Reynolds numbers) was carried out. It is shown that energy transfer from the rotational to the longitudinal motion takes place. The problem of a laminar jet flow with pressure difference between outer and inner ambient area was formulated additionally.

It is shown that equations which describe this phenomenon can be reduced to Cauchy's problem for the system of ODE. This equation set was solved by the *method of successive approximation*. The solution results are in good agreement with test.

Nonlinear analysis of instability shows that the film instability in meridional cross section is developed due to pressure difference. At positive pressure inner–outer difference, $\Delta p > 0$, two positive modes, $k = 4,\ and 16$ are stable while all negative modes, $k < 0$, are unstable. In the case $\Delta p < 0$ all negative modes are stable and all positive modes are unstable. These instabilities in meridional cross section developed due to pressure drop cannot be stabilized by rotation; only two modes, $k = 14$ and 16, become stable due to rotation.

7 Appendix

In the case of $\varepsilon \equiv Re^{-1} = 0$, Eu $= 0$, $Fr \to \infty$ the problem (1.5)–(1.7) leads to the system

$$\begin{array}{ll} \frac{dV_\tau}{ds} = \frac{V_\theta^2}{V_\tau R}\sin\psi, & \frac{dV_\theta}{ds} = -\frac{V_\theta}{R}\sin\psi, \\ \frac{d\psi}{ds} = \frac{\cos\psi}{R}\cdot\frac{V_\theta^2/V_\tau - We^{-1}R}{V_\tau - We^{-1}R}, & \frac{dR}{ds} = \sin\psi, \quad \frac{dx}{ds} = \cos\psi \end{array} \tag{7.1a}$$

with the boundary conditions

$$x|_{s=0} = 0, \quad R|_{s=0} = 1, \quad V_\tau|_{s=0} = 1, \quad V_\theta|_{s=0} = V_\Omega, \quad \psi|_{s=0} = \psi_0. \tag{7.1b}$$

Substituting the fourth equation into the first and the second ones gives

$$V_\tau\frac{dV_\tau}{ds} = \frac{V_\theta^2}{R}\frac{dR}{ds}, \quad \frac{1}{V_\theta}\frac{dV_\theta}{ds} = -\frac{1}{R}\frac{dR}{ds}, \quad \frac{d\psi}{ds}$$
$$= \frac{\cos\psi}{R}\cdot\frac{V_\theta^2/V_\tau - We^{-1}R}{V_\tau - We^{-1}R}, \quad \frac{dR}{ds} = \sin\psi, \quad \frac{dx}{ds} = \cos\psi.$$

Hence

$$V_\tau = \sqrt{C_1 - C^2/(2\,R^4)},\ \ V_\theta = C/R,\ \ \frac{d\psi}{ds} = \frac{\cos\psi}{R}\cdot\frac{V_\theta^2/\,V_\tau - We^{-1}R}{V_\tau - We^{-1}R},\ \ \frac{dR}{ds} = \sin\psi,\ \ \frac{dx}{ds} = \cos\psi.$$

Using boundary conditions one obtains

$$\begin{aligned} &V_\tau = \sqrt{\tfrac{1}{2}V_\Omega^2\,(1 - R^{-4}) + 1}, \quad V_\theta = V_\Omega/R, \\ &\tfrac{d\psi}{ds} = \tfrac{\cos\psi}{R}\cdot\tfrac{(V_\Omega^2/R^2)/[\frac{1}{2}V_\Omega^2(1-R^{-4})+1] - We^{-1}R}{\frac{1}{2}V_\Omega^2(1-R^{-4})+1-We^{-1}R}, \quad \tfrac{dR}{ds} = \sin\psi, \quad \tfrac{dx}{ds} = \cos\psi. \end{aligned} \tag{7.2}$$

Proposition 1. *When* $R = const$, $\psi = const$, $V_\tau = const$, *and* $V_\theta = const$ *are the solution to* (7.2), (7.1b) *it is necessary and sufficient to have the boundary values:* $\psi_0 = 0,\ V_\Omega = We^{-1/2}$.

Proof. (1) Necessity. If R = const, ψ = const, then follow the initial conditions (7.1b) $R \equiv 1$, $\psi \equiv 0$, and so the system (7.2) has a simple view:

$$\begin{aligned} &R = \text{const} = 1, \quad \psi = \text{const} = 0, \quad V_\tau = \text{const} = 1, \quad V_\theta = \text{const} = V_\Omega, \\ &\tfrac{d\psi}{ds} = -\cos\psi\,\tfrac{V_\Omega^2 - We^{-1}}{1 - We^{-1}} = 0. \end{aligned}$$

The result $V_\Omega = We^{-1/2}$ proves the necessity.

(2) Sufficiency. Let V_Ω and ψ_0 be equal to $V_\Omega = We^{-1/2}$, $\psi_0 = 0$, and we assume that $R \neq$ const, $V_\tau \neq$ const, $V_\theta \neq$ const, and $\psi \neq$ const are the solution to (7.2) and (7.1a,b). Then the constant functions $R = 1$, $\psi = 0$, $V_\tau = 1$, and $V_\theta = We^{-1/2}$ also satisfy the problem as it was proven in the previous point 1. Hence, because of the uniqueness theorem for Cauchy's problem, only the single solution, namely, $R \equiv 1$, $\psi \equiv 0$, $V_\tau \equiv 1$, $V_\theta \equiv We^{-1/2}$, satisfies to (7.2) and (7.1a,b). □

Acknowledgements The first, second, and fourth authors are grateful to Israel Science Foundation (ISF) for the support (ISF Grant No 661/11).

References

1. Chuech S. G., Numerical simulations of non-swirling and swirling annular liquid jets, *AIAA Journal*, 1993, Vol. 31, N. 6, 1022–1027
2. Epikhin V. E., Forms of annular streams of trickling liquid, *Fluid Dynamics*, 1977, Vol. 12, N. 1, 6–10
3. Epikhin V. E., On the shapes of swirling annular jets of a liquid, *Fluid Dynamics*, 1979, Vol. 14, N. 5, 751–755

4. Epikhin V. E. and Shkadov V.Ya., On the length of annular jets interacting with an ambient medium, *Fluid Dynamics*, 1983, Vol. 18, N. 6, 831–838
5. Epikhin V. E. and Shkadov V. Ya., Numerical simulation of the nonuniform breakup of capillary jets, *Fluid Dynamics*, 1993, Vol. 28, N. 2, 166–170
6. Gaissinski I., Kelis O., and Rovenski V., *Hydrodynamic Instability Analysis*, Verlag VDM, 2010
7. Gaissinski I. M., Kalinin A. V., and Stepanov A. E., The influence of the magnetic field on the α-particles deceleration, *Sov. Journal of Applied Mechanics and Technical Physics*, 1979, N. 6, pp. 40–46
8. Gaissinski I. and Rovenski V., *Nonlinear Models in Mechanics: Instabilities and Turbulence*, Sec. 4.1.7, Lambert, 2010
9. Kazennov A. K., Kallistov Yu. N., Karlikov V. P., and Sholomovich G. I., Investigation of thin annular jets of incompressible liquids, *Nauchn. Tr. Inst. Mekh. Moscow Univ.*, 1970, No. 1, p. 21
10. Kihm K. D. and Chigier N. A., Experimental investigations of annular liquid curtains, *J. of Fluid Engineering*, 1990, Vol. 112, N. 3, 61–66
11. Lefebvre A. H., *Atomization and Sprays*, N.Y.: Combustion (Hemisphere Pub. Corp.), 1989
12. Nayfeh A. H., *Perturbation Methods*, Wiley Classic Library, Wiley–Interscience, 2004
13. Yarin A. L., *Free Liquid Jets and Films: Hydrodynamics and Rheology*, Longman Scientific & Technical and Wiley & Sons, Harlow, New York, 1993

On Cycles and Other Geometric Phenomena in Phase Portraits of Some Nonlinear Dynamical Systems

Gaidov Yu. A. and Golubyatnikov V.P.

Abstract We show existence of cycles in some special nonlinear 4-D and 5-D dynamical systems and construct in their phase portraits invariant surfaces containing these cycles. In the 5D case, we demonstrate non-uniqueness of the cycles. Some possible mechanisms of this non-uniqueness are described as well.

Keywords Dynamical system • Cycle

Mathematics Subject Classifications (2010): 34C05, 34C25, 92C45, 37E99

We study geometric properties of phase portraits of nonlinear dissipative dynamical systems considered as models of gene networks functioning. In some simple cases, when the gene network is regulated by negative feedbacks only, all its n biochemical species, such as proteins, RNA, etc., can be ordered cyclically: $x_1, x_2, \dots x_n, x_{n+1} = x_1$ (see [8, 11, 13]). Here, $x_j \geq 0$ denotes concentration of the jth species in the gene network, $j = 1, 2, \dots n$. Usually, these negative feedbacks are described by monotonically decreasing functions $f_j \geq 0$ in the right-hand sides of the chemical kinetics equations:

$$\frac{dx_j}{dt} \equiv \dot{x}_j = f_j(x_{j-1}) - k_j \cdot x_j, \qquad j = 1, \dots, n. \tag{1}$$

Gaidov Yu. A. (✉)
State pedagogical University, Novosibirsk, Russia
e-mail: gja822@narod.ru

Golubyatnikov V.P.
Sobolev institute of mathematics, Novosibirsk, Russia
e-mail: glbtn@math.nsc.ru

V. Rovenski and P. Walczak (eds.), *Geometry and its Applications*, Springer Proceedings in Mathematics & Statistics 72, DOI 10.1007/978-3-319-04675-4_10,

The nonnegative function f_j describes here the rate of synthesis of the jth species, and the negative term $-k_j \cdot x_j$ corresponds to the natural degradation of this species. Just for simplicity of exposition, we consider below the *dimensionless* form of the system (1), i.e., when $k_j = 1$ for all $j = 1, \dots, n$.

In theoretical considerations, and in numerical experiments (see, e.g., [6, 11, 13, 14]), the monotonically decreasing functions f_j usually have either the form $f_j(x) = \alpha_j \cdot (1 + x^{\gamma_j})^{-1}$, i.e., they are the Hill's functions, or they have the threshold piecewise constant form

$$L_j(w) = A_j > 2 \quad \text{for} \quad 0 \le w < 1; \quad L_j(w) = 0 \quad \text{for} \quad 1 \le w;$$

here $A_j = \text{const}$. Note that such a threshold regulation corresponds to the value $\gamma_j = \infty$ in the Hill's function case.

In our previous publications cited above, we have obtained some conditions of existence, stability, and non-uniqueness of cycles in the cases of odd-dimensional dynamical systems (1), i.e., if $n = 2m + 1$. It should be noted that the phase portraits of even-dimensional dynamical systems of the type (1) have quite different structure; see below.

Consider symmetric 5-dimensional dynamical system

$$\frac{dx_1}{dt} = f(x_5) - x_1; \ \frac{dx_2}{dt} = f(x_1) - x_2; \ \dots \ \frac{dx_5}{dt} = f(x_4) - x_5. \tag{2}$$

Here, the functions $f_j(x) \equiv f(x)$ in all these equations coincide. Let $\alpha = f(0)$ be their maximal value. It was shown in [7] that the cube

$$Q = [0, \alpha] \times [0, \alpha] \times [0, \alpha] \times [0, \alpha] \times [0, \alpha] \subset \mathbb{R}^5_+$$

is an invariant set of the system (2), and there is exactly one stationary point M^0 of this system in the cube Q. Our main aim is to construct integral surfaces in the phase portraits of the symmetric system (2) in the case of threshold functions $f(x)$ and to find periodic trajectories on these surfaces. But we start our exposition from the smooth case.

For geometric description of phase portraits of this system, consider the partition of Q by 5 hyperplanes containing the stationary point $M^0 \in Q$ and parallel to the coordinate hyperplanes. It is easy to see that for symmetric system (2) all coordinates of the point M^0 are equal: $x^* := x_1^0 = x_2^0 = x_3^0 = x_4^0 = x_5^0$. So, we get a collection of 32 small blocks, which can be enumerated by binary indices:

$$\{\varepsilon_1 \varepsilon_2 \varepsilon_3 \dots \varepsilon_5\} = \{\mathbf{X} \in Q \mid x_1 \gtrless_{\varepsilon_1} x^*, \, x_2 \gtrless_{\varepsilon_2} x^*, \dots, \, x_5 \gtrless_{\varepsilon_5} x^*\}, \tag{3}$$

here $\mathbf{X} = (x_1, x_2, \dots, x_5)$, $\varepsilon_1, \varepsilon_2, \dots \varepsilon_5 \in \{0, 1\}$, and the relations are defined as follows: the symbol $\gtrless_0$ means $\le$, and the symbol $\gtrless_1$ means $\ge$. The faces of these blocks contained in the interior of Q (i.e., in the hyperplanes $x_j = x^*$) will be called *the interior faces*.

Similar constructions can be realized for higher-dimensional dynamical systems and for threshold functions $L(x)$ as well. In this piecewise constant case, the role of the stationary point plays the point E_5 with coordinates $\{1,1,1,1,1\}$; thus, the partition (3) here is determined by the inequalities $x_j \gtrless_{\varepsilon_j} 1$.

Direct calculations of the signs of the right-hand sides of the system (2) show that all its trajectories which start in the block $\{10101\}$ travel through the blocks (3) according to the cyclic diagram

$$\{10101\} \to \{00101\} \to \{01101\} \to \{01001\} \to \{01011\} \to \{01010\} \to$$
$$\to \{11010\} \to \{10010\} \to \{10110\} \to \{10100\} \to \{10101\} \ldots \quad (4)$$

Here, each trajectory of the system (2) passes through the common face of two adjacent blocks according to the corresponding arrow in this diagram. This construction implies that the union Q_{10} of 10 blocks listed in (4) is an invariant domain of this system. In terms of [2], this domain belongs to the *first potential level* of the phase portrait of the system (2). If this system is smooth, then its linearization at the stationary point M^0 has the eigenvalues

$$\lambda_k = -1 - p \cdot \left(\cos\frac{2\pi}{5}(k-1) + i \cdot \sin\frac{2\pi}{5}(k-1)\right), \quad k = 1,2,3,4,5.$$

Here, $p \equiv -df(w)/dw > 0$; all derivatives are calculated at the point M^0.

In the sequel we consider another enumeration of these eigenvalues; it corresponds to the values of their real parts: $\mathbb{R}e\lambda_j \le \mathbb{R}e\lambda_{j+1}$. It is easy to see that for the system (2) $\mathbb{R}e\lambda_1 < \mathbb{R}e\lambda_2 = \mathbb{R}e\lambda_3 < 0$.

Denote by $P_{4,5}$ the 2-dimensional plane containing the point M^0 and parallel to the eigenvectors corresponding to the eigenvalues λ_4, λ_5 with maximal real parts. Some open disks of this plane centered at the point M^0 are contained in the domain Q_{10}.

If $\mathbb{R}e\lambda_4 = \mathbb{R}e\lambda_5 > 0$, then the stationary point M^0 is hyperbolic (i.e., the real parts of all eigenvalues are either strictly positive or strictly negative). It was shown in [7] that if the point M^0 is hyperbolic, then the invariant domain Q_{10} contains at least one cycle of the system (2). From now on we assume that this point is hyperbolic.

According to the well-known Grobman–Hartman theorem (see, e.g., [9]), each nonlinear dynamical system can be linearized in some small neighborhood $U(M^0)$ of each of its hyperbolic stationary point.

Thus, the eigenvalues λ_2, λ_3 correspond to an invariant 2-dimensional surface $\Pi_{2,3}$ in the neighborhood $U(M^0)$. Let $P_{2,3}$ be its tangent plane at the point M^0. This surface becomes a part of corresponding 2-dimensional plane after the linearization of the system (2) in $U(M^0)$. On this surface $\Pi_{2,3}$, the trajectories of this system travel through the blocks (3) according to the diagram

$$\{11110\} \to \{11100\} \to \{11101\} \to \{11001\} \to \{11011\} \to \{10011\} \to$$
$$\to \{10111\} \to \{00111\} \to \{01111\} \to \{01110\} \to \{11110\} \ldots \quad (5)$$

Denote by $\hat{Q}_{10}$ the union of all 10 blocks listed in this diagram. Following the proofs of the main results of [3, 7, 8], one can verify that in contrast with (4), for each block of the diagram (5), the trajectories of the system (2) can leave it through three adjacent blocks. Only one of them is contained in $\hat{Q}_{10}$.

In particular, the trajectories which start in the third block $\{11101\}$ of the diagram (5) can pass to the blocks $\{01101\}, \{10101\}$, and $\{11001\}$ through the faces $x_1 = x^*$, $x_2 = x^*$, $x_3 = x^*$, respectively. Only the last of these 3 blocks is mentioned in the diagram (5), and the first two blocks listed here are contained in Q_{10}.

Hence, if the point M^0 is hyperbolic, then some trajectories of the system (2) with starting point in $\hat{Q}_{10}$ are attracted to the cycle, which is contained in Q_{10}. Thus, $\hat{Q}_{10}$ is not invariant. In terms of [2], this domain belongs to the *unstable potential level* 3 of the phase portrait of the system (2).

As in the case of Q_{10}, one can verify that the intersection $\hat{Q}_{10} \cap P_{2,3}$ contains some disk centered at the point M^0. Note that both these domains are non-convex polyhedra with disjoint interiors, and they are star-shaped with respect to their common vertex M^0.

All these considerations can be reproduced for any odd-dimensional symmetric dynamical system analogous to (2). For any pair of complex conjugate eigenvalues $\lambda_{2j}, \lambda_{2j+1}$ of linearization of such $(2m + 1)$-dimensional system at its stationary point, the diagrams analogous to (4) and (5) can be constructed as well. Here, for any $j \leq m$, the number of consecutive nonzero indices in $\{\varepsilon_1\varepsilon_2 \dots \varepsilon_{2m+1}\}$ in the jth diagrams is "proportional" to the angle $\arg \lambda_{2j}$ and corresponds to the potential level of all $4m + 2$ blocks of such a diagram.

Note that the right-hand sides of the equations in (2) define a vector field in Q, and the divergence of this field equals identically -5. Hence, the volume of any bounded domain $W \subset \mathbb{R}^5_+$ decreases exponentially when t tends to ∞:

$$Vol(W(t)) = Vol(W(0)) \cdot \exp(-5t).$$

See, for example [4]. Thus, each unstable cycle of the system (1) cannot repel its trajectories in all directions. Otherwise, the volume of some of its toric neighborhood will not decrease when $t \to \infty$.

It is possible to derive some lower bounds for the number of cycles in the phase portraits in the cases when the odd dimension of symmetric system of the type (2) is not a prime number; see [2]. Let $2m + 1 = p \cdot q$, $p \neq q$. It is not difficult to find invariant p-dimensional and q-dimensional planes P^p and P^q in the phase portraits of such systems. If the point M^0 is hyperbolic for restrictions of the "ambient" dynamical system to these planes, then, according to [8], each of these planes contains a cycle of this system, and there is one more cycle in Q_{2m+1} which does not intersect $P^q \cup P^q$, and has the potential level 1 in terms of [2].

For example, the phase portrait of symmetric 21-dimensional system of the type (2) contains 7-dimensional invariant plane P^7 defined by the equations $\{x_j = x_{j+3}\}$ and 3-dimensional invariant plane P^3 defined by the equations $\{x_m = x_{m+7}\}$. Here $1 \leq j \leq 18, 1 \leq m \leq 14$.

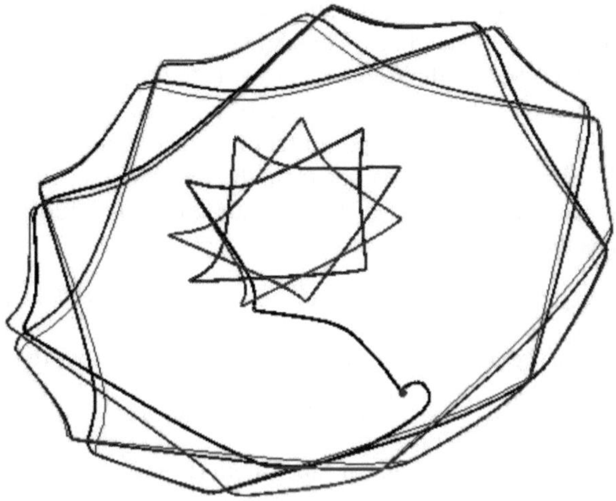

Fig. 1 Projections of two cycles of 11-D Hill's system onto 3-D plane

If the point M^0 is hyperbolic in P^3 and P^7, then according to [6], and [8], respectively, each of these planes contains a cycle of the system (2); hence, this system has at least 3 different cycles.

Actually, non-uniqueness of cycles is observed in numerical experiments in the case of prime dimensions as well. For example, Fig. 1 shows projections of two trajectories and their limit cycles of symmetric 11-dimensional dynamical system of the Hill's type onto 3-dimensional plane corresponding to the eigenvalues λ_1, λ_8, and λ_9. Here $f(x) = 130 \cdot (1 + x^6)^{-1}$.

For smooth odd-dimensional dimensionless asymmetric dynamical systems of the type (1), we have already obtained conditions of existence of cycles in the invariant domains similar to Q_{10}; see [7, 8].

Consider now the symmetric threshold dynamical systems of the type (1). Making corresponding change of the variables, we can assume that $k_j = 1$ for all j, $1 \leq j \leq n$. In this case, existence of symmetric cycles in their phase portraits follows from direct calculation, which we reproduce here for 4-dimensional dynamical system

$$\dot{x}_1 = L(x_4) - x_1;\ \dot{x}_2 = L(x_1) - x_2;\ \dot{x}_3 = L(x_2) - x_3;\ \dot{x}_4 = L(x_3) - x_4; \qquad (6)$$

analogous to (2). Similar 5-dimensional threshold dynamical system will be denoted by (6_5).

Let $A = L(0)$. As in the 5-dimensional case, we construct 4-dimensional invariant cube $k^4 = [0, A] \times [0, A] \times [0, A] \times [0, A] \subset \mathbb{R}^4_+$ in the phase portrait of the system (6). Consider the diagram

$$\ldots \to \{1110\} \to \{1100\} \to \{1101\} \to \{1001\} \to \{1011\} \to$$
$$\to \{0011\} \to \{0111\} \to \{0110\} \to \{1110\} \to \ldots \qquad (7)$$

As in the previous diagrams, let us denote by Q_8 the union of all eight blocks in the cube k^4 listed here. This union is a non-convex polyhedron which is star-shaped with respect to the point $E_4 = (1, 1, 1, 1)$.

Let us describe construction of a cycle $\mathcal{C}_8$ of the system (6) which is symmetric with respect to cyclic permutations of the variables x_j.

Since this cycle passes from the first block of the diagram (7) to the second one through the face $x_3 = 1$, we assume that its intersection point $X^{(0)}$ with this face has the coordinates

$$X^{(0)} = \{x_1^{(0)} > 1,\ x_2^{(0)} > 1,\ x_3^{(0)} = 1,\ x_4^{(0)} < 1\}. \tag{8}$$

In the second block of the diagram (7) the system (6) has the form

$$\dot{x}_1 = A - x_1;\ \dot{x}_2 = -x_2;\ \dot{x}_3 = -x_3;\ \dot{x}_4 = A - x_4,$$

hence, the trajectories of the system (6) in this block are described by the equations

$$x_1(t) = A + (x_1^{(0)} - A)e^{-t};\ x_2(t) = 0 + (x_2^{(0)} - 0)e^{-t};$$
$$x_3(t) = 0 + (x_3^{(0)} - 0)e^{-t};\ x_4(t) = A + (x_1^{(0)} - A)e^{-t},$$

or in the vector form: $X(t) = B_0 + (X^{(0)} - B_0)$, where $B_0 = (A, 0, 0, A)$, and $X(t) = (x_1(t), x_2(t), x_3(t), x_4(t))$.

Hence, in this block all these trajectories are contained in the rays with the end-point B_0, and the shifts along these trajectories can be represented as homotheties with the center B_0 and coefficients e^{-t}.

The following proposition can be easily verified for any dimension n in a similar way.

Lemma. *The trajectories of a threshold dynamical system analogous to* (6) *are rectilinear inside the interior of each block* $\{\varepsilon_1\varepsilon_2\dots\varepsilon_n\}$.

For $t = t_1 = \ln(A - x_4^0) - \ln(A - 1)$ the cycle $\mathcal{C}_8$ intersects the face $x_4 = 1$ of the second block at the point $X^{(1)}$ with the coordinates

$$x_1^{(1)} > 1, \quad x_2^{(1)} > 1, \quad x_3^{(1)} < 1, \quad x_4^{(1)} = 1,$$

which can be easily calculated in terms of coordinates of the point $X^{(0)}$.

Now, according to the diagram (7), the cycle $\mathcal{C}_8$ passes to its third block, where the system (6) has the form

$$\dot{x}_1 = -x_1;\ \dot{x}_2 = -x_2;\ \dot{x}_3 = -x_3;\ \dot{x}_4 = A - x_4,$$

and the shifts along its trajectories are represented by the homotheties with the center $B_1 = (0, 0, 0, A)$.

According to the diagram (7), the cycle $\mathcal{C}_8$ leaves the third block through the face $x_2 = 1$ at some point $X^{(2)} = \{x_1^{(2)}, x_2^{(2)} = 1, x_3^{(2)}, x_4^{(2)}\}$. Its coordinates can be easily expressed in terms of coordinates of the point $X^{(0)}$. Since the cycle $\mathcal{C}_8$ is symmetric, we get

$$x_3^{(0)} = x_2^{(2)} = 1, \quad x_4^{(0)} = x_3^{(2)} = \frac{1}{x_2^{(0)}}, \quad x_1^{(0)} = x_4^{(2)} = A - \frac{A - x_4^{(0)}}{x_2^{(0)}},$$

$$x_2^{(0)} = x_1^{(2)} = \frac{A(A - x_4^{(0)})}{x_2^{(0)}(A-1)} - \frac{A - x_1^{(0)}}{x_2^{(0)}}.$$

For $A > 2$, this algebraic system with unknowns $x_1^{(0)}, x_2^{(0)}, x_4^{(0)}$ has a unique solution which satisfies the initial conditions (8). Actually, we have to verify that the cubic equation

$$z^3 + z^2 + z + 1 - z^2 \cdot \frac{A^2}{A-1} = 0$$

has a unique solution $z = x_4^{(0)}$ on the segment $[0, 1]$, if $A > 2$. Let $\mathbf{X}^{(0)}$ be such point that its coordinates satisfy the algebraic system above.

Thus, after eight steps along the diagram (7), the trajectory which starts at this point $\mathbf{X}^{(0)}$ returns to $\mathbf{X}^{(0)}$, and so we have constructed the required cycle $\mathcal{C}_8 \subset Q_8$ of the dynamical system (6).

Since the trajectories of the system (6) are represented by homotheties inside each block of the partition of the cube P^4, the intersection of each ray $E_4X^{(j)}$, $j = 0, 1, 2,$ etc. with corresponding block is shifted along these trajectories to the ray $E_4X^{(j+1)}$. It is assumed here that $E_4X^{(8)} = E_4\mathbf{X}^{(0)}$, since $\mathcal{C}_8$ is a cycle.

So, these linear shifts allow us to construct an invariant piecewise linear surface $M^2 \subset k^4$ containing the cycle $\mathcal{C}_8$ in its interior. This surface is composed by 8 triangles with common vertex E_4 which is incident to their edges contained in adjacent blocks of the diagram (7). It should be noted that the classical theorems on central manifolds of smooth dynamical systems state that these manifolds are either small, or are bounded by corresponding cycles, and, in general, cannot be extended beyond these limits; see, for example, [1, 10, 12].

The phase portrait of dynamical system (6) has two stable stationary points $Z_1 = (0, A, 0, A)$ and $Z_2 = (A, 0, A, 0)$ contained in the blocks $\{0101\}$ and $\{1010\}$, respectively. Considerations of the homothety shifts in these blocks show that $\{0101\}$ is contained in the attraction basin U_1 of the point Z_1 and $\{1010\}$ is contained in the attraction basin U_2 of the point Z_2. Similar stable stationary points $Z_1 = (0, A, \dots, 0, A)$, $Z_2 = (A, 0, \dots, A, 0)$ do exist in phase portrait of any even-dimensional symmetric threshold dynamical system. Asymmetric systems of this type have also two stable stationary points.

These attraction basins U_1 and U_2 are separated by 3-dimensional invariant surface M^3, such that $M^2 \subset M^3$. The diagonal Δ of the cube k^4 which joins the origin with the points E_4 and (A, A, A, A) is contained in M^3 as well. This diagonal is an invariant 1-dimensional manifold of the symmetric dynamical system (6) and all its analogues odd-dimensional and even-dimensional. Thus, we have obtained the theorem:

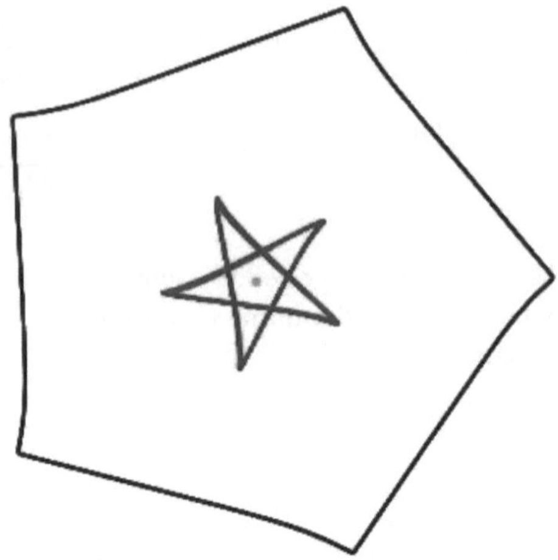

Fig. 2 Projections of two cycles of the system (2) onto 2-D plane

Theorem 1. *Let $A > 2$; then the dynamical system* (6) *has exactly one piecewise linear cycle $\mathcal{C}_8$. This cycle passes through the blocks of the diagram* (7) *and is contained in the piecewise linear surface $M^2 \subset \mathcal{Q}_8$ which is composed by* 8 *triangles with common vertex E_4.*

Note that the odd-dimensional dynamical systems of this type do not have stable stationary points. For all these higher-dimensional threshold dynamical systems, similar constructions and considerations can be reproduced as well; cf. [3]. For example, in the case of 5-dimensional dynamical system (6_5), one can verify the existence of a piecewise linear cycle $\hat{\mathcal{C}}_{10}$ which passes through the blocks (3) according to the diagram (5).

So, we can deduce the following result:

Theorem 2. *Let $A > \frac{5+\sqrt{5}}{2}$; then the 5-dimensional dynamical system* (6_5) *analogous to* (6) *has at least two piecewise linear cycles $\mathcal{C}_{10}$ and $\hat{\mathcal{C}}_{10}$. These cycles pass through the blocks of the diagrams* (4) *and* (5), *respectively. Each of these two cycles is contained in an invariant piecewise linear surface: $\mathcal{C}_{10} \subset M_1^2 \subset \mathcal{Q}_{10}$ and $\hat{\mathcal{C}}_{10} \subset M_2^2 \subset \hat{\mathcal{Q}}_{10}$. Each of these invariant surfaces is composed by* 10 *triangles with common vertex E_5.*

Figure 2 shows projections of two cycles of "almost threshold" symmetric smooth Hill's system with $\alpha = 2050$, $\gamma = 10$ onto 2-dimensional plane corresponding to the eigenvalues λ_4, λ_5 with positive real parts. The large pentagon is the projection of the cycle which follows the diagram (4), and the star inside is the projection of the cycle which follows the diagram (5). The point in the center is the projection of the stationary point M^0.

Much more difficult is the problem of classification of all cycles, stable and unstable, for asymmetric dynamical systems (1).

More complicated models of gene networks regulated by combinations of negative and positive feedbacks were considered in [5, 11]. Each of the corresponding nonlinear dynamical systems contains several stationary points in its phase portrait. The neighborhoods of stationary points with negative topological index can be decomposed to small blocks, as it was done in the diagrams above, and conditions of existence of cycles (and stable cycles) can be formulated for these dynamical systems as in the case of absence of positive feedbacks.

Acknowledgements The authors are indebted to A.A.Akinshin, I.V.Golubyatnikov, A.E.Gutman, and V.A.Likhoshvai for useful discussions and helpful assistance. The work was supported by RFBF, grant 12-01-00074, and by interdisciplinary grant 80 of SB RAS.

References

1. Abraham R.H., Robbin J. *Transversal mappings and flows*. New York: W.A.Benjamin. 1967. — 169 pp.
2. Akinshin A.A., Golubyatnikov V.P. *On cycles in symmetric dynamical systems* (Russian). Bulletin of Novosibirsk state university. 2012, v. 12, N 2, 3–12.
3. Akinshin A.A., Golubyatnikov V.P., Golubyatnikov I.V. *On some multidimensional models of gene networks functioning* (Russian). Siberian journ. of industrial mathematics. 2013, v. 16, N 1, 3–9.
4. Arnold V.I. *Mathematical methods of classical mechanics*. New York: Springer. 1989. — 508 pp.
5. Gaidov Yu.A., Golubyatnikov V.P. *On the Existence and Stability of Cycles in Gene Networks with Variable Feedbacks*. Contemporary mathematics. 2011, v. 553, 61–74.
6. Golubyatnikov V.P. V.A.Likhoshvai V.A., A.V.Ratushny A.V. *Existence of Closed Trajectories in 3-D Gene Networks*. The journ. of 3-dimensional images 3D Forum. 2004, v. 18, N 4, 96–101.
7. Golubyatnikov V.P., Golubyatnikov I.V., Likhoshvai V.A. *On the Existence and Stability of Cycles in Five-Dimensional Models of Gene Networks*. Numerical analysis and applic. 2010, v. 3, N 4, 329–335.
8. Golubyatnikov V.P., Golubyatnikov I.V. *On periodic trajectories in odd-dimensional gene networks models*. Russian journal of numerical analysis and mathematical modeling. 2011, v. 28, N 4, 379–412.
9. Grobman D.M. *Homeomorphisms of differential equations* (Russian). Dokl. Akad. Nauk SSSR. 1959, v. 128, N 5, 880–881.
10. Guckenheimer J., Holmes P. *Nonlinear Oscillations, Dynamical Systems, and Bifurcations of Vector Fields*. Applied Mathematical Sciences, v. 42, Berlin, New York: Springer. 1997. — 459 pp.
11. Likhoshvai V.A., Golubyatnikov V.P., Demidenko G.V., Evdokimov A.A., Fadeev S.I. *Gene networks theory*. In: Computational systems biology (Russian). Novosibirsk, SB RAS. 2008, 395–480.
12. Marsden J.E., McCracken M. *The Hopf bifurcation and its applications*. New York: Springer. 1981. — 406 pp.
13. Murray J.D. *Mathematical biology*, v. 1. *An introduction*. 3-rd ed. New York: Springer. 2002. — 551 pp.
14. Volokitin E.P., Treskov S.A. *The Andronov-Hopf bifurcation in a model of hypothetical gene regulatory network*. Journal of applied and industrial mathematics. 2007, v. 1, N 1, 127–136.

Remez-Type Inequality for Smooth Functions

Yosef Yomdin

Abstract The classical Remez inequality bounds the maximum of the absolute value of a polynomial $P(x)$ of degree d on $[-1,1]$ through the maximum of its absolute value on any subset Z of positive measure in $[-1,1]$. Similarly, in several variables the maximum of the absolute value of a polynomial $P(x)$ of degree d on the unit ball $B^n \subset \mathbb{R}^n$ can be bounded through the maximum of its absolute value on any subset $Z \subset Q_1^n$ of positive n-measure $m_n(Z)$. In [11] a stronger version of Remez inequality was obtained: the Lebesgue n-measure m_n was replaced by a certain geometric quantity $\omega_{n,d}(Z)$ satisfying $\omega_{n,d}(Z) \geq m_n(Z)$ for any measurable Z. The quantity $\omega_{n,d}(Z)$ can be effectively estimated in terms of the metric entropy of Z and it may be nonzero for discrete and even finite sets Z.

In the present paper we extend Remez inequality to functions of finite smoothness. This is done by combining the result of [11] with the Taylor polynomial approximation of smooth functions. As a consequence we obtain explicit lower bounds in some examples in the Whitney problem of a C^k-smooth extrapolation from a given set Z, in terms of the geometry of Z.

Keywords Remez inequality • Smooth functions • Polynomial approximation • Entropy

Mathematics Subject Classifications (2010): 26D10, 26C05

Y. Yomdin (✉)
Department of Mathematics, The Weizmann Institute of Science, Rehovot 76100, Israel
e-mail: yosef.yomdin@weizmann.ac.il

V. Rovenski and P. Walczak (eds.), *Geometry and its Applications*, Springer Proceedings in Mathematics & Statistics 72, DOI 10.1007/978-3-319-04675-4_11,

1 Introduction

The classical Remez inequality ([9], see also [5]) reads as follows:

Theorem 1.1. *Let $P(x)$ be a polynomial of degree d. Then for any measurable $Z \subset [-1,1]$*

$$\max_{[-1,1]}|P(x)| \le T_d\left(\frac{4-m}{m}\right)\max_Z|P(x)|, \qquad (1.1)$$

where $m = m_1(Z)$ is the Lebesgue measure of Z and $T_d(x) = cos(d\ arccos(x))$ is the dth Chebyshev polynomial.

In several variables a generalization of Theorem 1.1 was obtained in [2]:

Theorem 1.2. *Let $\mathcal{B} \subset \mathbb{R}^n$ be a convex body and let $\Omega \subset \mathcal{B}$ be a measurable set. Then for any real polynomial $P(x) = P(x_1,\dots,x_n)$ of degree d we have*

$$\sup_{\mathcal{B}}|P| \le T_d\left(\frac{1+(1-\lambda)^{\frac{1}{n}}}{1-(1-\lambda)^{\frac{1}{n}}}\right)\sup_{\Omega}|P|. \qquad (1.2)$$

Here $\lambda = \frac{m_n(\Omega)}{m_n(\mathcal{B})}$, with m_n being the Lebesgue measure on $\mathbb{R}^n$. This inequality is sharp and for $n = 1$ it coincides with the classical Remez inequality.

It is clear that Remez inequality of Theorems 1.1 and 1.2 cannot be verbally extended to smooth functions: such function f may be identically zero on any given closed set Z and nonzero elsewhere. In the present paper we show that adding a "remainder term" (expressible through the bounds on the derivatives of f) provides a generalization of the Remez inequality to smooth functions. Our main goal is to study the interplay between the geometry of the "sampling set" Z, the bounds on the derivatives of f, and the bounds for the extension of f from Z to the ball B^n of radius 1 centered at the origin in $\mathbb{R}^n$. To state our main "general" result we need some definitions:

Definition 1.1. For a set $Z \subset B^n \subset \mathbb{R}^n$ and for each $d \in \mathbb{N}$ the Remez constant $R_d(Z)$ is the minimal K for which the inequality $\sup_{B^n}|P| \le K \sup_Z |P|$ is valid for any real polynomial $P(x) = P(x_1,\dots,x_n)$ of degree d.

For some Z the Remez constant $R_d(Z)$ may be equal to ∞. In fact, $R_d(Z)$ is infinite if and only if Z is contained in the set of zeroes $Y_P = \{x \in \mathbb{R}^n, \mid P(x) = 0\}$ of a certain polynomial P of degree d. See [3] for a detailed discussion.

Definition 1.2. Let $f : B^n \to \mathbb{R}$ be a k times continuously differentiable function on B^n. For $d = 0,1,\dots,$ the approximation error $E_d(f)$ is the minimum over all the polynomials $P(x)$ of degree d of the absolute deviation $M_0(f-P) = \max_{x\in B^n}|f(x)-P(x)|$.

Theorem 1.3. *Let $f : B^n \to \mathbb{R}$ be a k times continuously differentiable function on B^n, and let a subset $Z \subset B^n$ be given. Put $L = \max_{x \in Z}|f(x)|$. Then*

$$\max_{x \in B^n}|f(x)| \leq \inf_d [R_d(Z)(L + E_d(f)) + E_d(f)]. \tag{1.3}$$

Proof. Let for a fixed d $P_d(x)$ be the polynomial of degree d for which the best approximation of f is achieved: $E_d(f) = \max_{x \in B^n}|f(x) - P(x)|$. Then $\max_{x \in Z}|P(x)| \leq L + E_d(f)$. By definition of the Remez constant $R_d(Z)$ we have $\max_{x \in B^n}|P(x)| \leq R_d(Z)(L + E_d(f))$. Returning to f we get $\max_{x \in B^n}|f(x)| \leq R_d(Z)(L + E_d(f)) + E_d(f)$. Since this is true for any d, we finally obtain $\max_{x \in B^n}|f(x)| \leq \inf_d[R_d(Z)(L + E_d(f)) + E_d(f)]$. □

In this paper we produce, based on Theorem 1.3, explicit Remez-type bounds for smooth functions in some typical situations.

2 Bounding $R_d(Z)$ via Metric Entropy

It is well known that the inequality of the form (1.1) or (1.2) may be true also for some sets Z of measure zero and even for certain discrete or finite sets Z. Let us mention here only a couple of the most relevant results in this direction: in [4, 8, 12] such inequalities are provided for Z being a regular grid in $[-1, 1]$. In [6] discrete sets $Z \subset [-1, 1]$ are studied. In this last paper the invariant $\phi_Z(d)$ is defined and estimated in some examples, which is the best constant in the Remez-type inequality of degree d for the couple $(Z \subset [-1, 1])$.

In [11] (see also [1]) a strengthening of Remez inequality was obtained: the Lebesgue n-measure m_n was replaced by a certain geometric quantity $\omega_{n,d}(Z)$, defined in terms of the metric entropy of Z, and satisfying $\omega_{n,d}(Z) \geq m_n(Z)$ for any measurable $Z \subset Q_1^n$. So we have the following proposition, which combines the result of Theorem 3.3 of [11] with the well-known bound for Chebyshev polynomials (see [5]):

Proposition 2.1. *For each $Z \subset B^n$ and for any d the Remez constant $R_{n,d}(Z)$ satisfies*

$$R_{n,d}(Z) \leq T_d\left(\frac{1 + (1-\lambda)^{\frac{1}{n}}}{1 - (1-\lambda)^{\frac{1}{n}}}\right) \leq \left(\frac{4n}{\lambda}\right)^d, \tag{2.1}$$

where $\lambda = \omega_{n,d}(Z)$.

In what follows we shall omit the dimension n from the notations for $\omega_d(Z) = \omega_{n,d}(Z)$. It was shown in [11] that in many cases (but not always!) the bound of Proposition 2.1 is pretty sharp. In the present paper we recall the definition of $\omega_d(Z)$ and estimate this quantity in several typical cases, stressing the setting where Z is fixed, while d changes.

2.1 Definition and Properties of $\omega_d(Z)$

To define $\omega_d(Z)$ let us recall that the covering number $M(\varepsilon, A)$ of a metric space A is the minimal number of closed ε-balls covering A. Below A will be a subset of $\mathbb{R}^n$ equipped with the l^∞ metric. So the ε-balls in this metric are the cubes Q^n_ε.

For a polynomial P on $\mathbb{R}^n$ let us consider the sublevel set $V_\rho(P)$ defined by $V_\rho(P) = \{x \in B^n, |P(x)| \le \rho\}$. The following result is proved in ([10]):

Theorem 2.1 (Vitushkin's bound). *For $V = V_\rho(P)$ as above*

$$M(\varepsilon, V) \le \sum_{i=0}^{n-1} C_i(n,d)\left(\frac{1}{\varepsilon}\right)^i + m_n(V)\left(\frac{1}{\varepsilon}\right)^n, \tag{2.2}$$

with $C_i(n,d) = C'_i(n)(2d)^{(n-i)}$. For $n = 1$ we have $M(\varepsilon, V) \le d + \mu_1(V)(\frac{1}{\varepsilon})$, and for $n = 2$ we have

$$M(\varepsilon, V) \le (2d-1)^2 + 8d\left(\frac{1}{\varepsilon}\right) + \mu_2(V)\left(\frac{1}{\varepsilon}\right)^2.$$

For $\varepsilon > 0$ we denote by $M_{n,d}(\varepsilon)$ (or shortly $M_d(\varepsilon)$) the polynomial of degree $n-1$ in $\frac{1}{\varepsilon}$ as appears in (2.2):

$$M_d(\varepsilon) = \sum_{i=0}^{n-1} C_i(n,d)\left(\frac{1}{\varepsilon}\right)^i. \tag{2.3}$$

In particular,

$$M_{1,d}(\varepsilon) = d,\ M_{2,d}(\varepsilon) = (2d-1)^2 + 8d\left(\frac{1}{\varepsilon}\right).$$

Now for each subset $Z \subset B^n$ (possibly discrete or finite) we introduce the quantity $\omega_d(Z)$ via the following definition:

Definition 2.1. Let Z be a subset in $B^n \subset \mathbb{R}^n$. Then $\omega_d(Z)$ is defined as

$$\omega_d(Z) = \sup_{\varepsilon>0} \varepsilon^n[M(\varepsilon, Z) - M_d(\varepsilon)]. \tag{2.4}$$

The following results are obtained in [11]:

Proposition 2.2. *The quantity $\omega_d(Z)$ for $Z \subset B^n$ has the following properties:*

1. *For a measurable Z $\omega_d(Z) \ge m_n(Z)$.*
2. *For any set $Z \subset B^n$ the quantities $\omega_d(Z)$ form a nonincreasing sequence in d.*
3. *For a set Z of Hausdorff dimension $n-1$, if the Hausdorff $n-1$ measure of Z is large enough with respect to d, then $\omega_d(Z)$ is positive.*

4. *Let* $G_s = \{x_1 = -1, x_2, \dots, x_s = 1\}$ *be a regular grid in* $[-1,1]$. *Then* $\omega_d(G_s) = \frac{2(s-d)}{s-1}$.
   Let $Z_r = \{1, \frac{1}{2^r}, \frac{1}{3^r}, \dots, \frac{1}{k^r}, \dots\}$. *In this case* $\omega_d(Z_r) \asymp \frac{r^r}{(r+1)^{r+1}} \frac{1}{d^r}$.
   Let $Z(q) = \{1, q, q^2, q^3, \dots, q^m, \dots\}$, $0 < q < 1$. *Then* $\omega_d(Z(q)) \asymp \frac{q^d}{\log(\frac{1}{q})}$.

We need the following result, which, although in the direction of the results in [11], was not proved there explicitly. Let S be a connected smooth curve in B^2 of the length σ. Define ε_0 as the maximal ε such that for each $\delta \leq \varepsilon$ we have $M(\delta, S) \geq \frac{l(S)}{2\delta}$. The parameter ε_0 is a kind of "injectivity radius" of the curve S, and for any curve of length σ inside the unit ball B^2 it cannot be larger than $\frac{1}{\sigma}$. Write ε_0 as $\varepsilon_0 = \frac{1}{l\sigma}$, $l \geq 1$. The computation below essentially compares the length of S with the maximal possible length of an algebraic curve of degree d inside B^2, which is of order d. So it is convenient for any given d to write σ as $\sigma = md$.

Proposition 2.3. *In the notations above,* $\omega_d(S)$ *satisfies*

$$\omega_d(S) \geq \frac{1}{2l}\left(1 - \frac{24}{m}\right). \tag{2.5}$$

In particular, for the length of S *larger than* $24d$, $\omega_d(S)$ *is strictly positive.*

Proof. By definition,

$$\omega_d(S) = \sup_{\varepsilon} \varepsilon^2[M(\varepsilon, S) - M_d(\varepsilon)] = \sup_{\varepsilon} \varepsilon^2 \left[M(\varepsilon, S) - (2d-1)^2 - 8d\left(\frac{1}{\varepsilon}\right)\right].$$

Substituting here $\varepsilon_0 = \frac{1}{l\sigma}$ we get

$$\omega_d(S) \geq \left(\frac{1}{lmd}\right)^2 \left[\frac{l(md)^2}{2} - (2d-1)^2 - 8lmd^2\right] =$$

$$= \frac{1}{2l}\left(1 - \frac{2}{m}\left[\left(\frac{2d-1}{d}\right)^2 + 8\right]\right) \geq \frac{1}{2l}\left(1 - \frac{24}{m}\right).$$

In particular, for $m > 24$, i.e., for the length of S larger than $24d$, the quantity $\omega_d(S)$ is strictly positive. □

3 Bounding Smooth Functions

Let $f : B^n \to \mathbb{R}$ be a k times continuously differentiable function on B^n. For $l = 0, 1, \dots, k$ put $M_l(f) = \max_{B^n} \|d^l f\|$, where the norm of the lth differential of f is defined as the sum of the absolute values of all the partial derivatives of

f of order l. To simplify notations, we shall not make specific assumptions on the continuity modulus of the last derivative $d^k f$. Now we use Taylor polynomials of an appropriate degree between 0 and $k-1$ in order to bound from above the approximation error $E_d(f)$, $d = 0, 1, \ldots, k$. Applying Theorem 1.3, we obtain the following result:

Proposition 3.1. *Let $f : B^n \to \mathbb{R}$ be a k times continuously differentiable function on B^n, with $M_l(f) = \max_{B^n} \|d^l f\|$, $l = 0, 1, \ldots, k$, and let a subset $Z \subset B^n$ be given. Put $L = \max_{x \in Z} |f(x)|$. Then*

$$M_0(f) = \max_{x \in B^n} |f(x)| \leq \min_{d=0,1,\ldots,k-1} [R_d(Z)(L + E_d^T(f)) + E_d^T(f)], \tag{3.1}$$

where $E_d^T(f) = \frac{1}{(d+1)!} M_{d+1}(f)$ is the Taylor remainder term of f of degree d on the unit ball B^n.

Proof. We restrict infimum in Theorem 1.3 to a smaller set of d's and replace $E_d(f)$ with a larger quantity $E_d^T(f)$. □

In general we cannot get an explicit answer for the minimum in Proposition 3.1, unless we add more specific assumptions on the set Z and the sequence $M_d(f)$. However, this proposition provides an explicit and rather sharp information in the case where the set Z is "small." Let us pose the following question: for a fixed $s = 1, \ldots, k-1$ and a given set $Z \subset B^n$, is it possible to bound $M_0(f) = \max_{x \in B^n} |f(x)|$ through $L = \max_{x \in Z} |f(x)|$ and $M_{s+1}(f)$ only, without knowing bounds on the derivatives $d^l(f)$, $l \leq s$?

Proposition 3.2. *If $R_s(Z) < \infty$, then $M_0(f) \leq R_s(Z)(L + E_s^T(f)) + E_s^T(f)$ with $E_s^T(f) = \frac{1}{(s+1)!} M_{s+1}(f)$. If $R_s(Z) = \infty$ then $M_0(f)$ cannot be bounded in terms of L and $M_l(f)$, $l \geq s+1$.*

Proof. In case $R_s(Z) < \infty$ the required bound is obtained by restricting the minimization in (3.1) to $d = s$ only. If $R_s(Z) = \infty$ then already polynomials of degree s vanishing on Z cannot be bounded on B^n. □

Now we can apply explicit calculations of $\omega_d(Z)$ in Sect. 2 to get explicit inequalities relating the geometry of Z, the values of f on this set, and the bounds on the derivatives of f. We shall restrict ourselves to the case of Z being a curve in the plane, as considered in Proposition 2.3. Other situations presented in Proposition 2.2 can be treated in the same way. Let S be a connected smooth curve in B^2 of the length σ and the injectivity radius ε_0. For $d \leq \frac{\sigma}{24} - 1$ put $\kappa_d = \frac{1}{2l}(1 - \frac{24}{m})$, in notations of Proposition 2.3.

Proposition 3.3. *Let $f : B^2 \to \mathbb{R}$ be a k times continuously differentiable function on B^2, with $M_l(f) = \max_{B^2} \|d^l f\|$, $l = 0, 1, \ldots, k$, and $S \subset B^2$ be a curve with the length σ and with the injectivity radius ε_0. Put $L = \max_{x \in S} |f(x)|$. Then for each $s \leq \frac{\sigma}{24} - 1$ we have*

$$M_0(f) \leq \left(\frac{8}{\kappa_s}\right)^s (L + E_s^T(f)) + E_s^T(f), \tag{3.2}$$

with $E_s^T(f) = \frac{1}{(s+1)!}M_{s+1}(f)$ *and* $\kappa_s = \frac{1}{2l}(1 - \frac{24}{m}) > 0$. *For each s there are curves* $S_s \subset B^2$ *of the length at least* $2s$ *such that* $M_0(f)$ *cannot be bounded in terms of L and* $M_l(f)$, $l \geq s+1$.

Proof. The bound follows directly from Propositions 3.2, 2.3, and 2.1. Now take as a curve S_s a zero set of a polynomial $y = T_s(x)$ inside B^2. Then for $f(x,y) = K(y - T_s(x))$ vanishing on S_s $M_0(f)$ cannot be bounded through $M_l(f)$, $l \geq s+1$. □

Another way to extract more explicit answer from Proposition 3.1 is to bound the norms $M_l(f)$ of the lth-order derivatives of f, for $l = 0, 1, \dots, k$, by their maximal value $M = M(f)$, to substitute M instead of $M_l(f)$ into the inequality 3.1, and to explicitly minimize the resulting expression in d.

We shall fix the smoothness k and consider sets $Z \subset B^n$ for which $\omega(Z) = \omega_{k-1}(Z) > 0$. In particular, let $Z \subset B^n$ be a measurable set with $m_n(Z) > 0$. Then $\omega_d(Z) \geq m_n(Z)$ for each d. Sets Z in the specific classes, discussed in Sect. 2 above, provide additional examples. Since $\omega_0(Z) \geq \omega_1(Z) \geq \dots \geq \omega_{k-1}(Z)$, by Proposition 2.1 for each $d = 0, \dots, k-1$ we have $R_d(Z) \leq (\frac{4n}{\omega(Z)})^d$. Let us denote $\frac{4n}{\omega(Z)} \geq 4n$ by $q = q(Z)$.

The following theorem provides one of possible forms of an explicit inequality, generalizing the Remez one to smooth functions:

Theorem 3.1. *Let* $f : B^n \to \mathbb{R}$ *be a k times continuously differentiable function on* B^n, *with* $M_l(f) = \max_{B^n} \|d^l f\| \leq M = M(f)$, $l = 0, 1, \dots, k$, *and let a subset* $Z \subset B^n$ *with* $\omega_{k-1}(Z) > 0$ *be given. Put* $L = \max_{x \in Z}|f(x)|$, $q = q(Z) \geq 4n$. *Then*

$$M_0(f) = \max_{x \in B^n}|f(x)| \leq 2q^{d_0}L + \frac{1}{(d_0+1)!}M, \tag{3.3}$$

where $d_0 = d_0(M, L)$, *satisfying* $1 \leq d_0 \leq k-1$, *is defined as follows:* $d_0 = 0$ *if* $L > M$, $d_0 = k-1$ *if* $L \leq \frac{1}{k!}M$, *and for* $\frac{1}{k!}M \leq L \leq M$ *the degree* d_0 *is defined by* $\frac{1}{(d_0+1)!}M \leq L \leq \frac{1}{d_0!}M$.

In particular, for $L > M$ *the inequality takes the form*

$$M_0(f) \leq L + 2M, \tag{3.4}$$

while for $L \leq \frac{1}{k!}M$ *we get*

$$M_0(f) \leq 2q^{k-1}L + \frac{1}{k!}M. \tag{3.5}$$

Proof. As above, $R_d(Z) \leq (\frac{4n}{m_n(Z)})^d = q^d$. By Theorem 1.3 we have

$$\max_{x \in B^n}|f(x)| \leq \inf_{d=0,1,\dots,k}[q^d(L + E_d^T(f)) + E_d^T(f)] \leq$$

$$\leq q^d\left(L + \frac{1}{(d+1)!}M\right) + \frac{1}{(d+1)!}M.$$

Now we guess the value of d which approximately minimizes the expression in the right-hand side: let $d_0 = d_0(M, L)$ be defined as follows:
$d_0 = 0$ if $L > M$, $d_0 = k - 1$ if $L \le \frac{1}{k!}M$, and for $\frac{1}{k!}M \le L \le M$ the degree d_0 is uniquely defined by the condition

$$\frac{1}{(d_0+1)!}M \le L \le \frac{1}{d_0!}M.$$

In each case we have $1 \le d_0 \le k - 1$. Substituting d_0 into the above expression we obtain for $L > M$ the inequality $M_0(f) = \max_{x \in B^n} |f(x)| \le L + 2M$, while for $L \le M$ we get $M_0(f) \le 2q^{d_0}L + \frac{1}{(d_0+1)!}M$. In the case $L \le \frac{1}{k!}M$ we get $d_0 = k - 1$, and the inequality takes the form $M_0(f) \le 2q^{k-1}L + \frac{1}{k!}M$. □

Remark. In the case $L > M$ in Theorem 3.1 we have $d_0 = 0$ and the resulting inequality (3.4) is rather straightforward. Indeed, we take one point $x_0 \in Z$. By the assumptions, $|f(x_0)| \le L$, while $||df|| \le M$ on B^n. For each $x \in B^n$ we have $||x - x_0|| \le 2$. Hence $|f(x)| \le L + 2M$. However, for smaller L, i.e., for larger d_0, the result apparently cannot be obtained by a similar direct calculation. Compare a discussion in the next section.

4 Whitney Extension of Smooth Functions

There is a classical problem of Whitney (see [7] and references therein) concerning extension of C^k-smooth functions from closed sets. Recently a major progress has been achieved in this problem. The following "finiteness principle" has been obtained, in its general form, by C. Fefferman in 2003: for a finite set $Z \subset B^n$ and for any real function f on Z denote by $||f||_{Z,k}$ the minimal C^k-norm of the C^k-extensions of f to B^n.

There are constants N and C depending on n and k only, such that for any finite set $Z \subset B^n$ and for any real function f on Z we have $||f||_{Z,k} \le C \max_{\tilde{Z}} ||f||_{\tilde{Z},k}$, with $\tilde{Z}$ consisting of at most N elements.

The original proof of this result, as well as its further developments in [7] and other publications, provides rich connections between the geometry of Z and the behavior of the C^k-extensions of F. Effective algorithms for the extension have been also investigated in [7]. Still, the problem of an explicit connecting the geometry of Z, the behavior of f on Z, and the analytic properties of the C^k-extensions of f to B^n for $n \ge 2$ remains widely open. In one variable divided finite differences provide a complete answer (Whitney). The following result illustrates the role of the Remez constant $R_d(Z)$ in the extension problem.

Theorem 4.1. *For a finite set $Z \subset B^n$ and for any $x \in B^n \setminus Z$ let $Z_x = Z \cup \{x\}$. Let $f_{Z,x}$ be zero on Z and 1 at x and let $\tilde{f}_{Z,x}$ be a C^k-extensions of $f_{Z,x}$ to B^n. Then for each $d = 0, \ldots, k-1$ we have $M_{d+1}(\tilde{f}_{Z,x}) \ge \frac{(d+1)!}{R_d(Z)+1}$.*

Proof. By Proposition 3.1 we have for the extension $\tilde{f}_{Z,x}$

$$M_0(\tilde{f}_{Z,x}) \leq \min_{d=0,1,\ldots,k-1}[R_d(Z)(L + E_d^T(f)) + E_d^T(f)],$$

where $E_d^T(\tilde{f}_{Z,x}) = \frac{1}{(d+1)!}M_{d+1}(\tilde{f}_{Z,x})$ is the Taylor remainder term of f of degree d on the unit ball B^n. In our case $M_0(\tilde{f}_{Z,x}) \geq 1$ while $L = 0$. So we obtain $1 \leq \min_{d=0,1,\ldots,k-1}(R_d(Z) + 1)\frac{1}{(d+1)!}M_{d+1}(\tilde{f}_{Z,x})$. We conclude that for each $d = 0, \ldots, k-1$ we have $M_{d+1}(\tilde{f}_{Z,x}) \geq \frac{(d+1)!}{R_d(Z)+1}$. □

The results of Sect. 3 can be translated into more results on extension from finite set, similar to that of Theorem 4.1. More importantly, Remez inequality for polynomials can be significantly improved, taking into account, in particular, a specific position of x with respect to Z. We plan to present these results separately.

Acknowledgements This research was supported by the ISF, Grant No. 639/09, and by the Minerva Foundation.

References

1. A. Brudnyi, On Covering Numbers of Sublevels Sets of Analytic Functions, *J. of Appr. Theory*, **162** (2010), no. 1, 72–93.
2. Yu. Brudnyi, and M. Ganzburg, On an extremal problem for polynomials of n variables, *Math. USSR Izv.* **37** (1973), 344–355.
3. A. Brudnyi, and Y. Yomdin, Remez Sets, in preparation.
4. D. Coppersmith, and T. J. Rivlin, The growth of polynomials bounded at equally spaced points, *SIAM J. Math. Anal.* **23** (1992), no. 4, 970–983.
5. T. Erdelyi, Remez-type inequalities and their applications, *J. Comp. Appl. Math.* **47** (1993) 167–209.
6. J. Favard, Sur l'interpolation, *Bull. de la S. M. F.*, **67** (1939), 103–113.
7. C. Fefferman, and B. Klartag, Fitting a C^m-smooth function to data. Part I, *Ann. of Math.* (2) 169 (2009), 315–346. Part II, *Rev. Mat. Iberoam.* 25 (2009), 49–273.
8. E. A. Rakhmanov, Bounds for polynomials with a unit discrete norm, *Ann. of Math.* (2) **165** (2007), no. 1, 55–88.
9. E. J. Remez, Sur une propriete des polynomes de Tchebycheff, *Comm. Inst. Sci. Kharkov* **13** (1936) 93–95.
10. A. G. Vitushkin, *O mnogomernyh Variaziyah*, Gostehisdat, Moskow, (1955)
11. Y. Yomdin, Remez-Type Inequality for Discrete Sets, *Isr. J. of Math.*, Vol 186 (Nov. 2011), 45–60.
12. Y. Yomdin, and G. Zahavi, High-Order processing of Singular Data, *Non-Commutativity and Singularities*, 173–207, Adv. Stud. Pure Math., 55, Math. Soc. Japan, Tokyo, 2009.

MIX
Papier aus verantwortungsvollen Quellen
Paper from responsible sources
FSC® C105338

If you have any concerns about our products,
you can contact us on
ProductSafety@springernature.com

In case Publisher is established outside the EU,
the EU authorized representative is:
Springer Nature Customer Service Center GmbH
Europaplatz 3, 69115 Heidelberg, Germany

Printed by Libri Plureos GmbH
in Hamburg, Germany